# Advances in Condition Monitoring of Railway Infrastructures

# Advances in Condition Monitoring of Railway Infrastructures

Editors

**Araliya Mosleh**
**Diogo Ribeiro**
**Abdollah Malekjafarian**
**Maria D. Martínez-Rodrigo**

MDPI

Basel • Beijing • Wuhan • Barcelona • Belgrade • Novi Sad • Cluj • Manchester

*Editors*
Araliya Mosleh
University of Porto
Porto
Portugal

Diogo Ribeiro
Polytechnic of Porto
Porto
Portugal

Abdollah Malekjafarian
University College Dublin
Dublin
Ireland

Maria D. Martínez-Rodrigo
Universitat Jaume I
Castellón
Spain

*Editorial Office*
MDPI
St. Alban-Anlage 66
4052 Basel, Switzerland

This is a reprint of articles from the Special Issue published online in the open access journal *Sensors* (ISSN 1424-8220) (available at: https://www.mdpi.com/journal/sensors/special_issues/Condition_Monitoring_Railway_Infrastructures).

For citation purposes, cite each article independently as indicated on the article page online and as indicated below:

Lastname, A.A.; Lastname, B.B. Article Title. *Journal Name* **Year**, *Volume Number*, Page Range.

**ISBN 978-3-7258-1269-1 (Hbk)**
**ISBN 978-3-7258-1270-7 (PDF)**
**doi.org/10.3390/books978-3-7258-1270-7**

# Contents

# About the Editors

**Araliya Mosleh**

Araliya Mosleh is a senior researcher at the Faculty of Civil Engineering, University of Porto. She obtained her PhD degree in 2016 from the University of Aveiro, Portugal. Since then she has actively engaged in 9 national and international projects in the field of railway infrastructure. She was a visiting researcher at Bundeswehr University (2015), Wollongong University (2017), and Evoleo Company (2019). Mosleh was/is supervising several PhD and Master students/researchers, co-author of more than 60 technical and scientific publications in peer-reviewed international journals and conferences, and was awarded in the CEEC-4th edition Junior Researcher individual call in 2022 (5th out of 72 applicants). Scopus H-Index: 12.

**Diogo Ribeiro**

Diogo Ribeiro is the professor at ISEP-IPP, Director of the Bachelor course in Civil Engineering of ISEP-IPP (since 2014) and Director of the Postgraduate Programs in BIM Coordination (since 2019) and Prefabrication in Concrete (since 2020). Regular invited teacher at University of São Paulo and Federal University of Ouro Preto. Integrated Member of the Institute of R&D in Structures and Construction (CONSTRUCT). Diogo Ribeiro was coordinator or researcher in more than 20 R&D projects funded by the industry, FCT and EU programs in the field of railway infrastructures and digital construction. He is the main editor of the Springer Book Series "Digital Innovations in Architecture, Engineering and Construction". Visiting researcher at Bauhaus Universität Weimar and University California San Diego. He was awarded with a Fulbright Grant for doctoral researchers/teachers by Fundação Luso-Americana (2016) and Ferry Borges prize (2022). He chaired of the IABSE Task Group TG5.9 - Remote inspection of bridges. In addition, he has authored of ¿300 scientific and technical publications in journals, books, and conferences. Scopus H-Index: 22.

**Abdollah Malekjafarian**

Dr. Abdollah Malekjafarian received his PhD in Civil Engineering from University College Dublin in 2016. He is currently an Assistant Professor and leader of "Structural Dynamics and Assessment Laboratory (SDA-Lab)" in the school of Civil Engineering at UCD. His main areas of research interest are in Structural Dynamics and Random Vibrations for Civil Infrastructure including "Transport Infrastructure" and "Offshore Wind Turbines". He is currently the Principal Investigator of the Di-Rail project, which is funded by the Science Foundation Ireland, under the Frontiers for the Future call. He is also the Coordinator of the WindLEDeRR project, funded by the Sustainable Energy Authority of Ireland, under the RD&D call. Abdollah is a member of the Young Academy of Ireland (YAI) selected by the Royal Irish Academy (RIA). He also received the "KB Broberg" medal in 2020 and Royal Irish Academy Charlemont award in 2018 for his outstanding contribution to his field of research. He is also a member of the editorial boards of "Journal of Shock and Vibration" and "Journal of Vibroengineering".

**María D. Martínez-Rodrigo**

Mª Dolores Martínez Rodrigo is a Full Professor in the Division of Continuum Mechanics and Structural Analysis at Jaume I University of Castellón, Spain. Her research falls within the field of Computational Mechanics and Structural Dynamics, applied to seismic- and railway-induced-vibrations related problems. Her Doctoral Thesis focused on vibration control of railway bridges, and led to the COMSA Railway Award from Polytechnical University of Catalonia, the Best Doctoral Thesis award of the Polytechnical University of Valencia and National Patent ES 2 372 095. She is currently the leader of the Computational Mechanics and Structural Analysis research group at UJI, and belongs to the inter-university group USUJI with the University of Seville, devoted to improving the safety, functionality, and sustainability of railway infrastructures based on predictions and experimentation in noise and vibration, performing vibratory studies for the construction of the Madrid-Galicia and León-Asturias High-Speed Lines. Prof. Martínez has participated in nine national projects, five regional, and four local ones for over EUR 1.4 M. Nowadays, she is UJI's representative in InBridge4EU project, funded by Europe's Rail Joint Undertaking. Scopus H-index 16.

# Preface

This Special Issue provides an overview of the scientific papers, focusing on advanced analytical and numerical simulation approaches, along with experimental contributions applied to railway infrastructures. Submissions for this issue have been received from China, Spain, Poland, Ireland, France, Portugal, the Netherlands, and Germany. Min et al. introduces a rail-surface defect-detection model, denoted as FS-RSDD, designed for the rail-surface-condition monitoring. Notably, it addresses the prevalent challenge of limited defect samples encountered with prior detection models. Arana et al. explore the condition monitoring of an overhead contact line (OCL) through the creation of a monitoring system tailored for a pantograph installed on electrical multiple units. Kinematic and dynamic modeling of the pantograph is undertaken to support the development of the monitoring system. Gosiewska et al. from Poland address object detection using computer vision in scenarios characterized by limited data by training YOLOv5 and MobileNet frameworks. It was demonstrated that a dataset comprising 120 observations is adequate for achieving high accuracy in the object detection task specific to railway infrastructure. In the research study conducted by Malekjafarian et al. from Ireland a novel approach for monitoring railway track conditions is presented, based on acceleration responses obtained from an operational train to detect alterations in the stiffness of underlying track sub-layers. An Artificial Neural Network (ANN) algorithm is formulated, operating on the energy content of the train's acceleration responses. Moya et al. from Spain present ongoing progress in the digitalization of freight wagons, encompassing the delineation, fabrication, and on-site trials conducted on a commercial rail line in Sweden. A diverse array of components and systems were installed in a freight wagon, envisaging the completion of an intelligent freight wagon. Mohammadi et al. from Portugal assess and compare the effectiveness of four distinct feature extraction methodologies, specifically, auto-regressive (AR), auto-regressive exogenous (ARX), principal component analysis (PCA), and continuous wavelet transform (CWT), in their capacity to autonomously differentiate between a defective wheel and a healthy one. Ton et al. investigate the applicability of three deep-learning-based models, namely, PointNet++, SuperPoint Graph, and Point Transformer, for the semantic segmentation of point clouds within the context of a practical real-world scenario. The study centers on a specific use case of catenary arches within the Dutch railway system, conducted in collaboration with Strukton Rail, a prominent contractor for rail projects. Tan et al. introduce an Ensemble Learning approach combining Improved MultiScale Retinex with Color Restoration (IMSRCR) and You Only Look Once (YOLO) based on acquired tunnel image data for the detection of corroded bolts in the lining. The features of the lining images are enhanced and strengthened using IMSRCR, mitigating the adverse effects of a dark environment in contrast to the existing MSRCR. Lorezen et al. from Germany propose a methodology for axle detection using accelerometers placed arbitrarily on a bridge structure. The model is implemented as a Fully Convolutional Network suitable for processing signals represented in the Continuous Wavelet Transforms format. This allows passages of any length to be processed in a single step with maximum efficiency while using multiple scales in a single evaluation. Xie et al. from China seek to comprehend the disparities in the impact of expressway bridges and subgrades on the near-surface blown sand environment. It examines variations in wind speed and profile, wind flow-field characteristics, and sand transport rates around bridges and subgrades.

**Araliya Mosleh, Diogo Ribeiro, Abdollah Malekjafarian, and Maria D. Martínez-Rodrigo**
*Editors*

sensors

Editorial

# Advances in Condition Monitoring of Railway Infrastructure

**Araliya Mosleh [1,*], Diogo Ribeiro [2], Abdollah Malekjafarian [3] and Maria D. Martínez-Rodrigo [4]**

1 CONSTRUCT—LESE, Faculty of Engineering, University of Porto, 4200-465 Porto, Portugal
2 CONSTRUCT—LESE, School of Engineering, Polytechnic of Porto, 4249-015 Porto, Portugal; drr@isep.ipp.pt
3 Structural Dynamics and Assessment Laboratory, School of Civil Engineering, University College Dublin, Belfield, D04V1W8 Dublin, Ireland; abdollah.malekjafarian@ucd.ie
4 Mechanical Engineering and Construction Department, Universitat Jaume I, 12006 Castellón, Spain; mrodrigo@uji.es
* Correspondence: amosleh@fe.up.pt

**Citation:** Mosleh, A.; Ribeiro, D.; Malekjafarian, A.; Martínez-Rodrigo, M.D. Advances in Condition Monitoring of Railway Infrastructure. *Sensors* **2024**, *24*, 830. https://doi.org/10.3390/s24030830

Received: 20 November 2023
Accepted: 19 January 2024
Published: 26 January 2024



In recent years, there has been a notable surge in investments directed towards developing new railway lines and revitalising existing ones, reflecting a global commitment to enhance transportation infrastructure. This wave of investment is essential in meeting the growing demands of modern societies for efficient and sustainable transportation options. These efforts encompass numerous critical infrastructures within the railway network, ranging from bridges and tunnels to tracks and signaling systems. Therefore, it is imperative to ensure the operational integrity and safety of these infrastructures throughout their life cycle, safeguarding against potential hazards and ensuring uninterrupted service [1,2].

This imperative has catalyzed significant advancements in the field of structural condition monitoring for railway infrastructures [3,4]. In addition, recent scientific and technological breakthroughs have transformed the way these infrastructures are monitored and maintained.

Moreover, integrating artificial intelligence (AI) and machine learning algorithms has been decisive in enhancing the accuracy and efficiency of structural condition assessment [5–7]. These technologies can process large amounts of data, identifying early anomalies and trends that may indicate forthcoming critical situations. Additionally, using unmanned aerial vehicles (UAVs) equipped with advanced imaging and sensing capabilities has paved the way for cost-effective and comprehensive inspections of hard-to-reach or hazardous areas, thus further supporting the capabilities of railway infrastructure monitoring.

The significance of these advancements in ensuring the longevity and safety of railway infrastructures cannot be overstated. They not only contribute to the overall efficiency and reliability of the transportation network but also play a crucial role in minimizing the environmental footprint associated with maintenance activities. The synergistic combination of strategic investments and innovative technological applications underscores a concerted effort toward building a resilient and sustainable railway infrastructure network capable of meeting future demands. Despite advancements in the railway industry in recent years, there remains a noticeable gap in the ability of science and technology to instigate transformative innovations in the railway industry at its core.

This Editorial refers to the Special Issue "Advances in Condition Monitoring of Railway Infrastructures", which serves as a compilation of the most recent research achievements in the scope of advanced planning, design, construction, monitoring, maintenance, and management of railway infrastructures. A total of 20 manuscripts were submitted for evaluation in the context of the Special Issue, with each manuscript undergoing a comprehensive and strict review process. Subsequently, 10 papers have been accepted for publication, and their respective contributions are detailed in the List of Contributions.

As indicated above, the contributions span diverse geographical regions, encompassing specific country cases such as China, Spain, Poland, Ireland, France, Portugal, Netherlands, and Germany. The topics covered include AI (1, 3, 4, 6), track monitoring approaches (1 and 4), wayside wheel defect detection (6), freight wagon digitalization (5), digital twins (7), Structural Health Monitoring (SHM) (9), and tunnels (10). Contributions 4, 6, and 9 entail numerical studies, while contributions 1, 2, 3, 5, 7, and 8 involve experimental field tests.

Contribution 1 introduces a rail surface defect detection model, denoted as FS-RSDD, designed for the rail surface condition monitoring. Notably, it addresses the prevalent challenge of limited defect samples encountered by prior detection models. The proposed model leverages a pre-trained framework to extract features from both normal and defective rail specimens. Subsequently, an unsupervised learning approach is employed to discern feature distributions and establish a feature prototype memory bank. Employing prototype learning strategies, FS-RSDD computes the likelihood of a test sample being associated with a defect at each pixel, guided by the information stored in the prototype memory bank. This methodology mitigates the constraints faced by deep learning algorithms based on supervised learning paradigms, which often deal with inadequate training samples and reduced reliability in validation. FS-RSDD attains noteworthy precision in defect detection and localization, even when trained with a restricted number of defective samples.

Contribution 2 explores the condition monitoring of an overhead contact line (OCL) by creating a monitoring system tailored for a pantograph installed on multiple electrical units. Kinematic and dynamic modeling of the pantograph is undertaken to support the development of the monitoring system. This modeling is validated through meticulous test-rig experiments, after which the proposed methodology is subjected to comprehensive field tests serving a dual purpose: firstly, to prove the efficiency of the monitoring system using benchmark measurements obtained from the tCat® trolley, and secondly, to evaluate the reproducibility of measurements under realistic operation scenarios.

Contribution 3 addresses object detection using computer vision in scenarios characterized by limited data by training YOLOv5 and MobileNet frameworks. A dataset comprising 120 observations was demonstrated to be adequate for achieving high accuracy in the object detection task specific to railway infrastructure. Additionally, a novel approach for the extraction of background images from railway imagery was introduced. To validate this method, the performance of YOLOv5 and MobileNet was evaluated on small datasets, both with and without background extraction. The experimental outcomes indicate that the application of background extraction reduces the sufficient data volume to 90.

In Contribution 4, a novel approach for monitoring railway track conditions is presented, based on acceleration responses obtained from an operational train to detect alterations in the stiffness of underlying track sub-layers. An artificial neural network (ANN) algorithm is formulated, operating on the energy content of the train's acceleration responses. A computational model of a half-car train interacting with a track profile is employed to simulate the vertical acceleration of the train. The induced damage is represented by reduced soil stiffness within the sub-ballast layer, representative of voided sleepers. Furthermore, a sensitivity analysis is conducted to evaluate the influence of signal noise, slice sizes, and the presence of multiple damaged locations on the performance of the damage index.

Contribution 5 presents ongoing progress in the digitalization of freight wagons, encompassing the delineation, fabrication, and on-site trials conducted on a commercial rail line in Sweden. A diverse array of components and systems were installed in a freight wagon, envisaging the completion of an intelligent freight wagon. The digitalization effort encompasses the seamless integration of sensors designed to provide various functions, including but not limited to train composition analysis, train integrity assessment, asset monitoring, and continuous wagon positioning. These strides herald the potential for real-time data analysis, anomaly detection, and the implementation of proactive maintenance strategies, envisaging the operational efficiency and safety of freight transportation.

Contribution 6 assesses and compares the effectiveness of four distinct feature extraction methodologies, specifically, the auto-regressive (AR) method, auto-regressive exogenous (ARX) method, principal component analysis (PCA), and continuous wavelet transform (CWT), in their capacity to autonomously differentiate between a defective wheel and a healthy one. The reference measurement employed in this investigation is the rail acceleration during the transit of freight vehicles. The study encompasses four sequential steps: (i) feature extraction; (ii) feature normalization; (iii) data fusion; and (iv) damage detection. The findings of this study underscore that the AR and ARX extraction methods exhibit superior efficiency in wheel flat damage detection compared to CWT and PCA techniques.

Contribution 7 investigates the applicability of three deep-learning-based models, namely, PointNet++, SuperPoint Graph, and Point Transformer, for the semantic segmentation of point clouds within the context of a practical, real-world scenario. The study centers on a specific use case of catenary arches within the Dutch railway system, conducted in collaboration with Strukton Rail, a prominent contractor for rail projects. A distinctive, complex, high-resolution, and annotated dataset is presented for the evaluation of point cloud segmentation models in railway environments. Comprising 14 individually labeled classes, this dataset represents the first of its kind to be openly accessible. The modified PointNet++ model emerges as the most effective, achieving a mean class Intersection over Union (IoU) of 71% for the semantic segmentation task.

Contribution 8 introduced an ensemble learning approach combining Improved MultiScale Retinex with Color Restoration (IMSRCR) and You Only Look Once (YOLO) based on acquired tunnel image data for the detection of corroded bolts in the lining. The features of the lining images are enhanced and strengthened by IMSRCR, mitigating the adverse effects of a dark environment in contrast to the existing MSRCR. Additionally, models with varying parameters, exhibiting diverse performance characteristics, are integrated using the ensemble learning method, resulting in a substantial improvement in accuracy. Sufficient comparisons based on a dataset collected from the tunnel are conducted to prove the superiority of the proposed algorithm.

Contribution 9 proposes a methodology for axle detection using accelerometers placed arbitrarily on a bridge structure. The model is implemented as a Fully Convolutional Network suitable for processing signals represented in the Continuous Wavelet Transforms format. This allows passages of any length to be processed in a single step with maximum efficiency while using multiple scales in a single evaluation. Consequently, the proposed method can effectively use acceleration signals from any location on the bridge structure, functioning as Virtual Axle Detectors (VADs) without constraint to specific bridge structural types. The efficiency of the proposed method is tested through the analysis of 3787 passages recorded on a steel railway bridge. The results derived from the measurement data indicate that the proposed model successfully detects 95% of the axles.

Contribution 10 seeks to comprehend the disparities in the impact of expressway bridges and subgrades on the near-surface blown sand environment. It examines wind speed and profile variations, wind flow-field characteristics, and sand transport rates around bridges and subgrades. The goal is to offer a scientific foundation for choosing expressway route forms in sandy regions. The study employs wind tunnel tests with models of a highway bridge and subgrade, comparing the environmental effects of wind-blown sand on both structures. The findings hold theoretical and practical importance for guiding expressway route selection in sandy areas.

The Guest Editors are pleased with the conclusive outcomes of the published papers in this Special Issue, anticipating their utility for researchers, engineers, designers, and other professionals engaged in diverse thematic aspects of advanced analytical and numerical simulation approaches, as well as experimental studies, applied to railway infrastructures. The Guest Editors extend their appreciation to all authors and reviewers for their crucial contributions and for the dissemination of scientific findings. Lastly, gratitude is extended to the Editorial Board of *Sensors* for their patience, support, and exceptional contributions.

**Conflicts of Interest:** The authors declare no conflict of interest.

**List of Contributions**

1. Min, Y.; Wang, Z.; Liu, Y.; Wang, Z. FS-RSDD: Few-Shot Rail Surface Defect Detection with Prototype Learning. *Sensors* **2023**, *23*, 7894.
2. Rodríguez-Arana, B.; Ciáurriz, P.; Gil-Negrete, N.; Alvarado, U. A Non-Intrusive Monitoring System on Train Pantographs for the Maintenance of Overhead Contact Lines. *Sensors* **2023**, *23*, 7890.
3. Gosiewska, A.; Baran, Z.; Baran, M.; Rutkowski, T. Seeking a Sufficient Data Volume for Railway Infrastructure Component Detection with Computer Vision Models. *Sensors* **2023**, *23*, 7776.
4. Malekjafarian, A.; Sarrabezolles, C.-A.; Khan, M.A.; Golpayegani, F. A Machine-Learning-Based Approach for Railway Track Monitoring Using Acceleration Measured on an In-Service Train. *Sensors* **2023**, *23*, 7568.
5. Moya, I.; Perez, A.; Zabalegui, P.; de Miguel, G.; Losada, M.; Amengual, J.; Adin, I.; Mendizabal, J. Freight Wagon Digitalization for Condition Monitoring and Advanced Operation. *Sensors* **2023**, *23*, 7448.
6. Mohammadi, M.; Mosleh, A.; Vale, C.; Ribeiro, D.; Montenegro, P.; Meixedo, A. An Unsupervised Learning Approach for Wayside Train Wheel Flat Detection. *Sensors* **2023**, *23*, 1910.
7. Ton, B.; Ahmed, F.; Linssen, J. Semantic Segmentation of Terrestrial Laser Scans of Railway Catenary Arches: A Use Case Perspective. *Sensors* **2022**, *23*, 222.
8. Tan, L.; Tang, T.; Yuan, D. An Ensemble Learning Aided Computer Vision Method with Advanced Color Enhancement for Corroded Bolt Detection in Tunnels. *Sensors* **2022**, *22*, 9715.
9. Lorenzen, S.R.; Riedel, H.; Rupp, M.M.; Schmeiser, L.; Berthold, H.; Firus, A.; Schneider, J. Virtual Axle Detector Based on Analysis of Bridge Acceleration Measurements by Fully Convolutional Network. *Sensors* **2022**, *22*, 8963.
10. Xie, S.; Zhang, X.; Pang, Y. Characteristic Differences of Wind-Blown Sand Flow Field of Expressway Bridge and Subgrade and Their Implications on Expressway Design. *Sensors* **2022**, *22*, 3988.

## References


1. Mosleh, A.; Meixedo, A.; Ribeiro, D.; Montenegro, P.; Calçada, R. Automatic clustering-based approach for train wheels condition monitoring. *Int. J. Rail Transp.* **2023**, *11*, 639–664. [CrossRef]
2. Neto, J.; Montenegro, P.A.; Vale, C.; Calçada, R. Evaluation of the train running safety under crosswinds—A numerical study on the influence of the wind speed and orientation considering the normative Chinese Hat Model. *Int. J. Rail Transp.* **2021**, *9*, 204–231. [CrossRef]
3. Figueiredo, E.; Park, G.; Farrar, C.R.; Worden, K.; Figueiras, J. Machine learning algorithms for damage detection under operational and environmental variability. *Struct. Health Monit.* **2011**, *10*, 559–572. [CrossRef]
4. Figueiredo, E.; Cross, E. Linear approaches to modeling nonlinearities in long-term monitoring of bridges. *J. Civ. Struct. Health Monit.* **2013**, *3*, 187–194. [CrossRef]
5. Jiang, H.; Lin, J. Fault diagnosis of wheel flat using empirical mode decomposition-Hilbert envelope spectrum. *Math. Probl. Eng.* **2018**, *2018*, 8909031. [CrossRef]
6. Amini, A.; Entezami, M.; Huang, Z.; Rowshandel, H.; Papaelias, M. Wayside detection of faults in railway axle bearings using time spectral kurtosis analysis on high-frequency acoustic emission signals. *Adv. Mech. Eng.* **2016**, *8*, 1687814016676000. [CrossRef]
7. Yonas, L.; Matthias, A.; Matti, R. Investigation of the Top-of-Rail Friction by Field Measurements on Swedish Iron Ore Line. *Int. J. COMADEM* **2015**, *18*, 17–20.

*sensors*

MDPI

*Article*

# FS-RSDD: Few-Shot Rail Surface Defect Detection with Prototype Learning

**Yongzhi Min [1,*], Ziwei Wang [1,*], Yang Liu [1] and Zheng Wang [2]**

1 School of Automation and Electrical Engineering, Lanzhou Jiaotong University, Lanzhou 730070, China; 12211463@stu.lzjtu.edu.cn
2 School of Mechanical Engineering, Lanzhou Jiaotong University, Lanzhou 730070, China; wangz@lzjtu.edu.cn
* Correspondence: minyongzhi@lzjtu.edu.cn (Y.M.); 11210387@stu.lzjtu.edu.cn (Z.W.)

**Abstract:** As an important component of the railway system, the surface damage that occurs on the rails due to daily operations can pose significant safety hazards. This paper proposes a simple yet effective rail surface defect detection model, FS-RSDD, for rail surface condition monitoring, which also aims to address the issue of insufficient defect samples faced by previous detection models. The model utilizes a pre-trained model to extract deep features of both normal rail samples and defect samples. Subsequently, an unsupervised learning method is employed to learn feature distributions and obtain a feature prototype memory bank. Using prototype learning techniques, FS-RSDD estimates the probability of a test sample belonging to a defect at each pixel based on the prototype memory bank. This approach overcomes the limitations of deep learning algorithms based on supervised learning techniques, which often suffer from insufficient training samples and low credibility in validation. FS-RSDD achieves high accuracy in defect detection and localization with only a small number of defect samples used for training. Surpassing benchmarked few-shot industrial defect detection algorithms, FS-RSDD achieves an ROC of 95.2% and 99.1% on RSDDS Type-I and Type-II rail defect data, respectively, and is on par with state-of-the-art unsupervised anomaly detection algorithms.

**Keywords:** rail surface defect detection; few-shot learning; prototype learning; transfer learning; unsupervised anomaly detection

**Citation:** Min, Y.; Wang, Z.; Liu, Y.; Wang, Z. FS-RSDD: Few-Shot Rail Surface Defect Detection with Prototype Learning. *Sensors* **2023**, *23*, 7894. https://doi.org/10.3390/s23187894

Academic Editor: Yi Qin

Received: 4 August 2023
Revised: 29 August 2023
Accepted: 5 September 2023
Published: 15 September 2023



## 1. Introduction

The rapid growth of railway operation mileage in recent years, due to the construction of numerous new railway lines in many countries, has significantly increased the pressure on maintenance. During the daily operation of railway systems, the interaction between wheels and rails inevitably leads to surface defects such as spalling, corrugation, and grinding, which pose serious hidden dangers to safe operation. Unlike internal defects in rails that can be detected using techniques such as ultrasound [1] and eddy current [2,3], traditional rail surface defect detection is mainly conducted through manual visual inspection, which is inefficient and heavily relies on human workers' experience [4]. In recent years, many researchers have focused on developing machine vision-based rail surface defect detection technologies that offer higher efficiency and accuracy to address the aforementioned issues. With the rapid development of artificial intelligence technology, deep-learning-based algorithms, specifically supervised learning-based defect detection algorithms, are being widely applied in rail surface defect detection [5–8].

However, defect samples are difficult to obtain in practical work; thus, defect detection methods based on supervised learning face two important challenges due to insufficient defect samples. One of them is the risk of overfitting caused by the limited training data, which may not adequately represent the distribution of defect; additionally, supervised learning methods typically require the use of a portion of the defect data for training, leading to a reduction in the number of testing samples available for validation, which affects the credibility of the validation results. Inspired by the concept of anomaly detection (AD), some researchers have turned their attention to utilizing unsupervised learning techniques to address the aforementioned issues in the field of defect detection [9–11]. However, these unsupervised learning-based methods rely completely on modeling the distribution of normal samples, lacking an understanding of defect data, which may lead to poor classification performance and a potentially high false-positive/negative rate [12].

To tackle the aforementioned shortcomings of supervised learning-based defect detection methods, this paper proposes a few-shot rail surface defect detection model called FS-RSDD (few-shot rail surface defect detection). Inspired by the prototype learning and feature-embedding-based unsupervised AD (anomaly detection algorithms), FS-RSDD uses a pre-trained neural network as a feature extractor for both normal and defective rail images. Global average pooling and mask average pooling are used to embed features for normal and defective samples, respectively, which aim to compress the feature maps into feature vectors to obtain a compact feature memory bank. Subsequently, an unsupervised learning algorithm is used to obtain the feature prototypes of normal samples. Finally, the detection of rail defects is accomplished through the similarity computation between input features and prototypes. In summary, our main contributions are as follows:

1. To overcome the challenges associated with using supervised learning-based defect detection algorithms when there is insufficient defect data available, we have introduced a simple yet effective few-shot rail surface defect detection method called FS-RSDD, which combines unsupervised anomaly detection with prototype learning. By effectively integrating the feature prototypes of normal rail images and defect rail images, we have achieved high accuracy in detecting rail surface defects with very little defect samples used for training.
2. By avoiding the partitioning of normal rail backgrounds into small image patches and individually modeling the feature distribution of each image patch, FS-RSDD achieves a compact feature memory bank for normal rail samples, alleviating the issue of memory bank redundancy in feature-embedding-based unsupervised anomaly detection algorithms.
3. FS-RSDD extensively leverages the fusion of multi-scale features to improve prediction accuracy. Furthermore, due to the integration of both normal background feature prototypes and defect feature prototypes for defect detection, the performance of the FS-RSDD model remains stable and robust compared to other few-shot industrial defect detection algorithms, even when the quality of the defect samples used for training is relatively low.
4. Through extensive experiments, our method outperformed most existing few-shot supervised defect detection algorithms under the same number of defect samples used for training and achieved comparable performance to existing unsupervised anomaly detection algorithms which assume the availability of normal training samples only.

## 2. Related Works

### 2.1. Rail Surface Defect Detection

Previous research on rail surface defect detection often utilizes traditional image processing techniques to extract features from defect images and trains detection models using corresponding machine learning methods [13–16]. However, the performance of these methods is limited by the design of feature extraction, and the detection results can easily be affected by factors such as lighting, noise, and other factors. With the rapid development of deep learning technology, an increasing number of researchers have started studying rail surface defect detection methods based on deep learning, especially supervised learning methods. Wang Hao et al. integrated the improved pyramid feature fusion and modified loss function into the Mask-RCNN algorithm for the purpose of detecting rail surface defects [4]. Meng Si et al. proposed a multi-task architecture for rail surface defect detection, which includes two branch models for rail detection and defect segmentation [17]. Zhang

Hui et al. cascaded the one-stage object detection algorithms SSD and YOLOv3, integrating the detection results from both networks to improve the accuracy of rail surface defect detection [18]. However, these approaches neglected the fact that defect samples are scarce and difficult to obtain in practical work.

Due to the limited number of defect samples in the field of defect detection, supervised algorithms-based defect detection models often face issues of overfitting and low validation credibility. To address these problems, many researchers have proposed corresponding solutions. D. Zhang et al. partitioned the rail image data into multiple segments and trained the defect detection model. However, this approach did not fundamentally solve the problem [19], and more researchers have recently started studying steel rail surface defect algorithms based on unsupervised anomaly detection algorithms. Q. Zhang et al. implemented the detection of rail surface defects using the multi-scale cross FastFlow model [20], while Menghui Niu et al. proposed an unsupervised stereoscopic saliency detection method for detecting rail surface defects and achieved good detection results [21]. However, some studies have pointed out that unsupervised anomaly defect detection algorithms often lead to a higher false detection rate [22,23] due to the lack of knowledge about defect samples during the training process. In this paper, we propose a simple yet effective few-shot rail surface defect detection algorithm that fully utilizes the feature information of normal steel rail samples and defect sample information to achieve defect detection.

### 2.2. Unsupervised Anomaly Detection for Industrial Images

Deep-learning-based algorithms are being widely used in industrial defect detection research in recent years due to their high efficiency and accuracy. Many researchers have devoted themselves to researching industrial defect detection algorithms based on supervised learning algorithms, which significantly depends on labeled defect data [24–29]. However, due to the hardship of collecting defective samples, it is extremely hard to obtain enough defect data for a deep model to learn its distribution. Furthermore, supervised learning-based methods require defect data for training, which further restricts the quantity of test datasets and affects the credibility of validation performance. In recent years, unsupervised-based anomaly detection (AD) algorithms have become the mainstream paradigm for industrial defect detection, which can be categorized as reconstruction-based and feature-embedding-based [30–32].

Reconstruction-based methods aim to train a deep network such as an adversarial generative network (GAN) or auto encoder (AE) to reconstruct normal images. When defective images are fed into the network, the defective parts cannot be reconstructed well, allowing for the detection of defects. However, sometimes the model can also yield a good reconstruction for the defective parts due to the powerful ability of the deep model [30].

Feature-embedding-based methods became the prevalent architecture in recent years, which typically consisted of a feature extractor and a feature estimator. A feature extractor is a deep network, typically a ResNet [33], that is pre-trained on ImageNet datasets. It is used to extract features from normal images, which are then stored into a memory bank. A feature estimator is used to estimate the distribution for normal features, which can be a multidimensional Gaussian distribution [34], clustering methods [35], or flow-based methods [36]. To avoid the deviation caused by different data distribution between industrial images and ImageNet datasets, only features from shallow layers are used. After distribution estimation, a distance metric is typically used to detect defects, since defects should be far from the center of the estimated distribution. One major drawback of embedding-based anomaly detection algorithms is that they estimate the distribution separately for each patch of the feature map, resulting in a massive and redundant feature memory bank to restore features from each patch. Many researchers have tried different methods to alleviate the problem: Padim experimentally studied the possibility to reduce redundancy of the memory bank and eventually chose to randomly discard a portion of the extracted features [30]; Patchcore utilized a coreset subsampling method to select representative features [32], thereby compressing the size of the feature memory bank.

This paper introduces a feature representation method widely used in few-shot learning, which obtains a representative and compact feature memory bank and alleviates the aforementioned redundancy problem of the memory bank for rail surface defect detection.

### 2.3. Few-Shot Learning

In recent years, deep learning algorithms based on supervised learning have garnered significant attention from researchers due to the remarkable ability of deep models and large-scale datasets with high-quality labels. However, it is well known that supervised algorithms fail to acquire strong generalization ability when trained on a dataset with a small amount of data. Moreover, in many fields such as industrial defect detection, collecting a large-scale dataset with high-quality annotations proves to be challenging. This realization has prompted many researchers to shift their focus to the field of few-shot learning, with the aim of enabling the model to obtain strong generalization ability with only a few samples, akin to human beings.

Within the domain of few-shot learning in computer vision, image classification tasks are a prominent focal point. These tasks can be broadly categorized into three distinct classes: data-augmentation-based methods, parameter-optimization-based methods, and metric-learning-based methods.

Data-augmentation-based methods aim to address the challenge of limited samples in few-shot learning indirectly by enhancing the intricacy of the dataset through data augmentation. Trinet [37] employs autoencoders to map the features to the semantic space, followed by mapping the augmented features back to the sample space via semantic nearest neighbor search. Moreover, Patchmix [38] resolves the issue of distribution shift by substituting a specific region of the query image with random gallery images from diverse categories.

Parameter-optimization-based methods generally first train a meta-learner to learn common features (prior knowledge) of different tasks and then apply the obtained meta-knowledge to fine-tune the base learner on the query set. The model-agnostic meta-learning (MAML) [39], which first trains the model on a large number of task sets to obtain an adaptable weight and then fine-tunes the model on the target task to obtain the final classifier.

Metric-learning-based methods leverage pre-trained neural networks to extract features from training data. These extracted features are then utilized to measure similarity between the training data and test data using a metric. Representative methods include Siamese networks [40] and matching networks [41]. The former inputs two samples into the neural network and compares the similarity of the output feature vectors, while the latter uses attention mechanisms to obtain information about the correlation between feature vectors.

A typical embedding-based approach to few-shot image classification is the prototypical network [42], which utilizes a pre-trained model to extract features from a limited amount of labeled data and learns corresponding feature prototypes from them. The network then produces a distribution over classes for an input feature based on a softmax function over distances to the prototypes in the embedding space.

The prototypical network approach, combined with the utilization of mask average pooling, has been widely adopted in few-shot semantic segmentation methods. In addition, the idea of prototype features in prototypical networks has also been widely applied in many unsupervised anomaly detection algorithms [43,44].

## 3. Methods

This paper proposes an approach for rail surface defect detection called FS-RSDD. It aims to tackle the challenge of detecting surface defects with a limited number of defect samples. The proposed model combines defect feature prototypes and background feature prototypes to enable few-shot learning in this task. The architecture of the model is depicted in Figure 1, illustrating the integration of the proposed approach.

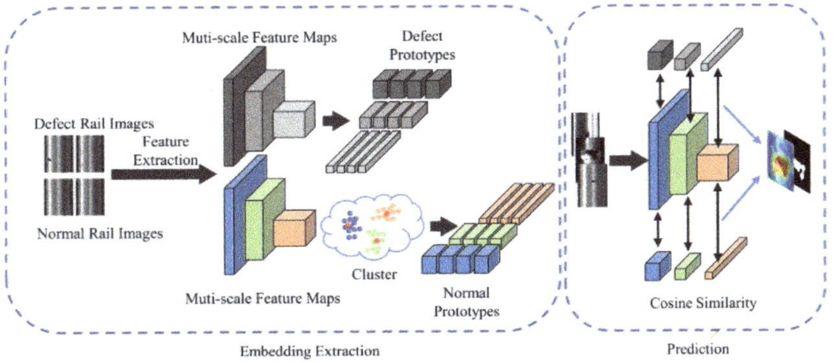


**Figure 1.** The architecture of proposed model.

Figure 1 depicts the proposed method, which consists of two parts: embedding extraction and prediction. In the embedding extraction phase, the approach is inspired by feature-embedding-based anomaly detection techniques. A pre-trained model is employed for extracting multi-scale features from the training set images. These extracted features are then processed to generate a compact memory bank.

During the prediction phase, the feature prototypes obtained from the embedding extraction phase are utilized to calculate the multi-scale similarity feature maps with the feature map of test images. These similarity feature maps of normal and defect samples are then synthesized at each scale to generate a segmentation probability map. Finally, the probability map is smoothed to obtain the final prediction result. This process enables the detection of rail surface defects with high accuracy with limited defect samples.

### 3.1. Embedding Extraction

In this paper, a ResNet $g(\cdot)$ pre-trained on the public dataset ImageNet is employed as a feature extractor, and k is defined as a layer index of ResNet. In order to avoid the deviation caused by different data distribution between industrial images and ImageNet datasets, only features from first three layers are used; thus, $k \in \{1,2,3\}$.

First, the pre-trained model weights are fixed, and then the training set images are passed through the feature extractor. Next, the feature maps are extracted from the shallow layer of the network. Specifically, we are presented with $\{N_{train}, D_{train}\}$, in which subset $N_{train} = \{x_1, x_2, \cdots, x_N\}$ only contains normal samples and subset $D_{train} = \{x_{N+1}, x_{N+2}, \cdots, x_{N+M}\}$ only contains defect samples with $N \gg M$. As shown in Equations (1) and (2), $F_D^k$ and $F_N^k$ refer to the defect feature maps and normal feature maps, respectively. They are obtained from the $k$-th layer of the feature extractor, which is denoted as $g_k(\cdot)$. $M$ and $N$ refer to number of defect samples and normal samples respectively. $C_k$, $H_k$, $W_k$ refer to channels, height, and width of feature map from layer $k$.

$$F_N^k = g_k(N_{train}),\ F_N^k \in R^{N\times C_k\times H_k\times W_k} \quad (1)$$

$$F_D^k = g_k(N_{train}),\ F_D^k \in R^{M\times C_k\times H_k\times W_k} \quad (2)$$

After obtaining the feature representations from defective and normal rail images, the corresponding feature memory bank can be created by the proposed process.

### 3.2. Compact Multi-Scale Memory Bank

After obtaining the corresponding feature maps, the global average pooling (GAP) operation is applied to the feature maps of normal rail images. This operation fuses the global information of normal samples into a feature vector. On the other hand, for defective rail images, since the defective parts only occupy a small portion of the entire image, the mask average pooling (MAP) operation is used. This operation, as shown in Figure 2, is widely employed in few-shot semantic segmentation. It eliminates the features of normal parts in the feature map and only preserves the defect-specific features by element-wise production between feature map and mask, and then global average pooling is applied to obtain the prototype of defects.

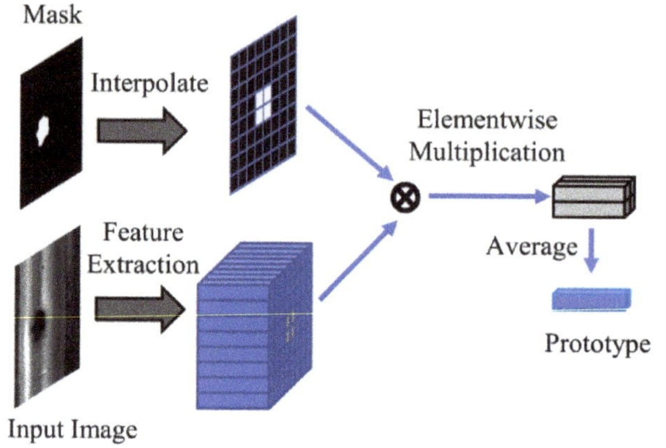


**Figure 2.** Mask average pooling.

In Equations (3) and (4), *GAP* represents the global average pooling operation, $mask_{Dj}^{k}$ represents the ground truth mask, $p_D^k(x_j)$ represents the feature prototype, both $mask_{Dj}^{k}$ and $p_D^k(x_j)$ correspond to a certain defective sample $x_j$, and $p_n^k(x_i)$ represents the feature prototype of the normal sample $x_i$. Additionally, $f_D^k(x_j)$ indicates a feature map corresponding to a certain image $x_j$, and $f_N^k(x_i)$ indicates a feature map corresponding to $x_i$. $\odot$ indicates the Hadamard product.

$$p_D^k(x_j) = GAP\left(f_D^k(x_j) \odot mask_{Dj}^k\right),\ p_D^k \in R^{C_k}, x_j \in D_{train} \tag{3}$$

$$p_N^k(x_i) = GAP\left(f_N^k(x_i)\right),\ p_N^k \in R^{C_k}, x_i \in D_{train} \tag{4}$$

The global average pooling operation is shown in Equation (5), where $p_N^k(x_i)$ represents the normal feature prototype obtained by applying global average pooling to a certain normal feature in the layer $k$, and $f_N^k(x_i)(h,w)$ represents the value of feature map $f_N^k(x_i)$ at position $(h,w)$.

$$p_N^k(x_i) = \frac{1}{H_k \cdot W_k} \sum_{h=1}^{H_k} \sum_{w=1}^{W_k} f_N^k(x_i)(h,w),\ p_N^k \in R^{C_k} \tag{5}$$

As the number of normal samples used is significantly higher than the number of defect samples, which is distinct from the few-shot learning scenario, unsupervised algorithms can be used to obtain the distribution of normal sample features. Instead of estimating the feature prototype using the mean of sample features, as carried out in the prototypical network, this study adopts a widely used clustering algorithm, K-Means, to cluster the normal sample features. The cluster centers are then used as the final feature prototypes of the normal samples.

For the normal sample feature prototype, which consists of a set of feature vectors, clustering is performed with a predetermined number of clusters denoted as n. In this study, a cluster center number of 30 is chosen to cluster the normal samples, and the resulting cluster centers are utilized as the final feature prototypes. Since the number of defect sample features is relatively small, no clustering is conducted, and they are directly used as feature prototypes. All prototypes will be stored as a memory bank.

### 3.3. Pixel-Level Defect Detection

After completing the construction of memory bank, the detection process involves several steps as illustrated in Figure 3. First, the test image is fed into the corresponding feature extractor, which is then used to extract multi-scale intermediate features of the image. Next, the obtained intermediate features are then compared to the feature prototypes obtained during the model construction stage, and based on their similarity, corresponding similarity feature maps are calculated.

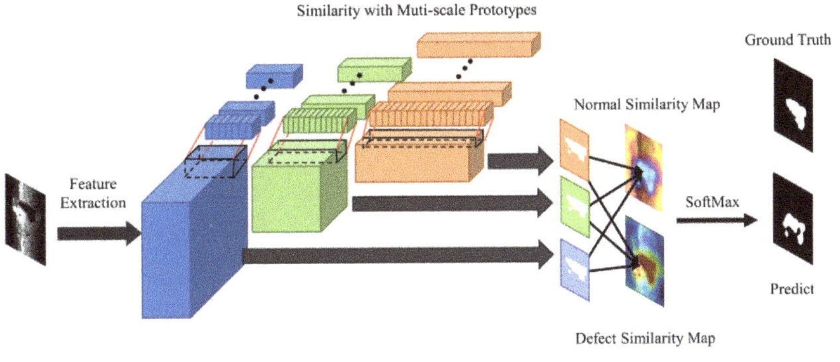


**Figure 3.** Detection procedure of FS-RSDD.

The features obtained from the test images are compared to the corresponding multi-scale normal and defect prototypes at each position using a similarity calculation $s(\cdot)$. The similarity calculation between input and prototypes is shown in Equations (6) and (7). $S_D^k(x)(h,w)$ refers to the similarity between defect feature prototypes and input image feature map at position (*h*,*w*), similarly $S_N^k(x)(h,w)$ refers to the similarity between normal feature prototypes and input image feature map. Specifically, $f_{img}^k$ denotes feature map of a input image.

$$S_D^k(x)(h,w) = \frac{1}{n}\sum_{i=1}^{n} s\left(p_D^k(x_i), f_{img}^k(h,w)\right), S_D^k(x)(h,w) \in R \tag{6}$$

$$S_N^k(x)(h,w) = \frac{1}{n}\sum_{i=1}^{n} s\left(p_N^k(x_i), f_{img}^k(h,w)\right), S_N^k(x)(h,w) \in R \tag{7}$$

In this study, cosine similarity was chosen for similarity calculation. The calculation process for the similarity feature map is demonstrated in Equation (8), where the defect prototype $p_D^k(x_j)$ and input image feature map $f_{img}^k(h,w)$ are both vectors of length $C_k$.

$$s\left(p_D^k(x_j), f_{img}^k(h,w)\right) = \frac{p_D^k(x_j)\cdot f_{img}^k(h,w)}{\|p_D^k(x_j)\|_2 \times \left\|f_{img}^k(h,w)\right\|_2} \tag{8}$$

After performing similarity calculations between all feature prototypes and the input image features, a probability distribution over defects for each position in the image is established using softmax. This allows us to obtain the probability of each position being a defect, as shown in Equation (9), where $q(y = defect|x)$ represents the conditional probability that y belongs to defect under the premise of given input $x$:

$$q(y = defect|\, x) = \frac{1}{3}\sum_{k=1}^{3} \frac{exp\left(S_D^k(x)(h,w)\right)}{exp(S_D^k(x)(h,w) + exp(S_N^k(x)(h,w))} \tag{9}$$

By combining Equation (9), we can observe that the essence of FS-RSDD is to evaluate the similarity between input samples and defect prototypes, as well as the dissimilarity between input samples and normal prototypes in three feature spaces (obtained from three layers of the feature extractor), as illustrated in Figure 4. Finally, defect detection is performed by integrating the prediction results from the three feature spaces, as shown in Equation (9).

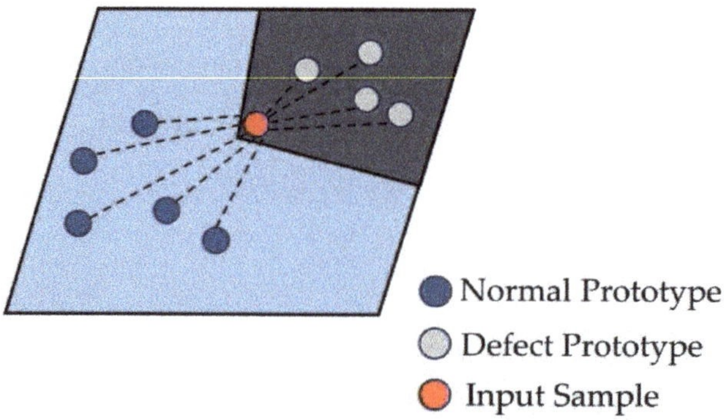


**Figure 4.** The similarity calculation of the FS-RSDD, blue and black areas represent the distribution bound of normal and defective samples, respectively.

### 3.4. *Image-Level Defect Detection*

Image-level defect detection aims to perform image-level binary classification between normal rail images and rail images containing defects. By processing the predicted results in Section 3.3 accordingly, we can obtain the corresponding image-level prediction results.

Our approach is based on a simple idea. If we define $q(y = defect|\, x)$ in Section 3.3 as the defect score of a certain pixel, we can represent the probability of an image containing defects by considering the defect score of the pixel with the highest defect score in the predicted image. However, this approach leads to poor performance, as it only considers individual pixels and lacks consideration for the local neighborhood pixels. In order to further improve the detection accuracy, we decided to use a simple Gaussian blur to fuse information from the local neighborhood of pixels. The process of Gaussian blur on an image is the convolution of the image with a two-dimensional Gaussian distribution that has been discretely sampled, as shown in Figure 5. Subsequently, we performed image-level defect detection. This approach significantly improved the performance of our model, as demonstrated in Section 4.3.

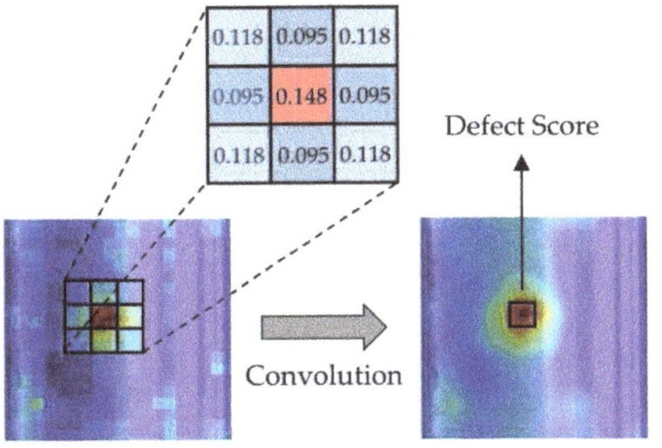


**Figure 5.** The schematic of the process of two-dimensional Gaussian blur, in heat map, the depth of red color represents the probability of the presence of defects in the area.

## 4. Experiments and Results

### *4.1. Evaluation Metrics*

This article focuses on the detection and localization of rail surface defects, which involves binary classification tasks at both image and pixel levels for defect rail images and normal rail images. The receiver operating characteristic (ROC) and precision recall (PR) are used as the evaluation metrics for the model.

These two performance metrics have different emphases, which enable this study to comprehensively evaluate the performance of the model during the experimental process. Additionally, we assessed the classification performance at both the image level and the pixel level. These two metrics, respectively, represent the algorithm's ability to classify defects and accurately locate them. By evaluating performance at both levels, a more comprehensive analysis of the algorithm's effectiveness can be obtained.

As defined in Equations (10) and (11), the $x$-axis of the ROC curve represents the false-positive rate ($FPR$), and the $y$-axis represents the true-positive rate ($TPR$), in which $FP$ denotes false positives (negative samples falsely predicted as positive), $TN$ denotes true negatives (negative samples correctly predicted as negative), $TP$ denotes true positives (positive samples correctly predicted as positive), and $FN$ denotes false negatives (positive samples falsely predicted as negative). A larger area under the ROC curve indicates better performance of the classifier. In this article, the model evaluation metrics are divided into image-level ROCs and pixel-level ROCs, which correspond to evaluation metrics for images and individual pixels, respectively.

$$FPR = \frac{FP}{FP + TN} \tag{10}$$

$$TPR = \frac{TP}{TP + FN} \tag{11}$$

The *recall* rate is represented on the $x$-axis of the $PR$ curve, while the accuracy precision is depicted on the vertical axis. The definitions of recall and *precision* are provided in Equations (12) and (13), respectively. The area under the $PR$ curve corresponds to the average accuracy (AP). A larger area under the $PR$ curve indicates better performance of the classifier.

$$Recall = \frac{TP}{TP + FN} \tag{12}$$

$$Precision = \frac{TP}{TP + FP} \tag{13}$$

### 4.2. Experiment Setup

#### 4.2.1. Dataset Setup

This article uses a dataset from the open-source Rail Surface Defect Detection dataset (RSDDS) [45]. RSDDS consists of two types of rail defect data: Type-I and Type-II. Type-I defects were obtained from 67 defect images collected from high-speed train tracks. Type-II defects, on the other hand, were collected from 128 defect images obtained from regular/heavy-duty transportation tracks. In this article, the two types of defect images are first divided into normal samples and defect samples through fixed ratio image cropping. During the cropping process, images that have a too small defect area are discarded. The image processing process is shown in Figure 6.

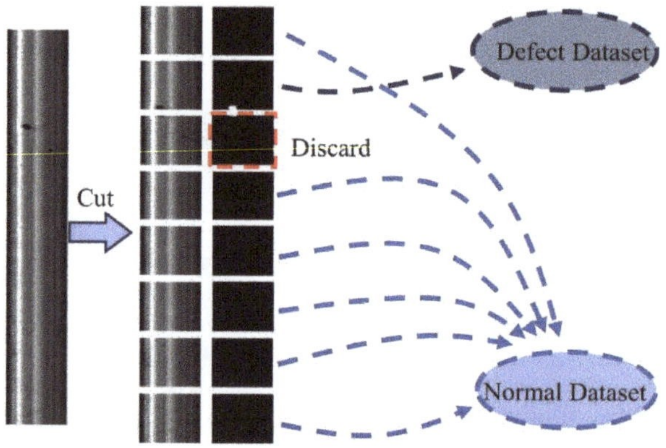


**Figure 6.** The process of dataset creation.

After the aforementioned process, there are 113 defect samples in Type-I dataset and 230 defect samples in Type-II dataset. To ensure a balanced representation of positive and negative samples in the test set and to provide a more accurate evaluation of the performance of the proposed method, we randomly selected normal rail samples for the test set, ensuring that the quantity was consistent with the number of defect samples.

Finally, the Type-I dataset consisted of 302 normal samples for the training set, 113 defect samples, and 113 normal samples for the testing set. Meanwhile, the Type-II dataset comprised 2071 normal samples for the training set, 230 defect samples, and 230 normal samples for the testing set. Additionally, the model necessitates a limited number of defect samples during the training phase, which will be randomly selected from the test set. After being partitioned and resized, the resolution of Type-I rail images is $160 \times 160$, while Type-II rail images have a resolution of $64 \times 64$.

#### 4.2.2. Comparison Experiment Setup

The proposed method in this article is compared with mainstream unsupervised industrial defect detection algorithms and existing few-shot supervised industrial defect detection algorithms in terms of classification evaluation metrics on the RSDDS dataset.

As there may be variations in the defect samples extracted during each training process, a random selection of a small subset of defect samples is employed for training during the experimental process. To ensure robustness, multiple experiments are conducted, and the average value is considered as the validation result of the model.

In the comparison experiments with unsupervised methods, since the defect samples for training are randomly selected in each experiment, the test set may not include the exact same defect samples in each experiment. Therefore, to maintain consistency, multiple tests are also conducted on the unsupervised industrial defect detection algorithms, and the defect samples utilized for our method are excluded from the test set to ensure a fair evaluation of both methods on the same test set, ensuring that the test set used aligns consistently with the test set employed in each experiment of the proposed method in this article.

Similarly, when comparing the performance with few-shot supervised industrial defect detection algorithms, multiple experiments are conducted, and the average test results are used as the final performance metric. Additionally, in each experiment, the defect samples utilized for training the few-shot supervised industrial defect detection algorithm are consistent with the defect samples randomly selected for training in the proposed method in this article. Furthermore, the default values were maintained for all other settings of the comparative models in the code. All the comparative models that were involved with the gradient decent process are trained to convergence to guarantee the impartiality of performance comparisons.

### *4.3. Comparison with Unsupervised-Based Algorithm*

The performance comparison results with unsupervised methods are presented in Tables 1–3. Table 1 displays the average image-level ROC, Table 2 shows the average image-level AP, and Table 3 presents the average pixel-level ROC. All of these metrics were obtained from 20 random sampling validations. In the training process, "m" refers to the defect sample used. It is worth mentioning that the unsupervised algorithms were implemented using the open-source industrial defect detection library anomalib [46].

**Table 1.** Image-level ROC of our proposed FS-RSDD and other unsupervised anomaly detection models.

| Dataset | Model | | | | | |
|---|---|---|---|---|---|---|
| | **FS-RSDD** | **Padim [30]** | **PatchCore [32]** | **stfpm [31]** | **cflow [9]** | **fastclow [10]** |
| RSDDS Type-I m = 5 | 0.941 | 0.950 | 0.951 | 0.899 | 0.866 | 0.895 |
| RSDDS Type-I m = 10 | 0.952 | 0.950 | 0.951 | 0.905 | 0.903 | 0.894 |
| RSDDS Type-II m = 5 | 0.989 | 0.976 | 0.996 | 0.990 | 0.875 | 0.983 |
| RSDDS Type-II m = 10 | 0.991 | 0.976 | 0.996 | 0.990 | 0.881 | 0.984 |

**Table 2.** Image-level AP of our proposed FS-RSDD and other unsupervised anomaly detection models.

| Dataset | Model | | | | | |
|---|---|---|---|---|---|---|
| | **FS-RSDD** | **Padim** | **PatchCore** | **stfpm** | **cflow** | **fastclow** |
| RSDDS Type-I m = 5 | 0.943 | 0.981 | 0.982 | 0.947 | 0.939 | 0.956 |
| RSDDS Type-I m = 10 | 0.953 | 0.980 | 0.981 | 0.951 | 0.950 | 0.953 |
| RSDDS Type-II m = 5 | 0.986 | 0.970 | 0.995 | 0.988 | 0.890 | 0.980 |
| RSDDS Type-II m = 10 | 0.986 | 0.970 | 0.995 | 0.988 | 0.886 | 0.981 |

**Table 3.** Pixel-level ROC of our proposed FS-RSDD and other unsupervised anomaly detection models.

| Dataset | Model | | | | | |
|---|---|---|---|---|---|---|
| | **FS-RSDD** | **Padim** | **PatchCore** | **stfpm** | **cflow** | **fastclow** |
| RSDDS Type-I m = 5 | 0.987 | 0.976 | 0.974 | 0.980 | 0.970 | 0.953 |
| RSDDS Type-I m = 10 | 0.991 | 0.977 | 0.975 | 0.981 | 0.971 | 0.954 |
| RSDDS Type-II m = 5 | 0.961 | 0.920 | 0.919 | 0.948 | 0.852 | 0.919 |
| RSDDS Type-II m = 10 | 0.962 | 0.920 | 0.920 | 0.948 | 0.855 | 0.919 |

Combining the data from Tables 1 and 2, it can be observed that our proposed method outperforms other unsupervised industrial defect detection algorithms in terms of image-level classification ROC, except for PatchCore. However, it does not show significant advantage over other unsupervised algorithms in terms of image-level AP.

The reason behind this result lies in the fact that AP is more inclined towards the detection of positive instances, i.e., defect samples, while ROC is a relatively balanced evaluation metric. The better performance of our method in ROC compared to AP may be attributed to the fact that, while maintaining a high precision, our method has a lower false-positive rate for defect detection. However, it has a higher false-negative rate compared to some algorithms, while under the same conditions, some unsupervised defect detection algorithms have a lower false-negative rate but a higher false-positive rate.

We further analyzed the defects that were not successfully detected by our method. Figure 7 shows the heatmap of the undetected defect samples and the successfully classified normal samples by our method, under the given defect detection threshold.

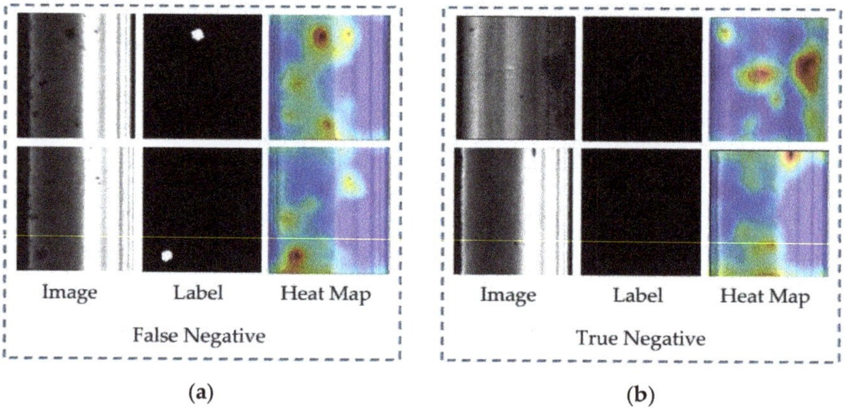


**Figure 7.** The heatmap visualization of false negatives and true negatives in the prediction results: (**a**) heatmap of false negatives, showing high defect scores in the actual defective regions; (**b**) heatmap of true negatives, showing high anomaly scores for noise or stains that are similar to defects.

By observing the heatmap of false-negative samples, we can visually see that the defective parts in the rail images are actually represented by darker colors. This means that our proposed method can accurately distinguish the defect foreground from the normal rail background. The reason why these defects were not detected can be further observed from the predicted results of true-negative samples. We can see that the reason for the lower AP in our method is that for those stains or noises that are difficult to distinguish from defects in the images, our method also considers them as potential defects. Although the probability of these noises belonging to defects may not be significantly higher than true defects, this ambiguous discrimination leads to our method's inability to provide clear judgments for some challenging cases. In other words, the trade-off of our method rarely misclassifying normal samples as defect samples is that some defect samples are also considered as normal samples. As a result, we have a higher ROC but a relatively lower AP.

Another thing we can observe from Tables 1 and 2 is that, regardless of the algorithm used, there is a significantly better performance on Type-II data compared to Type-I data. The reason behind this phenomenon is consistent with our previous analysis on the difference in performance between the two metrics, which is the presence of noise and interference in the images. As shown in Figure 8, it can be seen that, perhaps due to better image acquisition conditions, the Type-II rail images contain much less noise compared to Type-I data.

In Figure 8, the red curve indicates the defective area, while the green curve indicates the noise that is similar to the defect. It can be clearly seen that Type-I data contain much more noise that interferes with defect detection compared to Type-II data.

Furthermore, according to the data in Table 3, we can also observe that our method outperforms most unsupervised AD methods except Patchcore in terms of pixel-level ROC, indicating that our algorithm achieves more precise segmentation for the same defect.

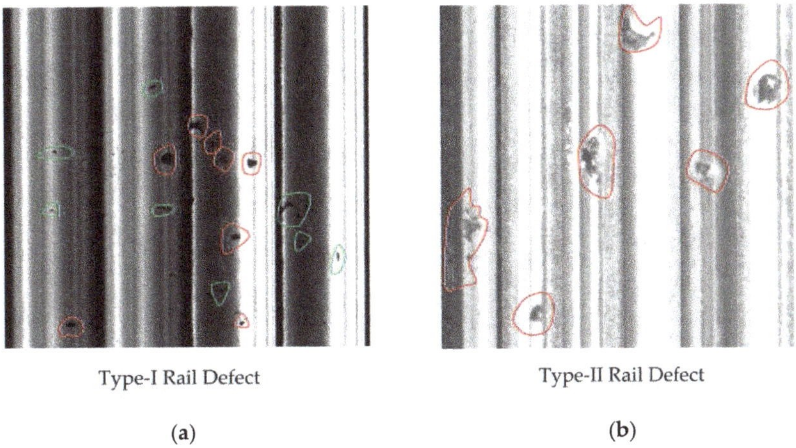


**Figure 8.** Type-I rail surface data and Type-II rail surface data. The true defects are circled in red, while the noise or stains similar to defects are circled in green. (**a**) Type-I data, where more noise and stains are visible; (**b**) for Type-II data, it is visually evident that there is not much noise interference.

In Figure 9, the segmentation results of different algorithms for the same defect sample are displayed. It is evident from this that our proposed algorithm exhibits more precise prediction and is less prone to generating false predictions on the background when compared to other algorithms.

**Figure 9.** Comparison of FS-RSDD with other unsupervised AD models in terms of prediction results.

### *4.4. Comparison with Few-Shot Supervised-Based AD Algorithms*

We also conducted comparative experiments with few-shot industrial defect detection algorithms. The experimental setting was similar to the unsupervised algorithm comparison experiment. We conducted 20 experiments, each time randomly selecting m defect samples for model training. The defect samples used for training in the comparative methods remained consistent with our proposed method. The average ROC and average AP were then calculated for performance comparison, as shown in Table 4. According to the results in the table, considering both the ROC and AP metrics, our method demonstrates advantages compared to DevNet [23] and DRA [22].

We not only compared the average performance but also recorded the performance of the model for each experiment in order to observe the impact of different training samples on the model's performance.

Figure 10 shows the changes in the model's ROC after training with randomly sampled defect data from Type-I and Type-II datasets, respectively.

**Table 4.** Comparison between FS-RSDD and other few-shot industrial defect detection models.

| | ROC | | | AP | | |
|---|---|---|---|---|---|---|
| Dataset | FS-RSDD | DevNet | DRA | FS-RSDD | DevNet | DRA |
| Type-I m = 5 | 0.941 | 0.905 | 0.888 | 0.943 | 0.967 | 0.958 |
| Type-I m = 10 | 0.952 | 0.930 | 0.927 | 0.953 | 0.976 | 0.973 |
| Type-II m = 5 | 0.989 | 0.858 | 0.955 | 0.986 | 0.901 | 0.963 |
| Type-II m = 10 | 0.991 | 0.911 | 0.974 | 0.986 | 0.939 | 0.978 |

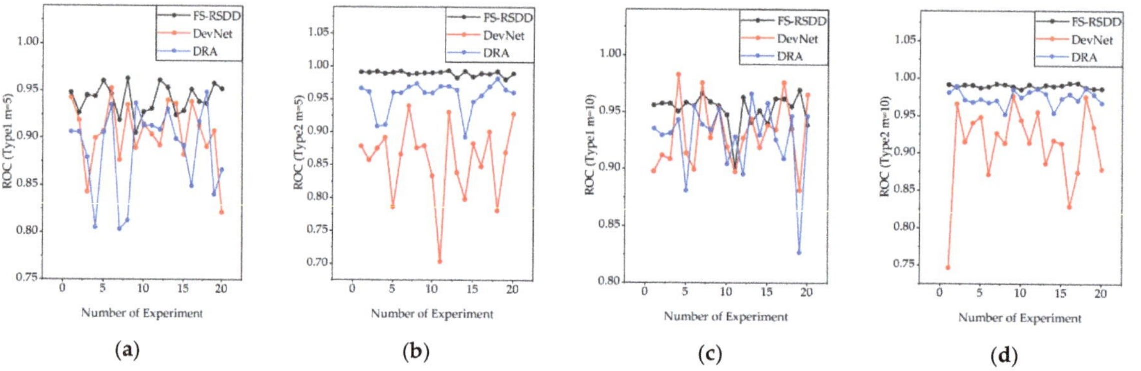


**Figure 10.** Performance fluctuations of FS-RSDD compared to the benchmarked few-shot supervised-based AD algorithms with different defect data for training: (**a**) Type-I, m = 5; (**b**) Type-II, m = 5; (**c**) Type-I, m = 10; (**d**) Type-II, m = 10.

From the observation of Figure 10, it can be noticed that FS-RSDD exhibits more stable and robust performance compared to other models when different defect data are used for training. Furthermore, although FS-RSDD shows more fluctuations on Type-I data compared to Type-II data, it shows better performance compared to the rest of the few-shot supervised-based AD models. This is mainly attributed to the fact that FS-RSDD not only utilizes defect features but also fully utilizes the features of a normal rail for defect detection. On the other hand, other algorithms tend to focus more on extracting information from defect samples, which can lead to lower accuracy when the quality of defect samples is poor.

### 4.5. *Ablative Studies*

In this section, we conducted ablation experiments to explore the impact of different settings on the performance of FS-RSDD. These experiments included comparative experiments on the model's performance using features from different semantic levels of the feature extractor, whether using Gaussian blur or not, and extracting features using different feature extractors. The experiments were conducted by extracting defect samples and training the model 20 times and then comparing the average performance of the model. The number of samples extracted was m = 10, and the defect samples used for training were consistent with those used in the experiments in Sections 4.1 and 4.2.

We first conducted comparative experiments on the performance of the FS-RSDD model using different feature extractors, both with and without Gaussian blur. Table 5 presents the performance of FS-RSDD on Type-I and Type-II rail surface defect datasets when using ResNet18, ResNet50, and WideResNet50 [47] as feature extractors.

From the experiment data in Table 5, we can clearly observe the significant impact of different feature extractors and the use of Gaussian blur on the detection performance of the model. From this, we can observe that when using ResNet18 as the feature extractor, the model has lower accuracy but faster speed. This is evident due to ResNet18 having fewer model parameters and faster inference speed but correspondingly poorer feature extraction capability. On the other hand, unlike ResNet18, WideResNet50, with its wider feature channels, can achieve better performance when used as a feature extractor, albeit with relatively slower detection speed.

**Table 5.** The impact of different feature extractors and the use of Gaussian blur on the performance of FS-RSDD.

| | Feature Extractor | Image-Level ROC | Image-Level AP | Pixel-Level ROC | FPS |
|---|---|---|---|---|---|
| Type-I m = 10 | ResNet18 | 0.896 | 0.889 | 0.964 | 130.890 |
| | ResNet50 | 0.906 | 0.901 | 0.976 | 70.403 |
| | WideResNet50 | 0.930 | 0.927 | 0.985 | 64.664 |
| | ResNet18 + Gaussian blur | 0.934 | 0.935 | 0.980 | 130.052 |
| | ResNet50 + Gaussian blur | 0.937 | 0.935 | 0.986 | 70.837 |
| | WideResNet50 + Gaussian blur | 0.952 | 0.953 | 0.991 | 63.519 |
| Type-II m = 10 | ResNet18 | 0.968 | 0.959 | 0.939 | 299.114 |
| | ResNet50 | 0.984 | 0.978 | 0.948 | 231.154 |
| | WideResNet50 | 0.985 | 0.979 | 0.952 | 204.306 |
| | ResNet18 + Gaussian blur | 0.986 | 0.983 | 0.955 | 265.375 |
| | ResNet50 + Gaussian blur | 0.990 | 0.986 | 0.959 | 215.554 |
| | WideResNet50 + Gaussian blur | 0.990 | 0.987 | 0.959 | 211.447 |

Additionally, by comparing the performance of each feature extractor with and without Gaussian blur, we can easily observe the extent of improvement that Gaussian blur brings to the model's performance. This demonstrates the enhancement of predictive performance through the fusion of pixel neighborhood features.

Comparative experiments were also conducted on the performance of the FS-RSDD model by utilizing features from different semantic levels to construct the memory bank, as presented in Table 6. In the experiment, the WideResNet50 is employed as the feature extractor, and Gaussian blur is applied to improve the performance. It can be clearly seen that the use of different combinations of semantic level features has an impact on the performance of FS-RSDD. When only using single- or two-level features for model construction, the performance of the model is suboptimal. However, when using multi-level features from a shallow layer, FS-RSDD exhibits the best performance. Furthermore, we conducted experiments with the utilization of features from deeper semantic levels. However, we observed no significant enhancement in the performance of FS-RSDD but a decrease in frames per second (FPS) due to the increased number of feature similarity calculations.

**Table 6.** The impact of constructing a feature memory bank using different hierarchical features on the performance of FS-RSDD.

| Dataset | Layer | Image-Level ROC | Pixel-Level ROC | Image-Level AP | FPS |
|---|---|---|---|---|---|
| Type-I m = 10 | Layer1 | 0.677 | 0.885 | 0.658 | 71.145 |
| | Layer1 + 2 | 0.920 | 0.976 | 0.911 | 63.552 |
| | Layer1 + 3 | 0.935 | 0.988 | 0.943 | 66.293 |
| | Layer2 + 3 | 0.954 | 0.991 | 0.956 | 75.039 |
| | Layer1 + 2 + 3 | 0.952 | 0.991 | 0.953 | 63.519 |
| | Layer1 + 2 + 3 + 4 | 0.949 | 0.988 | 0.948 | 61.557 |
| Type-II m = 10 | Layer1 | 0.846 | 0.925 | 0.868 | 235.983 |
| | Layer1 + 2 | 0.981 | 0.964 | 0.976 | 221.524 |
| | Layer1 + 3 | 0.990 | 0.957 | 0.986 | 222.655 |
| | Layer2 + 3 | 0.990 | 0.959 | 0.987 | 240.008 |
| | Layer1 + 2 + 3 | 0.991 | 0.962 | 0.987 | 211.447 |
| | Layer1 + 2 + 3 + 4 | 0.992 | 0.955 | 0.988 | 207.422 |

### 4.6. Time Complexity and the Size of Memory Bank

In this section, we conducted a comparative analysis of the computational complexity and size of memory banks among different models. The Type-I image data have a resolution of 160 × 160, while the Type-II data have a resolution of 64 × 64. All models employed the same network, WideResNet50, as the feature extractor. It is evident from Table 7 that FS-RSDD, benefiting from its compact feature memory bank that models the entire normal rail background, outperforms unsupervised anomaly detection and few-shot defect detection algorithms in terms of time complexity. Moreover, this advantage is more significant on the low-resolution Type-II dataset.

**Table 7.** The time complexity comparison between FS-RSDD and other benchmarked models.

| | FS-RSDD | Padim | Patchcore | stfpm | cflow | fastflow | devnet |
|---|---|---|---|---|---|---|---|
| Type-I m = 10 | 63.519 | 57.434 | 60.378 | 47.500 | 21.310 | 44.954 | 25.306 |
| Type-II m = 10 | 211.447 | 101.045 | 96.905 | 84.843 | 61.526 | 58.613 | 39.192 |

We also conducted experiments regarding the size of the memory bank. We extracted the memory bank of our model and other methods and compared them to demonstrate the compactness of the memory bank obtained by our approach. We contrasted our method with memory-bank-based approaches [30,32]. The results are as shown in Table 8. In the experiment, each model utilizes the WideResNet50 feature extractor, with an input image resolution of 160.

**Table 8.** Memory bank comparison: each element is a floating-point number.

| | FS-RSDD | Patchcore | Padim |
|---|---|---|---|
| Number of elements | 71,680 | 184,320 | 950,364,800 |
| File size | 282 KB | 721 KB | 3.54 GB |

## 5. Discussion

The method proposed in this article mainly combines the idea of feature-embedding-based industrial defect detection algorithms and the prototypical network. By embedding features of defects and normal rails, corresponding feature memory banks are obtained. FS-RSDD estimates the similarity of the input samples to the defect prototype and the normal prototype in the feature space for defect detection. This simple and direct method can achieve quite good results on the rail surface defect dataset using only a few samples. However, there are still some shortcomings. After studying the experimental results in Section 4.3, it can be concluded that although this method can effectively distinguish the rail background and defect foreground, it cannot effectively discriminate between defects and noise.

To explore the possibility of improving the model's detection performance using traditional image-processing techniques, we conducted additional experiments. We performed another experiment on both the RSDDS Type-I dataset without processing and the dataset processed with image processing. The experiment was conducted only once, and the random seed was fixed. Following the method described in reference [45], we used gamma transform to improve the uneven lighting in the images and combined it with Gaussian blur for image denoising. In the end, we achieved a pixel-level ROC of 99.3% and an image-level ROC of 96.5% on the original dataset, while on the denoised dataset we achieved a pixel-level ROC of 99.3% and an image-level ROC of 96.2%. It can be seen that after image processing, the model's performance did not improve as expected. We believe this may be due to the fact that deep-learning-based feature extractors have strong feature extraction capabilities, and the noise and uneven lighting that traditional image processing techniques can handle can also be distinguished by the feature extractor. However, noise and interference that are difficult for the feature extractor to distinguish are equally challenging for traditional image processing techniques. Therefore, instead of traditional image processing methods, we will focus on enhancing our work through novel image processing techniques in the future. Additionally, our research will prioritize exploring deep learning methods in defect detection.

## 6. Conclusions

This paper proposes a few-shot rail surface defect detection model, FS-RSDD, to address the issue of insufficient defect samples in the field of rail surface defect detection. FS-RSDD combines the idea of feature-embedding-based industrial defect detection algorithms with the prototypical network. The method utilizes a pre-trained convolutional neural network to embed features of both defective and normal samples. It then uses clustering algorithms to learn the distribution of features of normal samples. Finally, through the prototype learning approach, softmax is used to estimate the probability of a test sample's feature belonging to a defect in the feature space.

The proposed method surpasses all comparative algorithms in terms of speed by achieving a compact feature memory bank, which models the overall feature distribution of normal rail backgrounds. Additionally, the proposed method outperforms comparative few-shot defect detection algorithms in terms of accuracy on the RSDDS public dataset and is on par with the current state-of-the-art unsupervised anomaly detection algorithms.

**Author Contributions:** Conceptualization, Y.M. and Z.W. (Ziwei Wang); data curation, Z.W. (Ziwei Wang); methodology, Z.W. (Ziwei Wang), Y.L. and Z.W. (Zheng Wang); writing—original draft, Z.W. (Ziwei Wang) and Y.L.; writing—review and editing, Y.M. and Z.W. (Zheng Wang). All authors have read and agreed to the published version of the manuscript.

**Funding:** This research was funded by the National Natural Science Foundation of China (Grant No. 62066024) and the National Natural Science Foundation of China (Grant No.12162019).

**Institutional Review Board Statement:** Not applicable.

**Informed Consent Statement:** Not applicable.

**Data Availability Statement:** Not applicable.

**Conflicts of Interest:** The authors declare no conflict of interest.

## References

1. Li, Y.; Trinh, H.; Haas, N.; Otto, C.; Pankanti, S. Rail Component Detection, Optimization, and Assessment for Automatic Rail Track Inspection. *IEEE Trans. Intell. Transp. Syst.* **2014**, *15*, 760–770. [CrossRef]
2. Gao, B.; Bai, L.; Woo, W.L.; Tian, G.Y.; Cheng, Y. Automatic Defect Identification of Eddy Current Pulsed Thermography Using Single Channel Blind Source Separation. *IEEE Trans. Instrum. Meas.* **2014**, *63*, 913–922. [CrossRef]
3. Alvarenga, T.A.; Carvalho, A.L.; Honorio, L.M.; Cerqueira, A.S.; Filho, L.M.A.; Nobrega, R.A. Detection and Classification System for Rail Surface Defects Based on Eddy Current. *Sensors* **2021**, *21*, 7937. [CrossRef] [PubMed]
4. Wang, H.; Li, M.; Wan, Z. Rail Surface Defect Detection Based on Improved Mask R-CNN. *Comput. Electr. Eng.* **2022**, *102*, 108269. [CrossRef]
5. Hsieh, C.-C.; Hsu, T.-Y.; Huang, W.-H. An Online Rail Track Fastener Classification System Based on YOLO Models. *Sensors* **2022**, *22*, 9970. [CrossRef]
6. Luo, H.; Cai, L.; Li, C. Rail Surface Defect Detection Based on An Improved YOLOv5s. *Appl. Sci.* **2023**, *13*, 7330. [CrossRef]
7. Zhang, C.; Xu, D.; Zhang, L.; Deng, W. Rail Surface Defect Detection Based on Image Enhancement and Improved YOLOX. *Electronics* **2023**, *12*, 2672. [CrossRef]
8. Hu, J.; Qiao, P.; Lv, H.; Yang, L.; Ouyang, A.; He, Y.; Liu, Y. High Speed Railway Fastener Defect Detection by Using Improved YoLoX-Nano Model. *Sensors* **2022**, *22*, 8399. [CrossRef]
9. Gudovskiy, D.; Ishizaka, S.; Kozuka, K. CFLOW-AD: Real-Time Unsupervised Anomaly Detection with Localization via Conditional Normalizing Flows. In Proceedings of the IEEE/CVF Winter Conference on Applications of Computer Vision (WACV), Waikoloa, HI, USA, 3–8 January 2021.
10. Yu, J.; Zheng, Y.; Wang, X.; Li, W.; Wu, Y.; Zhao, R.; Wu, L. FastFlow: Unsupervised Anomaly Detection and Localization via 2D Normalizing Flows. *arXiv* **2021**, arXiv:2111.07677.

11. Yang, M.; Wu, P.; Liu, J.; Feng, H. MemSeg: A Semi-Supervised Method for Image Surface Defect Detection Using Differences and Commonalities. *Eng. Appl. Artif. Intell.* **2022**, *119*, 105835. [CrossRef]
12. Pang, G.; Shen, C.; Cao, L.; Hengel, A.V.D. Deep Learning for Anomaly Detection: A Review. *ACM Comput. Surv.* **2022**, *54*, 1–38. [CrossRef]
13. Yu, H.; Li, Q.; Tan, Y.; Gan, J.; Wang, J.; Geng, Y.; Jia, L. A Coarse-to-Fine Model for Rail Surface Defect Detection. *IEEE Trans. Instrum. Meas.* **2019**, *68*, 656–666. [CrossRef]
14. Ghorai, S.; Mukherjee, A.; Gangadaran, M.; Dutta, P.K. Automatic Defect Detection on Hot-Rolled Flat Steel Products. *IEEE Trans. Instrum. Meas.* **2013**, *62*, 612–621. [CrossRef]
15. Zhang, H.; Jin, X.; Wu, Q.M.J.; Wang, Y.; He, Z.; Yang, Y. Automatic Visual Detection System of Railway Surface Defects with Curvature Filter and Improved Gaussian Mixture Model. *IEEE Trans. Instrum. Meas.* **2018**, *67*, 1593–1608. [CrossRef]
16. Liu, Z.; Wang, W.; Zhang, X.; Jia, W. Inspection of Rail Surface Defects Based on Image Processing. In Proceedings of the 2010 2nd International Asia Conference on Informatics in Control, Automation and Robotics (CAR 2010), Wuhan, China, 3–7 March 2010; Volume 1, pp. 472–475.
17. Meng, S.; Kuang, S.; Ma, Z.; Wu, Y. MtlrNet: An Effective Deep Multitask Learning Architecture for Rail Crack Detection. *IEEE Trans. Instrum. Meas.* **2022**, *71*, 1–10. [CrossRef]
18. Zhang, H.; Song, Y.; Chen, Y.; Zhong, H.; Liu, L.; Wang, Y.; Akilan, T.; Wu, Q.M.J. MRSDI-CNN: Multi-Model Rail Surface Defect Inspection System Based on Convolutional Neural Networks. *IEEE Trans. Intell. Transp. Syst.* **2022**, *23*, 11162–11177. [CrossRef]
19. Zhang, D.; Song, K.; Wang, Q.; He, Y.; Wen, X.; Yan, Y. Two Deep Learning Networks for Rail Surface Defect Inspection of Limited Samples With Line-Level Label. *IEEE Trans. Ind. Inform.* **2021**, *17*, 6731–6741. [CrossRef]
20. Zhang, Q.; Wu, B.; Shao, Y.; Ye, Z. Surface Defect Detection of Rails Based on Convolutional Neural Network Multi-Scale-Cross FastFlow. In Proceedings of the 2022 5th International Conference on Pattern Recognition and Artificial Intelligence (PRAI), Chengdu, China, 19–21 August 2022; pp. 405–411.
21. Niu, M.; Song, K.; Huang, L.; Wang, Q.; Yan, Y.; Meng, Q. Unsupervised Saliency Detection of Rail Surface Defects Using Stereoscopic Images. *IEEE Trans. Ind. Inform.* **2021**, *17*, 2271–2281. [CrossRef]
22. Ding, C.; Pang, G.; Shen, C. Catching Both Gray and Black Swans: Open-Set Supervised Anomaly Detection. In Proceedings of the IEEE/CVF Conference on Computer Vision and Pattern Recognition (CVPR), New Orleans, LA, USA, 18–24 June 2022.
23. Pang, G.; Ding, C.; Shen, C.; Hengel, A.V.D. Explainable Deep Few-Shot Anomaly Detection with Deviation Networks. *arXiv* **2021**, arXiv:2108.00462.
24. Liu, B.; Gao, F.; Li, Y. Cost-Sensitive YOLOv5 for Detecting Surface Defects of Industrial Products. *Sensors* **2023**, *23*, 2610. [CrossRef]
25. Li, B.; Gao, Q. Defect Detection for Metal Shaft Surfaces Based on an Improved YOLOv5 Algorithm and Transfer Learning. *Sensors* **2023**, *23*, 3761. [CrossRef] [PubMed]
26. Ahmed, K.R. DSTEELNet: A Real-Time Parallel Dilated CNN with Atrous Spatial Pyramid Pooling for Detecting and Classifying Defects in Surface Steel Strips. *Sensors* **2023**, *23*, 544. [CrossRef]
27. Han, G.; Li, T.; Li, Q.; Zhao, F.; Zhang, M.; Wang, R.; Yuan, Q.; Liu, K.; Qin, L. Improved Algorithm for Insulator and Its Defect Detection Based on YOLOX. *Sensors* **2022**, *22*, 6186. [CrossRef]
28. Zheng, J.; Wu, H.; Zhang, H.; Wang, Z.; Xu, W. Insulator-Defect Detection Algorithm Based on Improved YOLOv7. *Sensors* **2022**, *22*, 8801. [CrossRef] [PubMed]
29. Kou, L.; Sysyn, M.; Fischer, S.; Liu, J.; Nabochenko, O. Optical Rail Surface Crack Detection Method Based on Semantic Segmentation Replacement for Magnetic Particle Inspection. *Sensors* **2022**, *22*, 8214. [CrossRef] [PubMed]
30. Defard, T.; Setkov, A.; Loesch, A.; Audigier, R. PaDiM: A Patch Distribution Modeling Framework for Anomaly Detection and Localization. In *Pattern Recognition. ICPR International Workshops and Challenges. ICPR 2021*; Lecture Notes in Computer Science; Springer: Cham, Switzerland, 2020.
31. Wang, G.; Han, S.; Ding, E.; Huang, D. Student-Teacher Feature Pyramid Matching for Anomaly Detection. *arXiv* **2021**, arXiv:2103.04257.
32. Roth, K.; Pemula, L.; Zepeda, J.; Scholkopf, B.; Brox, T.; Gehler, P. Towards Total Recall in Industrial Anomaly Detection. In Proceedings of the IEEE/CVF Conference on Computer Vision and Pattern Recognition (CVPR), New Orleans, LA, USA, 18–24 June 2022; IEEE: Piscataway, NJ, USA, 2022; pp. 14298–14308.
33. He, K.; Zhang, X.; Ren, S.; Sun, J. Deep Residual Learning for Image Recognition. In Proceedings of the IEEE Conference on Computer Vision and Pattern Recognition (CVPR), Las Vegas, NV, USA, 27–30 June 2016.
34. Li, C.-L.; Sohn, K.; Yoon, J.; Pfister, T. CutPaste: Self-Supervised Learning for Anomaly Detection and Localization. *arXiv* **2021**, arXiv:2104.04015.
35. Reiss, T.; Cohen, N.; Bergman, L.; Hoshen, Y. PANDA: Adapting Pretrained Features for Anomaly Detection and Segmentation. In Proceedings of the 2021 IEEE/CVF Conference on Computer Vision and Pattern Recognition (CVPR), Nashville, TN, USA, 20–25 June 2021; pp. 2805–2813.
36. Rudolph, M.; Wandt, B.; Rosenhahn, B. Same Same But DifferNet: Semi-Supervised Defect Detection with Normalizing Flows. In Proceedings of the 2021 IEEE Winter Conference on Applications of Computer Vision (WACV), Waikoloa, HI, USA, 3–8 January 2021; pp. 1906–1915.

37. Chen, Z.; Fu, Y.; Zhang, Y.; Jiang, Y.-G.; Xue, X.; Sigal, L. Multi-Level Semantic Feature Augmentation for One-Shot Learning. *IEEE Trans. Image Process.* **2019**, *28*, 4594–4605. [CrossRef]
38. Xu, C.; Liu, C.; Sun, X.; Yang, S.; Wang, Y.; Wang, C.; Fu, Y. PatchMix Augmentation to Identify Causal Features in Few-Shot Learning 2022. *arXiv* **2022**, arXiv:2211.16019.
39. Finn, C.; Abbeel, P.; Levine, S. Model-Agnostic Meta-Learning for Fast Adaptation of Deep Networks. In Proceedings of the 34th International Conference on Machine Learning, Sydney, Australia, PMLR. 17 July 2017; pp. 1126–1135.
40. Koch, G.R. Siamese Neural Networks for One-Shot Image Recognition. Master's Thesis, University of Toronto, Toronto, ON, Canada, 2015.
41. Vinyals, O.; Blundell, C.; Lillicrap, T.; Wierstra, D. Matching Networks for One Shot Learning. In *Advances in Neural Information Processing Systems*; Curran Associates, Inc.: New York, NY, USA, 2016; Volume 29.
42. Snell, J.; Swersky, K.; Zemel, R.S. Prototypical Networks for Few-Shot Learning 2017. *arXiv* **2017**, arXiv:1703.05175v2.
43. Yang, B.; Liu, C.; Li, B.; Jiao, J.; Ye, Q. Prototype Mixture Models for Few-Shot Semantic Segmentation. *arXiv* **2020**, arXiv:2008.03898.
44. Dong, N.; Xing, E.P. Few-Shot Semantic Segmentation with Prototype Learning. *BMVC* **2018**, *3*. Available online: http://bmvc2018.org/contents/papers/0255.pdf (accessed on 29 August 2023).
45. Gan, J.; Li, Q.; Wang, J.; Yu, H. A Hierarchical Extractor-Based Visual Rail Surface Inspection System. *IEEE Sens. J.* **2017**, *17*, 7935–7944. [CrossRef]
46. Akcay, S.; Ameln, D.; Vaidya, A.; Lakshmanan, B.; Ahuja, N.; Genc, U. Anomalib: A Deep Learning Library for Anomaly Detection. In Proceedings of the 2022 IEEE International Conference on Image Processing (ICIP), Bordeaux, France, 16–19 October 2022; pp. 1706–1710.
47. Zagoruyko, S.; Komodakis, N. Wide Residual Networks 2017. *arXiv* **2017**, arXiv:1605.07146v4.

*sensors*

MDPI

*Article*

# A Non-Intrusive Monitoring System on Train Pantographs for the Maintenance of Overhead Contact Lines

**Borja Rodríguez-Arana [1,2,*], Pablo Ciáurriz [1,2], Nere Gil-Negrete [2] and Unai Alvarado [1,2]**

1 Ceit-Basque Research and Technology Alliance (BRTA), Manuel Lardizabal 15, 20018 Donostia/San Sebastián, Spain
2 Universidad de Navarra, Tecnun, Manuel Lardizabal 13, 20018 Donostia/San Sebastián, Spain
* Correspondence: brodriguez@ceit.es or brodriguez@unav.es

**Abstract:** The condition monitoring of an overhead contact line (OCL) is investigated by developing an innovative monitoring system for a pantograph on an electrical multiple unit of a regional line. Kinematic and dynamic modelling of the pantograph is conducted to support the designed monitoring system. The modelling is proved through rigorous test-rig experiments, while the proposed methodology is then validated through extensive field tests. The field tests serve a dual purpose: First, to validate the monitoring system using benchmark measurements of the tCat® trolley, and second, to assess the reproducibility of measurements in a realistic case. This paper presents the OCL monitoring system developed in the framework of the H2020 project SIA. The accuracy of our results is not far from that of other commercial systems, with just 12 mm of absolute error in the height measurement. Therefore, they provide reliable information about trends in various key performance indicators (KPIs) that facilitates the early detection of failures and the diagnosis of anomalies. The results highlight the importance of model calibration and validation in enabling novel health monitoring capabilities for the pantograph. By continuously monitoring the parameters and tracking their degradation trends, our approach allows for optimized scheduling of maintenance tasks for the OCL.

**Keywords:** pantograph–catenary interaction; infrastructure monitoring; railway; experimental results

**Citation:** Rodríguez-Arana, B.; Ciáurriz, P.; Gil-Negrete, N.; Alvarado, U. A Non-Intrusive Monitoring System on Train Pantographs for the Maintenance of Overhead Contact Lines. *Sensors* **2023**, *23*, 7890. https://doi.org/10.3390/s23187890

Academic Editors: Abdollah Malekjafarian, Diogo Ribeiro, Araliya Mosleh and Maria D. Martínez-Rodrigo

Received: 11 July 2023
Revised: 10 September 2023
Accepted: 11 September 2023
Published: 14 September 2023



## 1. Introduction

The overhead contact line (OCL) or catenary system is a key element of the railway infrastructure that provides the supply of energy to the electrical vehicles running through a line. The electric current is transferred from the contact wire (CW) to the vehicle through pantographs installed on the roof. The pantograph and OCL interaction depends on their mechanical characteristics [1] and contributes to the degradation of both systems with regard to their design configurations. Monitoring such interactions allows a good characterization of the health status of the OCL. Furthermore, maintenance operations are scheduled periodically to recover the original state of the OCL based on the data gathered from periodic (however scarce) inspections. These data are normally provided by expensive equipment installed in dedicated vehicles and processed by expert data analysts. However, this maintenance approach increases the life cycle cost (LCC) of the railway infrastructure. A preventive opportunistic maintenance strategy can reduce the number of maintenance schedules [2]; the maintenance of an OCL could, therefore, be optimised by monitoring some of their characteristics, such as the height and stagger of the CW (Figure 1), from sensors placed on pantographs and reducing the LCC. In addition to maintenance, the prediction of probabilistic risks is also being studied [3].

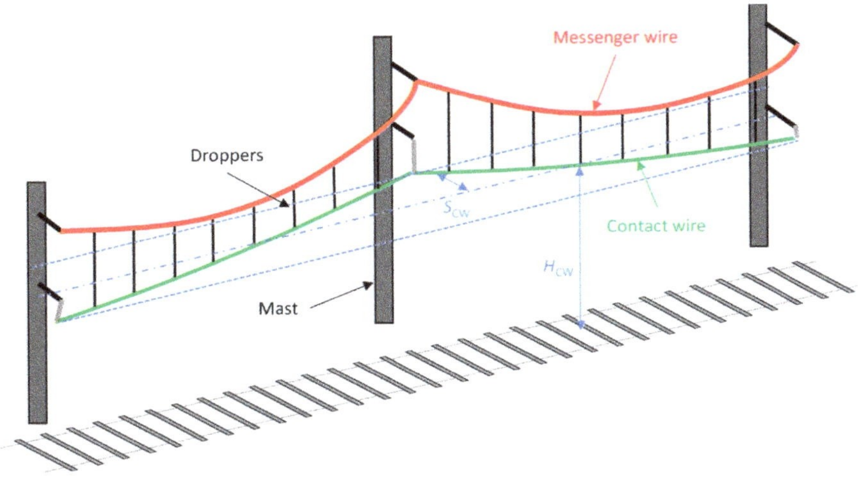


**Figure 1.** Schematic representation of the overhead contact line (OCL).

The geometrical study of the pantograph frame is described by Benet et al. [4], considering a three-dimensional model. Generally, the static force at the head of a pantograph remains constant regardless of the working height. This can be achieved through the accurate design of the geometry and the mechanism to lift the pantograph [5]. Not only the static lift force but also the tension of the CW have a significant influence on the dynamic performance of the pantograph and catenary system [6]. The deviation of the dropper length and the mast height from their nominal values is primarily responsible for tension changes [7]. Vesali et al. proposed a fast and accurate method for determining the static equilibrium configuration of a catenary [8]. Regarding the dynamic behaviour, wind loads and the irregularities of the CW also have a significant influence [9,10]. The dynamic performance due to local dropper defects has also been studied [11]. Most of the models employed are gathered together in the benchmark presented by Bruni et al. [12]. In these models, pantographs are usually modelled by two or three lumped masses, whose parameters can be obtained through experimental testing [13]. Accurate pantograph–catenary models provide a suitable approach for different investigations, such as assessing the current collection quality [14] or monitoring the stagger [15]. In addition to models, test facilities are developed to study the interaction between pantographs and OCL systems [16].

Inspection and monitoring are closely related concepts associated with railway infrastructure maintenance. Inspection entails systematic examination of the condition of different railway components, such as tracks, OCL, bridges, tunnels, and other structures, utilising sensors, cameras, and similar technologies. Typically conducted at regular intervals, such as annually or bi-annually, inspections aim to identify potential issues and defects that need to be repaired. On the other hand, monitoring involves the continual or regular gathering of data on the condition of railway components. These data are subsequently analysed to detect trends and patterns that may imply potential problems or concerns. The primary objective of monitoring is to enable early detection of potential issues and facilitate timely intervention before they escalate into major problems. Both inspection and monitoring serve as crucial instruments for ensuring the safety and reliability of the railway network. While inspections provide a detailed snapshot of the condition of railway components, monitoring allows a more continuous and real-time evaluation of their state.

The deployment and installation of infrastructure monitoring systems can vary, leading to different approaches [17]. Wayside monitoring and on-board monitoring are two distinct methods employed in assessing railway infrastructure. Wayside monitoring involves the utilization of sensors and stationary technologies positioned alongside the railway tracks, such as on the ground, on bridges, or within tunnels. These technologies gather data on the condition of vehicles and some network assets. Conversely, on-board monitoring entails the incorporation of sensors and embedded technologies directly on trains. These technologies collect data on the condition of the trains, their components, and the railway infrastructure's assets as the train traverses over them. On-board monitoring facilitates the real-time collection of data on the railway network's condition, enabling the identification of potential issues and defects in linear infrastructure, such as the OCL. In this case, different technologies can be used to assess the health status of the railway OCL. These technologies enable the detection of deformations such as sagging, twisting, or other anomalies that could impact the performance of the OCL system. Some examples of these technologies include laser scanners, which use lasers to precisely measure the position and alignment of the CW [18]; infrared cameras, capable of detecting temperature variations and other anomalies within the OCL system [19], aiding in identifying potential issues; other vision-based technologies for anomaly detection and failure diagnosis in the catenary equipment [20]; accelerometers, which measure the vibration of the wires and other related structures, providing valuable insights into their dynamic behaviour [21]; and distributed acoustic sensing (DAS) systems, which use fibre optic cables along the lines or in the pantograph to detect defects along the OCL system and its components [22].

The above-mentioned monitoring systems for OCL can pose challenges in terms of cost, installation, and maintenance, placing a significant burden on railway operators. To tackle this issue, there is an increasing demand for low-cost monitoring systems that offer ease of installation and maintenance, without the need for specialized tools or expertise. These systems should also have the capability to gather detailed and precise data regarding the condition of the OCL system. This enables early detection of potential issues and failures, facilitating timely repairs and maintenance interventions. By addressing the cost and complexity associated with monitoring, these low-cost systems aim to alleviate the burden on railway operators and enhance the overall efficiency of OCL maintenance.

Therefore, this work aims to develop a non-intrusive and low-cost monitoring system for an OCL that leverages the modelling of the pantograph. The system uses affordable MEMS (micro-electro-mechanical systems) sensors in the range of a few dollars, significantly reducing costs (up to 2–3 orders of magnitude) compared to traditional railway-specific and optic-based systems. While our system does not cover all the parameters covered by the EN50317 standard [23], it offers a crucial set of key performance indicators (KPIs). These KPIs include geometrical parameters (CW height and stagger) and an estimation of the static contact force, ensuring a comprehensive evaluation of the OCL condition. Initially, the modelling of a pantograph is implemented in MATLAB to consider the kinematics and static loads within the working range of the panhead height. Subsequently, experimental tests are conducted on a test rig to evaluate the actual characteristics of the pantograph. Based on the findings from these prior tasks, the design of a non-intrusive monitoring system is developed. Finally, the effectiveness of the developed system is validated through field tests, which yield results comparable to those obtained through conventional measuring methods.

## 2. System Modelling

In the current research, some models for simulating the mechanical behaviour of a pantograph are proposed, which enable the simulation of various pantograph details while accommodating changes in the characteristics of its components. The models have been successfully implemented using MATLAB and provide interesting results regarding contact height, static contact force and stagger of the CW. The proposed models are valid for any pantograph, but note that the torque calculation employed for the static contact force model is only for this specific design.

The model adopted for the kinematical assessment considers the structure of the pantograph to be a one-degree-of-freedom articulated quadrilateral mechanism. The symbolic solver of MATLAB allows calculating the different positions of the assembled structure, which comprises three primary parts: the push bar ($a$), lower arm ($c$) and connections on the upper arm ($b$), given an inclination value of the lower arm according to

$$a \cdot \cos(\alpha - \varphi) + b \cdot \cos(\beta - \varphi) - c \cdot \cos(\gamma - \varphi) - d = 0, \quad (1)$$

$$a \cdot \sin(\alpha - \varphi) + b \cdot \sin(\beta - \varphi) - c \cdot \sin(\gamma - \varphi) = 0, \quad (2)$$

where:

- $a$, $b$, $c$ are the length of the three bars of the articulated quadrilateral mechanism.
- $\alpha$, $\beta$, $\gamma$ are the inclination against the horizontal of $a$, $b$ and $c$ bars.
- $d$ is the distance between fastening points.
- $\varphi$ corresponds to the angle of fastening points against horizontal.

The solution obtained from the articulated quadrilateral mechanism defines the entire upper arm part of the pantograph. Figure 2 shows the downward movement of the pantograph head, where the head trajectory is defined by the dimensional characteristics of the structure. The parameters used in the articulated quadrangle are summarized in Table 1. The upper arm bar connected to $b$ has a length of 2.055 m and a relative inclination of 35.3 degrees.

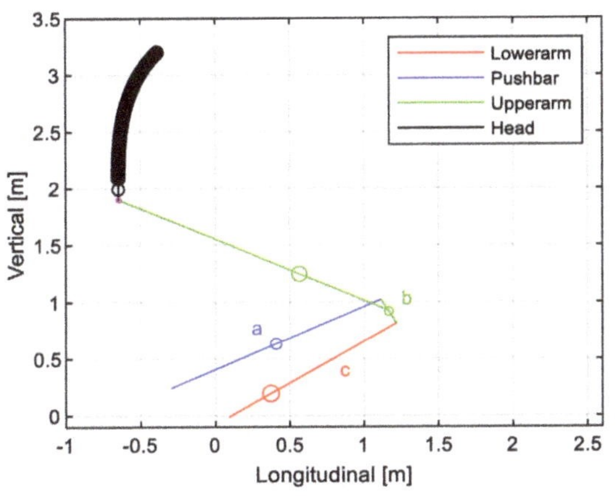


**Figure 2.** Simplified representation of a pantograph model and movement of the pantograph head.

**Table 1.** Parameters of the articulated quadrangle.

| Parameter | Value | Units |
|---|---|---|
| $a$ | 1.615 | m |
| $b$ | 0.235 | m |
| $c$ | 1.400 | m |
| $d$ | 0.4643 | m |
| $\varphi$ | 0.5814 | rad |

The uplift force of the pantograph originates from a torque applied to the lower arm. When the strips make contact with the CW, the pantograph reaches a steady state. This steady state can be evaluated for each position in a 2D analysis. The maximum vertical slope of a CW should be 2‰. Therefore, at the maximum speed of a regional EMU train (around 25 m/s), the vertical speed of the panhead could reach less than 0.1 m/s and dynamic effects on pantograph arms can be avoided. The supports for both the lower arm and the push bar are constrained in the vertical and longitudinal directions. As a result, the structure of the pantograph becomes hyperstatic, and a breakdown of its components is required to solve the problem. Figure 3 illustrates the unassembled structure, which facilitates the definition of a system comprising nine equations and nine unknowns, as follows:

$$\begin{pmatrix} 1 & 0 & 1 & 0 & 0 & 0 & 0 & 0 & 0 \\ 0 & 1 & 0 & 1 & 0 & 0 & 0 & 0 & 0 \\ a_{31} & a_{32} & 0 & 0 & 0 & 0 & 0 & 0 & 0 \\ 0 & 0 & 0 & 0 & 1 & 0 & 1 & 0 & 0 \\ 0 & 0 & 0 & 0 & 0 & 1 & 0 & 1 & 0 \\ 0 & 0 & 0 & 0 & a_{65} & a_{66} & 0 & 0 & 0 \\ 0 & 0 & -1 & 0 & 0 & 0 & -1 & 0 & 0 \\ 0 & 0 & 0 & -1 & 0 & 0 & 0 & -1 & -1 \\ 0 & 0 & 0 & 0 & 0 & 0 & a_{97} & a_{98} & a_{99} \end{pmatrix} \begin{Bmatrix} H_o \\ V_o \\ H_d \\ V_d \\ H_a \\ V_a \\ H_b \\ V_b \\ F \end{Bmatrix} = \left\{\overline{b_F}\right\}, \quad (3)$$

$$\begin{pmatrix} 1 & 0 & 1 & 0 & 0 & 0 & 0 & 0 & 0 \\ 0 & 1 & 0 & 1 & 0 & 0 & 0 & 0 & 0 \\ a_{31} & a_{32} & 0 & 0 & 0 & 0 & 0 & 0 & a_{39} \\ 0 & 0 & 0 & 0 & 1 & 0 & 1 & 0 & 0 \\ 0 & 0 & 0 & 0 & 0 & 1 & 0 & 1 & 0 \\ 0 & 0 & 0 & 0 & a_{65} & a_{66} & 0 & 0 & a_{69} \\ 0 & 0 & -1 & 0 & 0 & 0 & -1 & 0 & 0 \\ 0 & 0 & 0 & -1 & 0 & 0 & 0 & -1 & 0 \\ 0 & 0 & 0 & 0 & 0 & 0 & a_{97} & a_{98} & 0 \end{pmatrix} \begin{Bmatrix} H_o \\ V_o \\ H_d \\ V_d \\ H_a \\ V_a \\ H_b \\ V_b \\ T \end{Bmatrix} = \left\{\overline{b_T}\right\}. \quad (4)$$

where generally $a_{ij}$ terms depend on the geometrical design (distance units) of the structure and must be properly defined to obtain an equilibrium of forces and moments. The $a_{39}$ and $a_{69}$ non-dimensional terms, which are multiplied by a torque $T$, take 1 or 0 value as a function of where the torque is applied (joint a or o). The vectors $\overline{b_F}$ and $\overline{b_T}$ refer to the terms that are necessary to solve the contact force $F_c$ or applied torque $T$ respectively. These terms depend not only on the geometrical design but also on the mass of each bar. The terms $H_{o,d,a,b}$ and $V_{o,d,a,b}$ solve the horizontal and vertical reactions at the ends of the bars, while neglecting any bending effects. Hence, the terms of $\overline{b_F}$ and $\overline{b_T}$ vectors meet the equilibrium of force in the horizontal direction (mainly zero), the vertical direction (mainly mass of the bar multiplied by gravity, with force units) and torque (depending on mass, forces on the ends and their distance from the assessed point), respectively, for each bar of the structure.

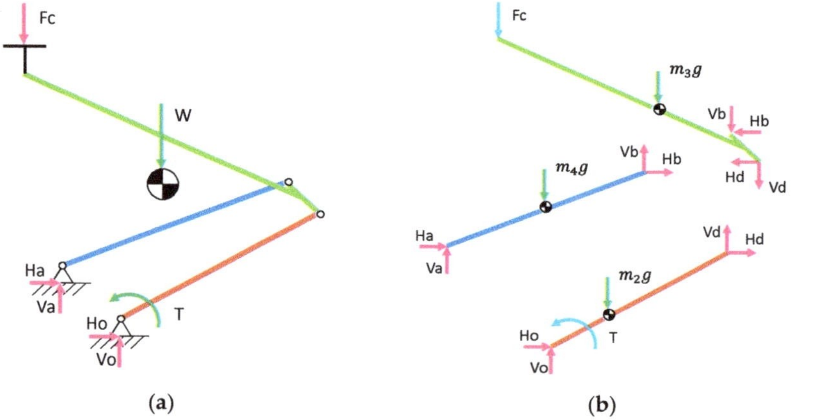


**Figure 3.** Simplified representation of a pantograph model. (**a**) Assembled and (**b**) breakdown structure.

As viscous and dry friction phenomena in articulated and prismatic joints are neglected in the model for the estimation of the static contact force, the calculated contact force for a known height should be given along with a confidence interval that can be obtained with previous tests.

The torque application mechanism on the lower arm can vary depending on the type of pantograph. In the case of the monitoring system here presented, the pantograph (of a regional train) is equipped with a passive torque actuator comprising two linear springs. However, the relationship between the applied torque and the inclination angle of the lower arm is non-linear. To achieve this non-linear torque characteristic, the pantograph incorporates a mechanical system that controls the distance between the spring and the rotation centre of the lower arm. This control is accomplished using two screws, which determine the position of a one-degree-of-freedom mechanism consisting of a helical spring and three bars.

The helical spring is elongated from its nominal length within the working range of the pantograph structure, exerting a force $F_s$. The applied torque $T$, opposing the rotation of the lower arm (denoted by $\gamma$), can be calculated using the stiffness of the linear spring $K_s$, its elongation $\delta_s$, and the distance to the rotation centre of the lower arm $d_s$, instead of a non-linear rotational stiffness $K_\gamma$ and the accumulated rotation, according to

$$T(\gamma) = K_\gamma \cdot \gamma = F_s \cdot d_s = K_s \cdot \delta_s \cdot d_s. \quad (5)$$

The mechanical system that introduces a non-linear torque to the pantograph structure has been modelled using MATLAB. This mechanism consists primarily of four bars, one of which (associated with helical springs) can change its length and is joined at one end to the baseplate. Another bar is fixed to the lower arm and the screws that control the mechanism, ensuring that their relative rotation with respect to the rotation centre is the same. The remaining two bars have circular-shaped ends, and their relative movement with the preceding bar becomes negligible once contact with the corresponding screw is established. Considering the lowering of the pantograph structure, before contact occurs, the bars with circular-shaped ends and the changeable bar are aligned. Figure 4a shows the representation of the computational model, where the elongation $\delta_s$ and distance $d_s$ are calculated for each inclination of the lower arm. Figure 4b provides a schematic representation of this model, overlaid on the actual mechanism.

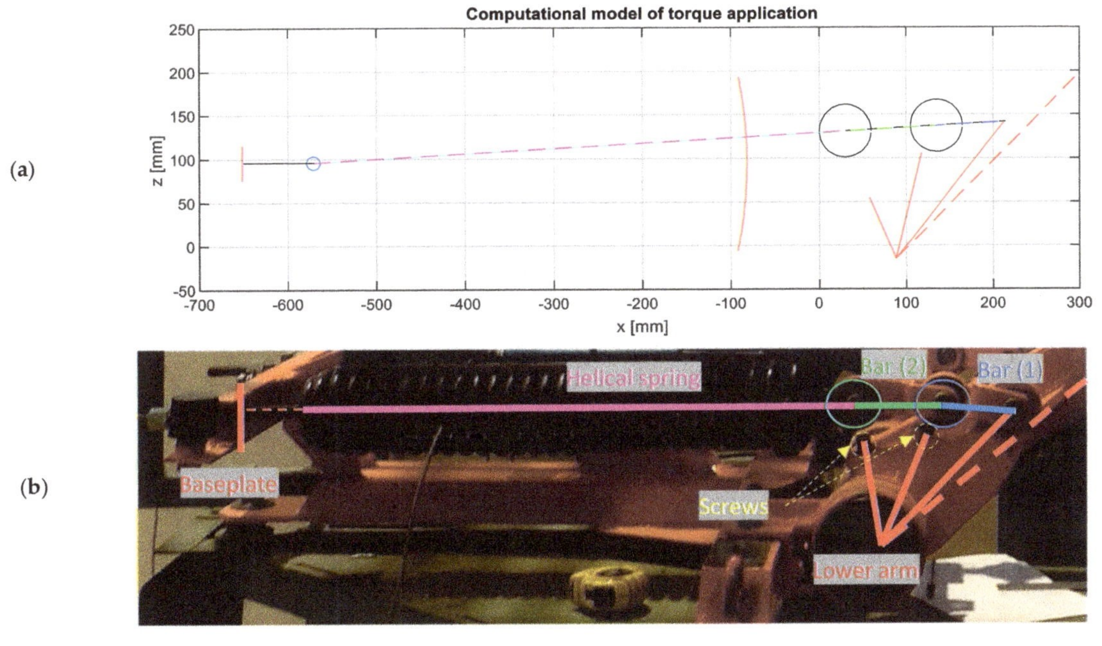


Figure 4. (a) Computational modelling of torque application and (b) schematic representation above the real mechanism.

Once the kinematical assessment of the mechanism of torque application is completed, the applied torque, as a function of the inclination of the lower arm, $T(\gamma)$, is found to be proportional to the stiffness of the helical spring. Using Equations (3) and (4), the static contact force $F_c$ on the top of the pantograph structure can be determined. Thus, the geometrical characteristics, the mass of the pantograph structure, the spring stiffness, and the distance of screws are sufficient to calculate the static contact force.

The system modelling approach employed in this study for the stagger assessment is based on the signal processing techniques used in the OCL model developed by Blanco et al. [15]. This model has a lumped mass representation of the pantograph, allowing for its interaction with a 2D OCL model. The model calculates panhead accelerations that, after SAWP signal processing, enable the monitoring of stagger in OCL systems. The calculation of the static force applied at the lowest mass of the lumped mass model, aimed at maintaining the contact force close to its nominal value, follows the guidelines outlined in the standard EN-50367 [24]. To ensure accurate results from this model, the lumped mass model should be representative of the pantograph installed on the train roof. For that purpose, instead of relying on parameter estimation from real tests, a linear system analysis (LSA) utilising multibody systems (MBS) can be employed [25].

In the system proposed here, the accelerations are directly obtained from sensors on the panhead, and only SAWP signal processing is employed without the need to use the dynamic model in [15].

## 3. Test Rig Experiments

To validate the developed model, test-rig experiments were conducted using an actual pantograph. The tests were carried out at the facilities of the Spanish train operating company FGC (Ferrocarrils de la Generalitat de Catalunya) on a working table designed for the calibration of pantographs. During these experiments, multiple acceleration and displacement sensors were strategically placed on the pantograph. An inclinometer was mounted on the lower arm to serve as a reference for the position of the structure. Furthermore, a load cell was utilised to measure the force at the panhead for different height positions. The configuration of the sensors on the pantograph is depicted in Figure 5.

The panhead of the pantograph was positioned at different heights within the operational range, from 0.5 to 2.5 metres. For each position, three measurements were considered:

- Shaft distance from baseplate
- Rotation angle of the lower arm from the horizontal
- Deformation of the helical spring relative to a previous reference.

Figure 6 presents a comparison between the measured shaft height and the inclination of the lower arm, as well as the kinematic assessment of the pantograph structure. The obtained results confirm the accuracy of the dimensions of the bars used in the system model and their links to the base frame. The computational assessment of the torque application mechanism provides the deformation of the helical springs and their distance to the articulation point between the lower arm and base frame. Figure 7a shows good agreement with the measured relative displacements. Figure 7b provides the distance obtained by simulation, which could not be directly measured but aligns well with the expected behaviour.

The deformation of helical springs and their distance to the articulated point enables the calculation of the non-linear torque provided by each spring. The total applied torque is obtained by summing the contributions from both mechanisms on each side of the pantograph, as depicted in Figure 8a. To validate the torque obtained from the kinematic assessment, a load cell is used to measure the force at different heights of the pantograph structure. By solving the system of nine equations and nine unknowns described in Section 2, it becomes possible to calculate the required torque for a known contact force at the panhead. Figure 8b illustrates that the torque calculated based on the kinematical assessment exhibits less than a 5% error compared to the torque derived from the measured force.

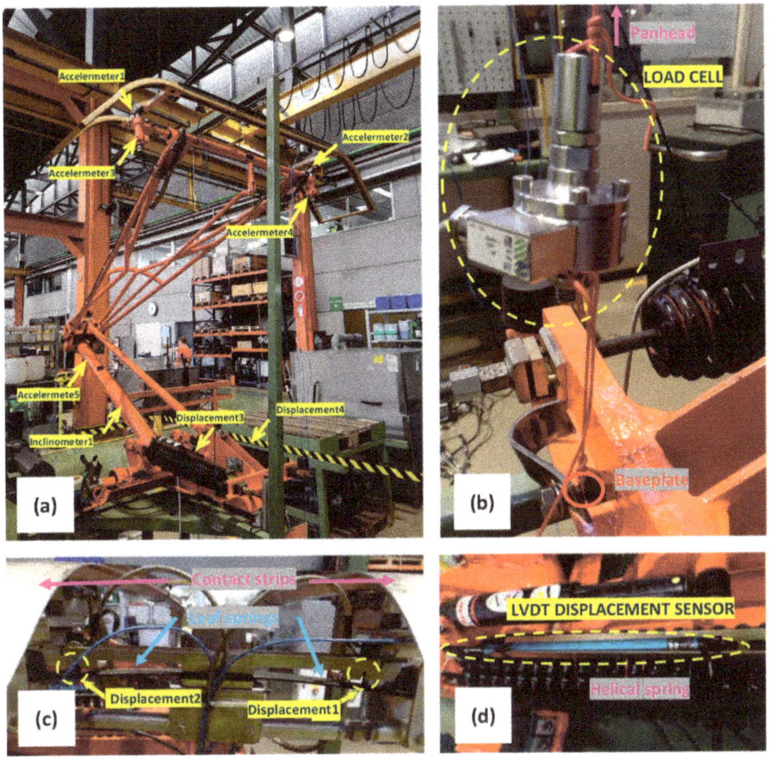


**Figure 5.** Pantograph instrumentation for test-rig experiments: (**a**) overall placement of accelerometers and displacement sensors; (**b**) load cell for the steady-state force at panhead; (**c**) displacement sensors on panhead leaf-springs and (**d**) displacement sensor for helical springs of structure.

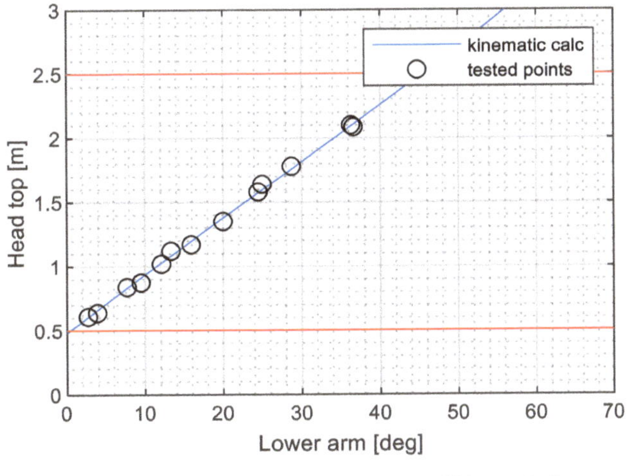


**Figure 6.** Comparison between measurements and kinematical assessment of the structure.

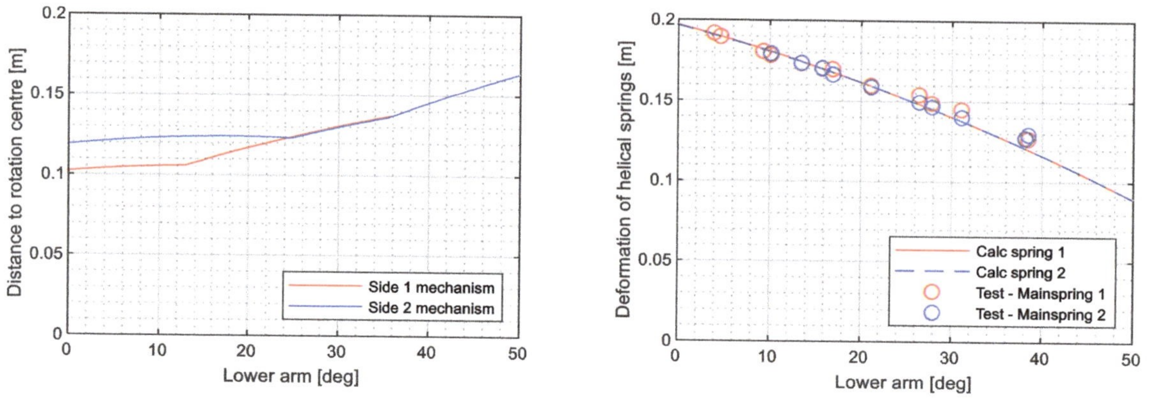


(a) (b)

**Figure 7.** Computational assessment of the torque mechanism by (**a**) distance of springs to rotation centre and (**b**) their deformation comparing with test results.

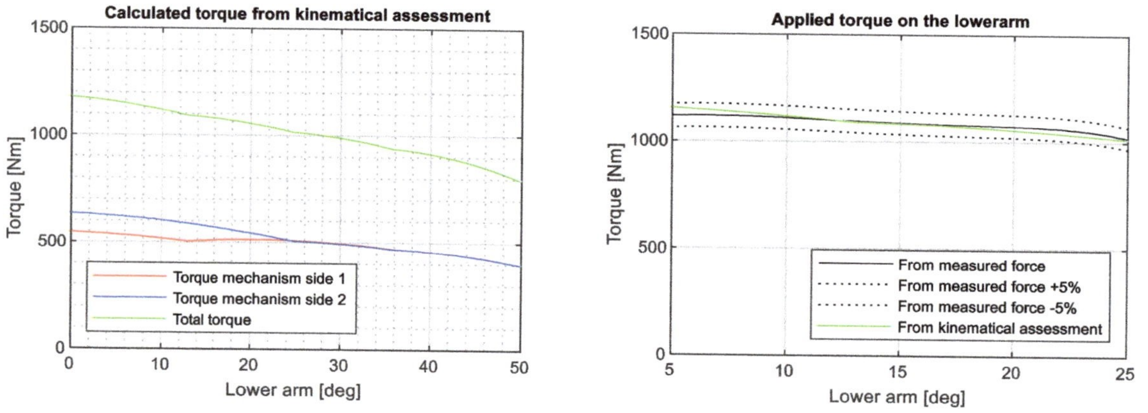


(a) (b)

**Figure 8.** (**a**) Calculation of applied torque from kinematical assessment and (**b**) comparison with calculated torque from the measured force at shaft.

Additionally, to validate the model, some extra tests were conducted in order to characterise the dynamic behaviour of the contact strips of the pantograph. The modal shapes of these elements were determined using a dynamometric hammer equipped with a dynamic force sensor (DYTRAN 1051V4). To perform this analysis, seven accelerometers were strategically positioned on one of the contact strips of the pantograph's head, as depicted in Figure 9. The structure of the pantograph was set in the lowest position and fixed to neglect additional movements. During the test, the strip was impacted vertically at the central position using the dynamometric hammer. A total of 5 impacts were recorded using a trigger for applied load and a time window of 5 s with a sampling rate of 800 Hz.

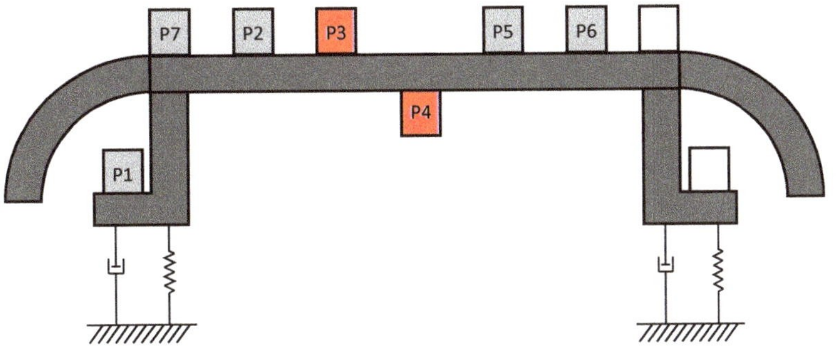


**Figure 9.** Sketch of accelerometers placed on the tested contact strip.

Afterwards, the recorded acceleration signals were post-processed using fast Fourier transforms (FFTs). Each signal is subjected to FFT analysis using a 3-s time window per impact and is normalised to its maximum value. It was necessary to discard the signals from channels corresponding to points P3 and P4 (coloured in red in Figure 9) due to acquisition errors. Figure 10 illustrates the resulting FFT for each point, representing the average of all impacts. Based on the identified peaks, the first and second bending modes are found to occur at frequencies of 57.78 Hz and 133.49 Hz, respectively. Additionally, the vertical mode of the rigid body can be observed at lower frequencies. It arises from the mass of the contact strip and the stiffness associated with the connection to the shaft of the pantograph structure.

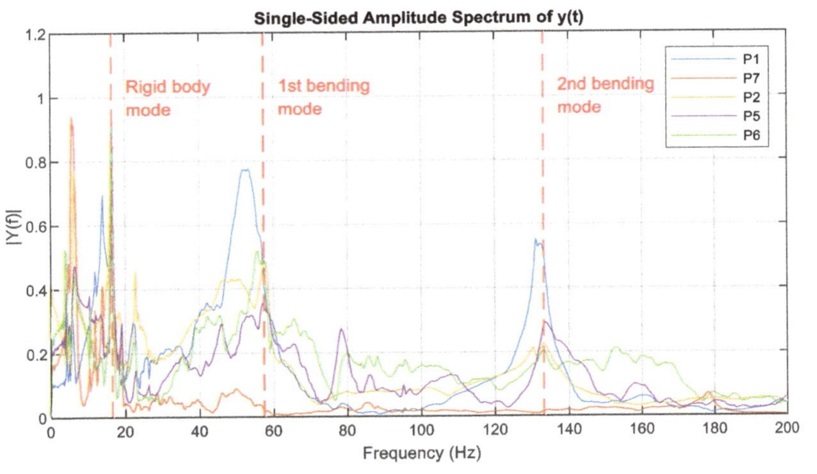


**Figure 10.** Averaged FFTs of measured points from five impacts with a time window of three seconds.

Table 2 shows the frequency obtained experimentally for each vibration mode. These values are similar to those in the literature [26]. The results obtained could be used for the implementation of the structural modes of both lead and rear strips on an OCL model.

**Table 2.** Experimental vertical vibration modes of a pantograph contact strip.

| Vibration Mode | Frequency (Hz) |
| --- | --- |
| Rigid body | 16.79 |
| 1st bending mode | 57.78 |
| 2nd bending mode | 133.49 |

## 4. Monitoring System

Based on conclusions extracted from simulations, a low-cost and non-invasive monitoring system has been developed, focusing on the dynamic interaction of the pantograph and OCL (in operational service trains). The developed system aims to monitor the necessary information to generate different KPIs, i.e., CW height, the estimation of the contact force and the CW stagger. The design of the system adheres to two primary constraints. Firstly, the system's hardware costs should be minimized to ensure affordability; a low-cost system is sought. Secondly, the installation process should be designed to be easy and non-invasive. These characteristics are crucial in enabling the monitoring system's suitability for use in service vehicles during commercial operations. Nevertheless, as a counterpart, the quality and accuracy of the data obtained may be lower compared to more sophisticated systems that require complex installation (e.g., disassembling the pantograph's head to accommodate force sensors to measure the contact force). The advantage of simpler and low-cost systems lies in their potential for installation in all trains operating within a network, which facilitates the generation of more frequent and pervasive data and offers additional value through further data analytics.

Based on that rationale, the developed monitoring system comprises several modules, namely sensors, positioning, processing and storage, communications, and power management. Each module contributes to the overall functionality of the system, ensuring efficient data collection, analysis, and transmission, while optimising power usage.

The analysis described in the preceding sections has allowed the identification of the most suitable sensors for capturing the required signals, which will be further processed to generate the desired KPIs. The system includes two piezoresistive accelerometers (with a range of $\pm 3$ g and a bandwidth spanning from 0 to 500 Hz), mounted at both ends of the pantograph's head. Furthermore, a dual-axis inclinometer/accelerometer ($\pm 90°$ and $\pm 1.7$ g) is placed at the lower arm (Figure 11).

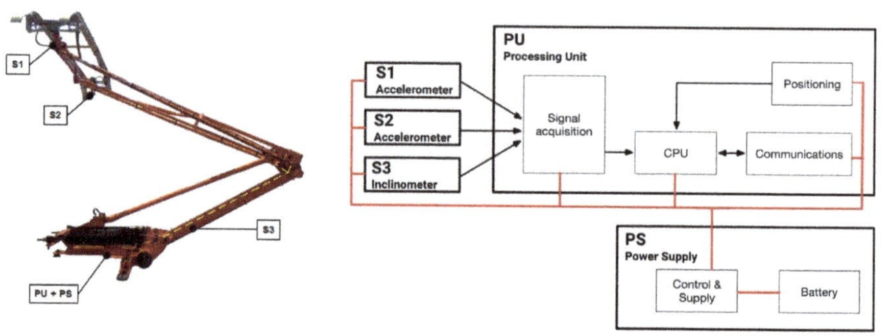


**Figure 11.** Architecture of the monitoring system and location of the different components in the pantograph.

Additionally, the monitoring system is equipped with a positioning module, which is necessary to generate both time and position stamps that are associated with the captured data. This module is based on a GNSS receiver that combines multi-constellation features (GPS and GALILEO) to enhance accuracy and availability. Providing an accurate (and available) position is key to geo-positioning the monitoring data of linear assets, enabling repeatability in the collected data. There are two key benefits to this approach. Firstly, it enables the precise localisation of defects along the railway line, facilitating the work of the maintenance staff in identifying and addressing issues. Secondly, when multiple trains provide large datasets that are accurately geo-positioned on the same line over an extended period (e.g., months or even years), it allows an advanced analysis to gain a better understanding of the health condition of the assets. This includes observing the evolution of various parameters over time at specific locations and contributing to comprehensive asset management and maintenance strategies.

The signals captured by the sensors are synchronised (time and position stamped) and processed onboard. The first step of the processing involves analysing the panhead accelerometries in order to detect impacts and shocks. When one of these events is detected, a real-time message is sent to a back-office application using an LTE wireless link. This message includes the severity and geo-localisation of the event, as well as additional contextual data such as train and service information. To facilitate the communication process, a publisher/subscriber system is implemented using an MQTT (message queuing telemetry transport) broker. The broker is subscribed to the events generated by the onboard monitoring system. Simultaneously, the raw data are processed to extract the target KPIs. The relevant information is then stored in binary files. These files are transferred via FTP (file transfer protocol). This transfer is carried out through the Wi-Fi installed in the stations. This approach ensures that there is sufficient bandwidth to transmit the files and minimises the risk of signal loss during the transmission process.

The power required to operate the electronics of the monitoring system is provided by a 12VDC GEL rechargeable battery. The negative terminal of the battery is connected to the pantograph structure and used as a reference voltage (i.e., virtual common ground) for the system, ensuring electrical isolation of the bodywork. To save energy during the pilot tests, a location-based trigger (using GNSS data) was implemented. This way, the recording of data can be controlled automatically and can be accessed and configured remotely. The architecture of the system and the placement of the different modules in the pantograph are depicted in Figure 11.

The installation of the monitoring system involves securely attaching the three sensors to the pantograph's structure using double-sided industrial tape. Prior to attaching the tape, cyanoacrylate-based glue is applied to the base of the sensor enclosure to ensure a firm and reliable bond. This installation method is non-invasive and does not hinder the normal functionality and free movement of the pantograph. The remaining electronic components and power supply share the same enclosure, which is mounted on the base frame of the pantograph using a specially designed base plate. The design is such that it does not interfere with the moving parts of the pantograph while maintaining proper clearance with the roof of the vehicle. The plate is attached to the frame using U-shaped metallic brackets, providing stability to the entire system. The sensor cables are neatly fixed and secured to the structure of the pantograph using zip ties. To accommodate the movement of the pantograph, sufficient space is provided for the cables to pass through any moving parts while adhering to the minimum flexion radius specified by relevant standards. Figure 12 provides details about the installation of the monitoring system in the pantograph.

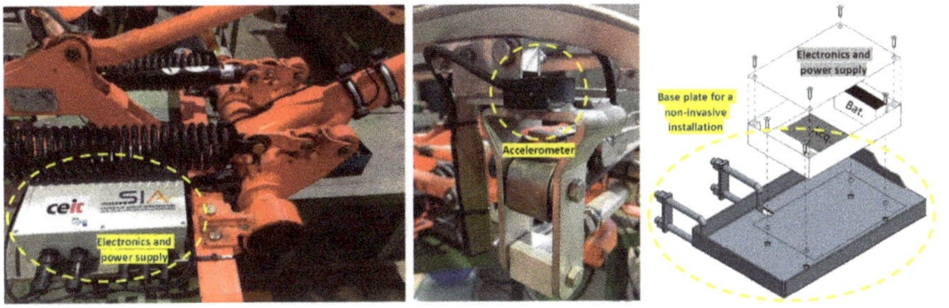


**Figure 12.** Installation of the monitoring system in the pantograph. Electronics and power supply (**left**); detail of an accelerometer (**centre**); detail of the base plate for a non-invasive installation (**right**).

Finally, regarding the results provided by the presented system, Table 3 lists a comparison with previous work where the employed methods are indicated.

**Table 3.** Comparison of provided results of CW monitoring with previous literature work and existing commercial systems.

| | | | Results of the CW Monitoring | | | |
|---|---|---|---|---|---|---|
| **Ref** | **Type** | **Methods** | **Wear** | **Tension** | **Stagger** | **Height** |
| Xu et al. [27] | On-board | Mechanics | Yes | No | No | No |
| Derosa et al. [28] | Wayside | Mechanics | No | Yes | No | No |
| Aydin et al. [29] | On-board | Vision | No | No | Yes | No |
| Chen et al. [30] | On-board | Vision | No | No | Yes | No |
| * tcat® [31] | Trolley | Vision | No | No | Yes | Yes |
| * Catenary Eye [32] | On-board | Vision | Yes | No | Yes | Yes |
| * Autocommute [33] | On-board | Vision | Yes | No | Yes | Yes |
| * CAT-T [34] | Trolley | Vision | No | No | Yes | Yes |
| * CAT-VW [35] | On-board | Vision | Yes | No | Yes | Yes |
| This work | On-board | Mechanics | No | No | Yes | Yes |

* Commercial devices.

## 5. Results of the Proposed System in Field Tests

This section focuses on assessing the measuring system through field tests conducted in real-world conditions. The accuracy and repeatability of the measuring system are crucial to ensuring the reliability and quality of the collected data. The results obtained from the field tests provide valuable insights regarding the ability of the system to accurately measure and record data, thus confirming its suitability for practical applications. The following subsections provide details about the case study conducted and present the significant results obtained, which contribute to the validation of the measuring system.

### *5.1. Description of the Case Study*

The monitoring system was installed in a pantograph of a train by the Spanish train operating company FGC during regular service in commercial operations. FGC, founded in 1979, features more than 92 million passengers per year and also provides freight transportation services. With a dedicated workforce of more than 1900 people and an extensive network spanning 290 km of track with international, narrow and Iberian gauges, FGC operates more than 1300 train circulations per day, with a peak headway of 150 s on the Barcelona–Vallès line.

To establish a baseline for the condition of the OCL, an initial testing campaign was conducted using the tCat® system [31], more precisely the tCat® 1435 system. This system consists of a measurement stroller pulled by an operator that collects several lines of information about the track and the OCL. The measurements were carried out in track 2, specifically between the stations of Rubí and Hospital General (milestones 20 + 124 and 18 + 377, respectively, from Plaza Catalunya station reference), as depicted in Figure 13. The height and stagger of the CW in static conditions were recorded at all poles along the track.

To validate the results obtained from the developed monitoring system and compare them with the measurements taken with the tCat®, the proposed monitoring system was installed in a regular pantograph mounted on a unit of the 112 series while running through the same section as the tCat®. The results of this validation stage against the baseline measurements are shown in Section 5.2.1.

Furthermore, to ensure the repeatability of the measurements and gather more data for analysis, an additional measurement campaign was conducted. During this campaign, two different sections of the line were analysed, with two circulations performed in each section. Each section had a length of approximately 500 m. To save energy during regular operations, the collection of data was automatically triggered using the GNSS subsystem.

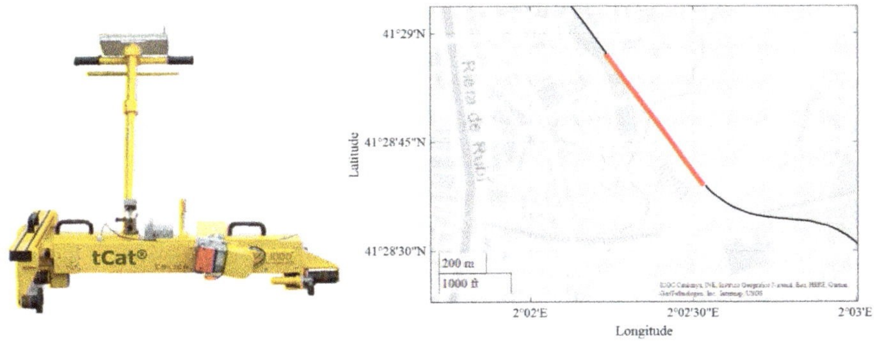


**Figure 13.** tCat® measuring system (**left**) and track section used in the baseline validation campaign (**right**).

Four parameters are analysed in each case: The stagger of the CW, the height of the pantograph, the height of the CW, and the estimated contact force. The contact force is estimated based on the experimental results from Section 3, particularly referring to Figure 8. The two height parameters are directly related and can be derived using the roof height. These relationships are graphically provided in Figure 14. The panhead height, and consequently the CW height, can be easily calculated using the kinematics of the mechanism and the angle of the lower arm ($\gamma$), as described by Equations (6) and (7). The parameters of these equations are obtained through linear fitting, as shown in Figure 6, and for a roof height of 3720 mm.

$$H_{\text{panhead}}(\gamma)\ [\text{mm}] = 493.953 + (\gamma \cdot 42.877), \tag{6}$$

$$H_{\text{CW}}(\gamma)\ [\text{mm}] = 3720 + H_{\text{panhead}}(\gamma). \tag{7}$$

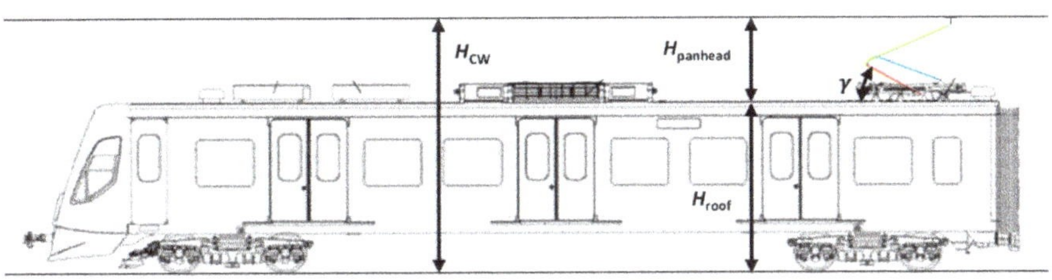


**Figure 14.** Scheme showing a graphical definition of important parameters of the height of the pantograph and CW.

### 5.2. Results

This section presents the results obtained from the field tests. The results were validated using a commercial measuring device (tCat®), and subsequently, their repeatability was studied through multiple runs in different sections of the track.

#### 5.2.1. Comparison with Commercial Measurement Device

The initial analysis seeks validation of the measurements taken by the system by comparing them with the tCat® commercial measurement system. Figure 15 presents the comparison of the CW height calculated by the monitoring system versus the reference values taken by the tCat®. In Figure 16, the results of the stagger of the CW are shown. Overall, the results show a very good correlation in the direct CW height measurements (Figure 15) and promising similarities in the CW stagger parameter (Figure 16), with some amplitude errors but good detection of the shape of the stagger.

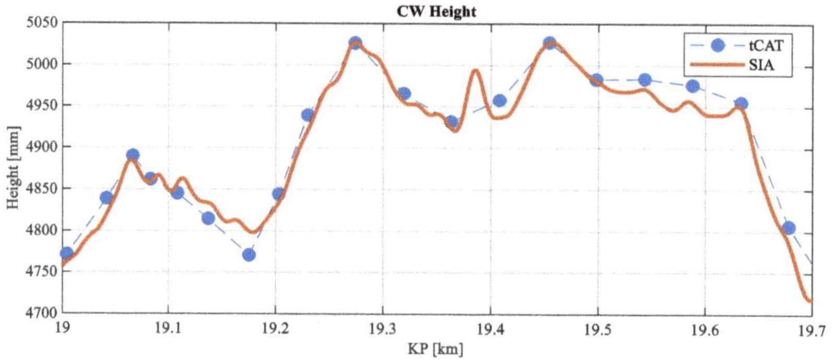


**Figure 15.** CW height comparison between tCat® and the monitoring system.

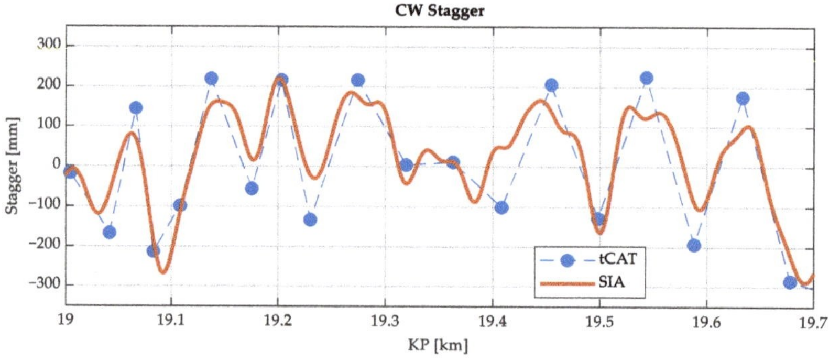


**Figure 16.** Calculated stagger with the monitoring system versus the reference values measured with the tCat® system.

It is important to note that the discrete reference measurements obtained with the tCat® device are taken statically, without direct contact with the CW, using vision equipment. In contrast, the results obtained with the monitoring system here presented are dynamic measurements captured during the actual circulation of the train, influenced by the dynamics of the interaction between the pantograph and the OCL. The uplift force of the pantograph with the proposed system, along with the vertical stiffness of the CW, changes the CW height slightly when compared with the steady-state situation. The speed was maintained constant during circulation through the section under study, with a reference value of 45 km/h.

Considering the same KP where there are measurements with the tCat® system, the accuracy of the developed SIA system has been calculated for the CW height. Table 4 lists the minimum, average and maximum errors.

**Table 4.** Comparison of error on measured CW height with SIA system against tCAT®.

| | **Absolute Error** | **Relative Error** |
|---|---|---|
| Minimum | 0.704 mm | 0.014% |
| Average | 11.951 mm | 0.245% |
| Maximum | 28.980 mm | 0.608% |

5.2.2. Reproducibility Analysis

After the validation stage, this subsection focuses on assessing the repeatability of measurements obtained by the monitoring system. For this study, two track sections are analysed, namely Section A and Section B, each with a length of 500 m. To evaluate the reproducibility of the measurements, two runs were conducted in each section, both in the same direction and at the same nominal circulation speeds, although in reality, both speeds were not exactly the same.

Figure 17 displays four different parameters (speed (a), CW height (b), CW stagger (c), and estimated contact force (d)) for each section under examination, comparing the two runs. The graph indicates that the measured speed values in Section A are similar between runs, whereas the differences in speed are bigger in the second case. Regarding CW height and stagger, both sections yield highly comparable results in both runs, particularly in the stagger KPI. Lastly, the lower section of the figure showcases the predicted contact force between the pantograph and the CW, displaying the estimated force along with the 95% confidence prediction intervals (represented as shadowed areas). These intervals are required because viscous and dry phenomena in joints are neglected in the calculation. Marques et al. [36] made a revision to the modelling and analysis of friction. It is evident that the estimated force remains consistent in both cases and falls well within the indicated red limits.

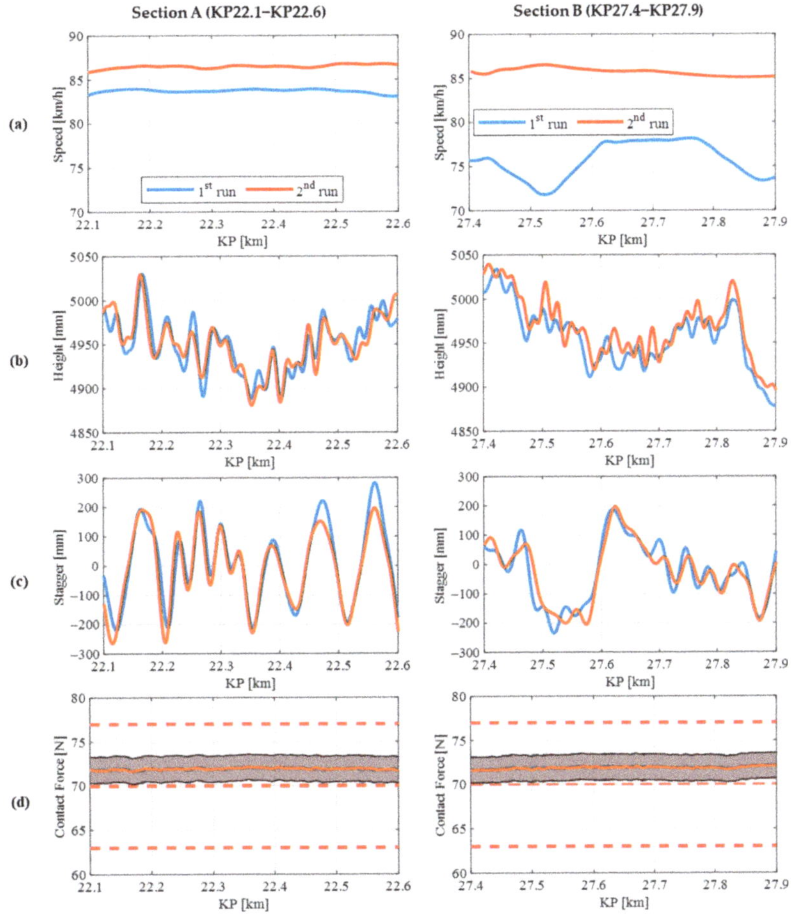


**Figure 17.** Results of the repeatability performance study of the monitoring system in different Section A (**left**) and Section B (**right**) for four different parameters: speed (**a**), CW height (**b**), CW stagger (**c**) and contact force (**d**).

Considering the same KP where there are measurements for the first run, the accuracy of the developed SIA system has been calculated for the CW height. Table 5 lists the minimum, average and maximum errors.

**Table 5.** Comparison of error on measured CW height with SIA system.

| | Absolute Error | Relative Error |
|---|---|---|
| Minimum | 0.010 mm | 0.001% |
| Average | 12.503 mm | 0.252% |
| Maximum | 42.051 mm | 0.847% |

## 6. Discussion

The monitoring of the OCL provides valuable information about the health status of this crucial asset in the railway system. This monitoring can be conducted continuously or periodically, serving different purposes. Continuous monitoring ensures that the analysed parameters remain consistent, and any deviations could indicate a fault in the system. In such cases, an alert can be sent to the infrastructure manager, enabling prompt action based on the reliability of the detected fault. On the other hand, periodic monitoring is sufficient for assessing the degradation trend over time. Figure 18 shows how a two-dimensional plot can effectively distinguish different failures in the height of the CW. Regions with the same measurement values are represented by flat-coloured surfaces. On the other hand, a continuous vertical drop of the CW shows a colour gradient for different measurements. In the case of an incorrect vertical slope, due, for example, to a broken dropper, the colour gradient intensifies with each measurement.

(a) (b) (c)

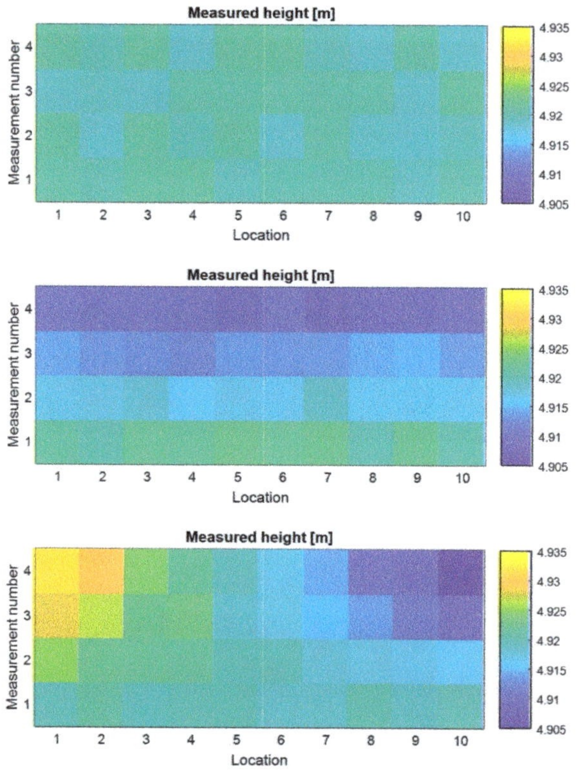


**Figure 18.** Degradation trend of the CW height. (**a**) Correct height, (**b**) vertical drop and (**c**) incorrect vertical slope.

In addition, conducting a statistical analysis of the measurements recorded at various positions along a track section and at different times allows for a quantitative understanding of the type of failure occurring. Figure 19 shows the mean and standard deviation (SD) values for three different states. In cases where the CW height is consistently correct, both the mean and SD values remain stable, albeit with slight variations between measurements due to the use of a monitoring device. On the contrary, when a failure is present, one of the parameters exhibits stability while the other shows a significant change. A vertical drop in the CW height leads to a shift in the mean value, while an incorrect vertical slope is characterised by a changing SD value. This quantitative method is essential for infrastructure digitalisation platforms to automatically show the locations of required interventions and the corresponding maintenance tasks throughout the entire rail line. As the monitoring system is not as accurate as other costly inspection methods, some differences can occur between measurements. However, assessing the trends of the statistical analysis confirms the feasibility of utilising low-cost devices for maintenance purposes.

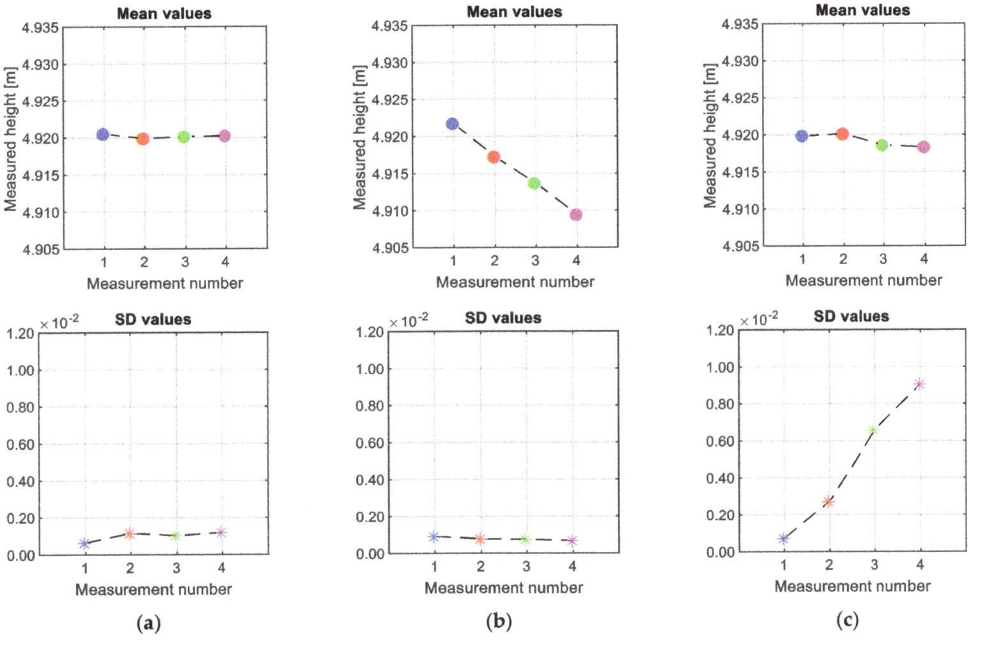


**Figure 19.** Statistical analysis of the CW height changes. (**a**) Correct height, (**b**) vertical drop and (**c**) incorrect vertical slope.

The static contact force is a parameter that can be directly obtained from the characterisation of the pantograph. It provides valuable information by quantifying the contact forces at different kilometric points along the railway line. Monitoring the evolution of this contact force can serve as an indicator of contact anomalies and potential issues [37]. It is worth noting that changes in force can be attributed to both the pantograph and the OCL, and using multiple instrumented pantographs can help detect and analyse these changes effectively. In addition to the static contact force, the dynamic contact force is also a valuable parameter for maintenance purposes. Models that describe the interaction between the pantograph and the OCL, such as those employed for stagger characterization using accelerometers, can be utilized to estimate dynamic forces. Machine learning techniques can be applied to further enhance the estimation of dynamic forces, although this area requires further development and research.

The stagger amplitude and the stagger central position, obtained from accelerations of both sides of the pantograph strips, show good repeatability. The reliability of these measurements is influenced by the running conditions of the train unit. In our study, we observed that the measurements exhibit consistent results across different runs and operating conditions, indicating a high level of reproducibility.

By analysing the measured data, the system can promptly detect changes in forces or OCL geometry and alert maintenance personnel in real time. This enables timely adjustments to the pantograph's force settings to prevent potential interaction problems and mitigate the risk of further failures in the OCL system.

## 7. Conclusions

In this research work, a non-intrusive monitoring system has been developed for the continuous monitoring of overhead contact lines that is adequate to be installed on a train pantograph. The proposed system utilises low-cost sensors and makes use of a set of formulae based on geometrical and mechanical parameters to ensure reliable measurements. The reliability of measurements is further enhanced through physical modelling, which allows for periodic updates of the pantograph's configuration parameters.

The system modelling is based on a mechanical design approach. The kinematic assessment of the pantograph structure and mechanism is accurately modelled, and the employed models are validated through test-rig experiments. As a result, the position of the pantograph head can be determined based on parameters such as the inclination of the lower arm or push bar. Additionally, by utilising an inclinometer and knowing the height of the carbody roof, the height of the overhead contact line can be monitored.

The static contact force can also be calculated using the inclinometer if the position of the screws controlling the applied torque mechanism is known. Although the calibration of this mechanism is not carried out on a daily basis and the contact force may change over time, variations in the force correspond to changes in the height of a specific pantograph. By employing multiple instrumented pantographs on the same overhead contact line, it is possible to detect whether the change in the pantograph height, and consequently the contact wire height, is due to the stress on the messenger wire or other factors.

The sensors chosen for the monitoring system have been carefully selected thanks to the modelling and laboratory tests conducted on the pantograph. The nature and location of these sensors are well suited to meet the required performance of the monitoring system, given the low-cost and non-intrusive nature of the application.

The field tests conducted have demonstrated the effectiveness of the monitoring system. The results obtained from the monitoring system were found to be comparable to those produced by a commercial device, indicating a high level of accuracy. Furthermore, the obtained results have exhibited repeatability across several runs conducted at different speeds. This consistency in performance and the reliable measurement of the target parameters highlight the robustness of the system and its suitability for real-world applications. The field tests provide confidence in the reliability and accuracy of the monitoring system, positioning it as a viable solution for measuring and monitoring the overhead contact line in operational environments.

**Author Contributions:** Conceptualization, B.R.-A. and U.A.; methodology, B.R.-A. and U.A.; validation, B.R.-A., P.C. and U.A.; investigation, B.R.-A., P.C. and U.A.; writing—original draft preparation, B.R.-A., P.C. and U.A.; writing—review and editing, B.R.-A., P.C., N.G.-N. and U.A.; visualization, B.R.-A. and P.C.; supervision, U.A.; funding acquisition, U.A. All authors have read and agreed to the published version of the manuscript.

**Funding:** This research was funded by HORIZON-ER-JU-2022-01, grant number 101101966.

**Institutional Review Board Statement:** Not applicable.

**Informed Consent Statement:** Not applicable.

**Data Availability Statement:** The data are not publicly available due to confidentiality.

**Conflicts of Interest:** The authors declare no conflict of interest.

## References


1. Pombo, J.; Ambrósio, J. Influence of pantograph suspension characteristics on the contact quality with the catenary for high speed trains. *Comput. Struct.* **2012**, *110–111*, 32–42. [CrossRef]
2. Cheng, H.; Cao, Y.; Wang, J.; Zhang, W. A preventive, opportunistic maintenance strategy for the catenary system of high-speed railways based on reliability. *J. Rail Rapid Transit* **2019**, *234*, 1149–1155. [CrossRef]
3. Wang, J.; Gao, S.; Yu, L.; Ma, C.; Zhang, D.; Kou, L. A data-driven integrated framework for predictive probabilistic risk analytics of overhead contact lines based on dynamic Bayesian network. *Reliab. Eng. Syst. Saf.* **2023**, *235*, 109266. [CrossRef]
4. Benet, J.; Cuartero, N.; Cuartero, F.; Rojo, T.; Tendero, P.; Arias, E. An advanced 3D-model for the study and simulation of the pantograph catenary system. *Transp. Res. Part C Emerg. Technol.* **2013**, *36*, 138–156. [CrossRef]
5. Gritti, M.G.; Giberti, H.; Collina, A. Optimal synthesis of a cam mechanism for train pantograph. In Proceedings of the 2013 IEEE International Conference on Mechatronics (ICM), Vicenza, Italy, 27 February–1 March 2013; pp. 406–411. [CrossRef]
6. Zhou, N.; Zhang, W. Investigation on dynamic performance and parameter optimization design of pantograph and catenary system. *Finite Elem. Anal. Des.* **2011**, *47*, 288–295. [CrossRef]
7. Gregori, S.; Tur, M.; Tarancón, J.E.; Fuenmayor, F.J. Stochastic Monte Carlo simulations of the pantograph–catenary dynamic interaction to allow for uncertainties introduced during catenary installation. *Veh. Syst. Dyn.* **2019**, *57*, 471–492. [CrossRef]
8. Vesali, F.; Rezvani, M.A.; Molatefi, H.; Hecht, M. Static form-finding of normal and defective catenaries based on the analytical exact solution of the tensile Euler–Bernoulli beam. *Proc. Inst. Mech. Eng. Part F J. Rail Rapid Transit* **2019**, *233*, 691–700. [CrossRef]
9. Song, Y.; Liu, Z.; Ronnquist, A.; Navik, P.; Liu, Z. Contact Wire Irregularity Stochastics and Effect on High-Speed Railway Pantograph-Catenary Interactions. *IEEE Trans. Instrum. Meas.* **2020**, *69*, 8196–8206. [CrossRef]
10. Duan, F.; Song, Y.; Gao, S.; Liu, Y.; Chu, W.; Lu, X.; Liu, Z. Study on Aerodynamic Instability and Galloping Response of Rail Overhead Contact Line Based on Wind Tunnel Tests. *IEEE Trans. Veh. Technol.* **2023**, *72*, 7211–7220. [CrossRef]
11. Song, Y.; Liu, Z.; Lu, X. Dynamic Performance of High-Speed Railway Overhead Contact Line Interacting with Pantograph Considering Local Dropper Defect. *IEEE Trans. Veh. Technol.* **2020**, *69*, 5958–5967. [CrossRef]
12. Bruni, S.; Ambrosio, J.; Carnicero, A.; Cho, Y.H.; Finner, L.; Ikeda, M.; Kwon, S.Y.; Massat, J.P.; Stichel, S.; Tur, M.; et al. The results of the pantograph-catenary interaction benchmark. *Veh. Syst. Dyn.* **2015**, *53*, 412–435. [CrossRef]
13. Pil Jung, S.; Guk Kim, Y.; Sung Paik, J.; Won Park, T. Estimation of dynamic contact force between a pantograph and catenary using the finite element method. *J. Comput. Nonlinear Dyn.* **2012**, *7*, 041006. [CrossRef]
14. Song, Y.; Wang, H.; Liu, Z. An Investigation on the Current Collection Quality of Railway Pantograph-Catenary Systems with Contact Wire Wear Degradations. *IEEE Trans. Instrum. Meas.* **2021**, *70*, 9003311. [CrossRef]
15. Blanco, B.; Errandonea, I.; Beltrán, S.; Arrizabalaga, S.; Alvarado, U. Panhead accelerations-based methodology for monitoring the stagger in overhead contact line systems. *Mech. Mach. Theory* **2022**, *171*, 104742. [CrossRef]
16. Koyama, T. Development of High-speed Test Facility for Pantograph/OCL Systems. *Q. Rep. RTRI Railw. Tech. Res. Inst.* **2022**, *63*, 128–132. [CrossRef] [PubMed]
17. Falamarzi, A.; Moridpour, S.; Nazem, M. A Review on Existing Sensors and Devices for Inspecting Railway Infrastructure. *J. Kejuruter.* **2019**, *31*, 1–10. [CrossRef]
18. Soilán, M.; Sánchez-Rodríguez, A.; Del Río-Barral, P.; Perez-Collazo, C.; Arias, P.; Riveiro, B. Review of laser scanning technologies and their applications for road and railway infrastructure monitoring. *Infrastructures* **2019**, *4*, 58. [CrossRef]
19. Huang, Z.; Chen, L.; Zhang, Y.; Yu, Z.; Fang, H.; Zhang, T. Robust contact-point detection from pantograph-catenary infrared images by employing horizontal-vertical enhancement operator. *Infrared Phys. Technol.* **2019**, *101*, 146–155. [CrossRef]
20. Zhang, F.; Tao, K.; Xie, X.; Liu, H.; Tian, L.; Yang, X.; Wang, M. Research on Fault Detection Method of Catenary Equipment Based on Deep Learning. In Proceedings of the 2021 IEEE 2nd International Conference on Information Technology, Big Data and Artificial Intelligence (ICIBA), Chongqing, China, 17–19 December 2021; Volume 2, pp. 478–482. [CrossRef]
21. Efanov, D.; Osadchy, G.; Sedykh, D.; Pristensky, D.; Barch, D. Monitoring system of vibration impacts on the structure of overhead catenary of high-speed railway lines. In Proceedings of the 2016 IEEE East-West Design & Test Symposium (EWDTS), Yerevan, Armenia, 14–17 October 2016; pp. 1–8. [CrossRef]
22. Boffi, P.; Cattaneo, G.; Amoriello, L.; Barberis, A.; Bucca, G.; Bocciolone, M.F.; Collina, A.; Martinelli, M. Optical fiber sensors to measure collector performance in the pantograph-catenary interaction. *IEEE Sens. J.* **2009**, *9*, 635–640. [CrossRef]
23. *EN50317:2012*; Railway Applications-Current Collection Systems-Requirements for and Validation of Measurements of the Dynamic Interaction between Pantograph and Overhead Contact Line. Aenor: Brussels, Belgium, 2012.
24. *EN50367:2022*; Railway Applications-Fixed Installations and Rolling Stock-Criteria to Achieve Technical Compatibility between Pantographs and Overhead Contact Line. Aenor: Brussels, Belgium, 2022.
25. Eberhard, P.; Schiehlen, W. Computational dynamics of multibody systems: History, formalisms, and applications. *J. Comput. Nonlinear Dyn.* **2006**, *1*, 3–12. [CrossRef]
26. Collina, A.; Bruni, S. Numerical simulation of pantograph-overhead equipment interaction. *Veh. Syst. Dyn.* **2002**, *38*, 261–291. [CrossRef]
27. Xu, F.T.; Duan, S.Y.; Wang, F.; Liu, G.R. A mechanics principle based inverse technique for real-time monitoring of wear-level of contact wire in pantograph-catenary systems. *Appl. Math. Sci. Eng.* **2022**, *30*, 75–93. [CrossRef]
28. Derosa, S.; Nåvik, P.; Collina, A.; Rønnquist, A. Railway catenary tension force monitoring via the analysis of wave propagation in cables. *Proc. Inst. Mech. Eng. Part F J. Rail Rapid Transit* **2021**, *235*, 494–504. [CrossRef]

29. Aydin, I.; Karakose, M.; Akin, E. Anomaly detection using a modified kernel-based tracking in the pantograph-catenary system. *Expert Syst. Appl.* **2015**, *42*, 938–948. [CrossRef]
30. Chen, R.; Lin, Y.; Jin, T. High-Speed Railway Pantograph-Catenary Anomaly Detection Method Based on Depth Vision Neural Network. *IEEE Trans. Instrum. Meas.* **2022**, *71*, 1502710. [CrossRef]
31. tCat OLE Mobile Mapping. Available online: https://tcat.es/ (accessed on 1 June 2023).
32. CATENARY EYE. Available online: https://www.meidensha.com/catalog/BA78-3054.pdf (accessed on 1 August 2023).
33. AUTOCOMMUTE. Available online: http://www.autocommute.com/ocsi.html (accessed on 1 August 2023).
34. CAT-T. Available online: http://www.selectravision.com/products-cat-t.php (accessed on 1 August 2023).
35. CAT-VW. Available online: http://www.selectravision.com/products-cat-vw.php (accessed on 1 August 2023).
36. Marques, F.; Flores, P.; Claro, J.C.P.; Lankarani, H.M. Modeling and analysis of friction including rolling effects in multibody dynamics: A review. *Multibody Syst. Dyn.* **2019**, *45*, 223–244. [CrossRef]
37. Harada, S.; Kusumi, S. Monitoring of overhead contact line based on contact force. In Proceedings of the 2006 IET International Conference on Railway Condition Monitoring, Birmingham, UK, 29–30 November 2006; pp. 188–193.

sensors

MDPI

Article

# Seeking a Sufficient Data Volume for Railway Infrastructure Component Detection with Computer Vision Models

**Alicja Gosiewska, Zuzanna Baran, Monika Baran and Tomasz Rutkowski ***

Nevomo IoT, 03-828 Warsaw, Poland; a.gosiewska@nevomo.tech (A.G.); z.baran@nevomo.tech (Z.B.); m.seniut@nevomo.tech (M.B.)
* Correspondence: t.rutkowski@nevomo.tech

**Abstract:** Railway infrastructure monitoring is crucial for transportation reliability and travelers' safety. However, it requires plenty of human resources that generate high costs and is limited to the efficiency of the human eye. Integrating machine learning into the railway monitoring process can overcome these problems. Since advanced algorithms perform equally to humans in many tasks, they can provide a faster, cost-effective, and reproducible evaluation of the infrastructure. The main issue with this approach is that training machine learning models involves acquiring a large amount of labeled data, which is unavailable for rail infrastructure. We trained YOLOv5 and MobileNet architectures to meet this challenge in low-data-volume scenarios. We established that 120 observations are enough to train an accurate model for the object-detection task for railway infrastructure. Moreover, we proposed a novel method for extracting background images from railway images. To test our method, we compared the performance of YOLOv5 and MobileNet on small datasets with and without background extraction. The results of the experiments show that background extraction reduces the sufficient data volume to 90 observations.

**Keywords:** object detection; computer vision; machine learning; railway

**Citation:** Gosiewska, A.; Baran, Z.; Baran, M.; Rutkowski, T. Seeking a Sufficient Data Volume for Railway Infrastructure Component Detection with Computer Vision Models. *Sensors* **2023**, *23*, 7776. https://doi.org/10.3390/s23187776

Academic Editor: Giovanni Betta, Abdollah Malekjafarian, Diogo Ribeiro, Araliya Mosleh and Maria D. Martínez-Rodrigo

Received: 20 July 2023
Revised: 29 August 2023
Accepted: 5 September 2023
Published: 9 September 2023



## 1. Introduction

Nowadays, rail transportation is widely accessible and is one of the most popular travel forms worldwide. An increasing amount of studies show that trains are more environmentally friendly than cars [1–3]. Additionally, high-speed trains are serious competitors for air transport [4]; therefore, railway has a good prospect of development ahead of it. However, with the rapid growth comes a new set of challenges for rail management. The increased popularity will put more strain on the infrastructure, which will need to be inspected more often; what is more, the expanding rail connections network will result in more and more kilometers of track to monitor. All of this implies an increase in the labor workload related to infrastructure maintenance.

The need for track inspections stems from the fact that the track is susceptible to weather conditions, such as extreme temperatures, high levels of humidity, and air pollution. Studies show that broken rails and welds were the leading derailment cause on tracks in the United States [5]. However, any defect on the railroad track can carry immense costs and even lead to catastrophic incidents such as train derailments. That is why monitoring railway infrastructure is crucial for the safety of travelers and the reliability of public transportation.

Currently, infrastructure inspections are manual. The specially trained staff, based on the visual evaluation and measurements from dedicated devices, assess the degradation of the track. Such a procedure requires much human labor, translating into a high maintenance cost. Moreover, the inspection speed is limited to the efficiency of the human eye; thus, it requires time and is prone to mistakes [6]. The solution for the highlighted issues could be incorporating computer vision techniques, especially machine learning, in infrastructure monitoring. Machine learning is a branch of artificial intelligence that consists of algorithms that can optimize themselves based on the provided data.

Nowadays, machine learning models achieve human-level performance in many areas, including medicine, finance, and technology . Therefore, artificial intelligence algorithms can be successfully used to support human decisions, including visual monitoring of rail infrastructure [7]. In recent years, much research has been conducted on the usefulness of computer vision for railway applications. A variety of methods were successfully applied, starting from traditional pattern recognition [8], through classical machine learning models, such as support vector machines [9], k-nearest neighbors [10], or random forest [11], ending with deep neural networks [12–15]. The latter methods usually yield the best results due to their capacity to solve complex problems. Therefore, deep learning has great potential for detecting railway defects [16].

Computer vision-based infrastructure monitoring for fault identification usually consists of two steps: (1) detection of railway components and (2) component-specific identification of defects. It is worth noting that in step (2), different rail components require different machine learning models for fault detection. This is due to the specific characteristics of faults for different components, and therefore, many studies are focusing on only one component of the rail infrastructure. Studies include the detection of wheel defects [17,18], the identification of bolt corrosion [19,20], assessing ballast support for sleepers [21], aiding in the design of prestressed concrete railway sleepers [22], the recognition of rail surface cracks [23], capturing fastener defect detection [24], and monitoring bridges' condition [25,26]. However, fault identification is impossible without accurate object detection (OD) in the previous step (1). For example, a crack on a sleeper would not be identified if the sleeper itself was not detected correctly. Due to its complexity and importance, separate studies often address the component-detection task, where deep neural networks detect track elements [27,28]. Over time, the need for rail object detection models will only increase. They will be in demand for various types of infrastructure elements and for different tracks, such as high-speed rail, maglev, or subway. Additionally, each country may need a different model due to the country-specific regulations and different ways the rail infrastructure is built. Therefore, it is important to study how to optimally build object-detection models.

This paper focuses on fast and accurate railway track component detection that can be used to support humans in monitoring rail infrastructure. We consider a scenario when only a small dataset is available, which is the most common case for railway data. The reason behind this stems from the low number of publicly available datasets with photos of the tracks. As a result, it is necessary to rely on a small number of public images or to gather new photos. The process of preparing new training datasets with railway images is costly and time-consuming since the images have to be labeled by experts with domain knowledge. This increases the need for precise upfront estimation of how many images are needed to obtain an accurate machine learning model. In this article, we show how much data are enough to train a neural network that detects track components and which architectures are best for this task. The key contributions of this paper are as follows:

- We conducted a benchmark to determine the sufficient data volume for railway component detection. We have shown how the YOLO and MobileNet neural network architectures perform for different sizes of datasets. We have used a completely new dataset with track images we collected and labeled. The results of the analysis will be valuable to anyone designing their own railway dataset, as we provide an estimate of the sufficient size of the data.
- We introduced a novel method of extracting background images (BIE) that can be used to enrich the datasets for the railway object detection task. We have shown that this method allows us to obtain better neural networks for really small datasets. BIE is useful to improve the performance of any models for railway track object detection.

This paper is organized as follows. Section 2.1 describes railway track and its components, Section 2.2 gives an overview of machine learning algorithms for OD, Section 2.3

introduces our novel Background Image Extraction method, and Section 2.4 exhibits the details of the OD benchmark. Section 3 outlines the main results of the benchmark. Section 4 summarizes the findings presented in the paper.

## 2. Materials and Methods

### 2.1. Railway Track

In this Section, we describe the railway track components that are detected in experiments in Section 2.4—rails, sleepers, and fasteners. The image areas without the mentioned components mostly contain track ballast. For the object-detection task, we consider ballast part of image background. The added examples are the images used in experiments, so they illustrate the data used in model training.

Rails are steel bars that are the surface on which trains can move. Figure 1 shows an example of rail used on a railway track. Sleepers serve as support for rails, fixing them in position. Figures 2 and 3 show examples of concrete and wooden sleepers. Fasteners are elements used to keep rails fastened to sleepers. Figure 4 shows examples of different types of fasteners on railway tracks. Track ballast is defined as a layer of crushed materials, usually rocks, placed around sleepers. Figure 5 shows a railway track with red arrows pointing to the ballast.

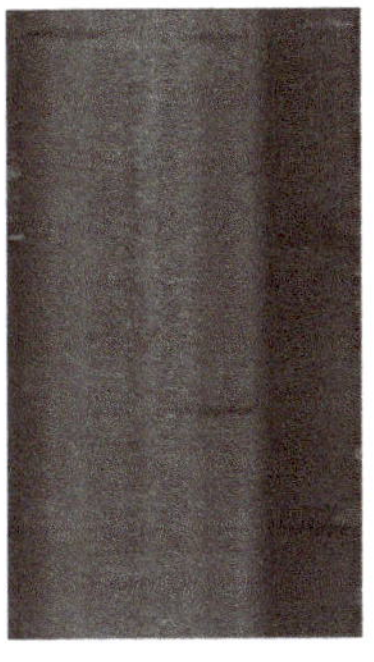

**Figure 1.** Example of rail.

**Figure 2.** Example of concrete sleeper.

**Figure 3.** Example of wooden sleeper.

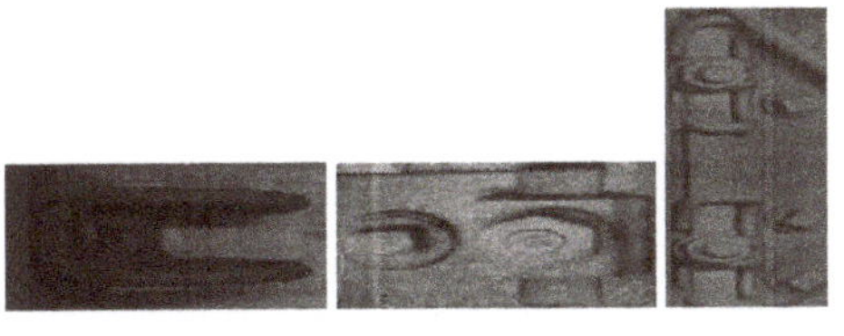

**Figure 4.** Examples of different types of fasteners.

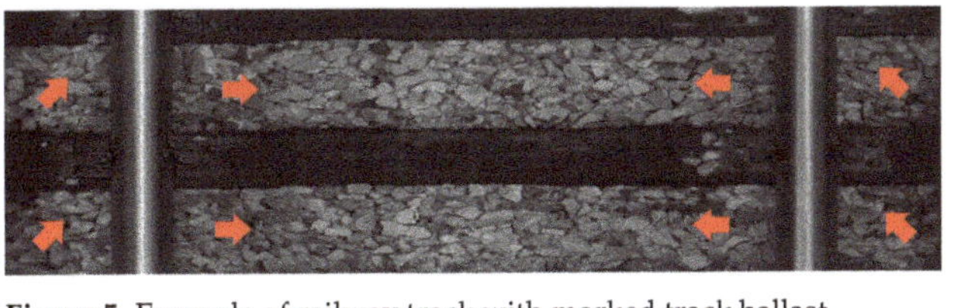

**Figure 5.** Example of railway track with marked track ballast.

### 2.2. Machine Learning Models for Object Detection

Object detection is a computer vision technique for locating objects with bounding boxes (bboxes). Nowadays, convolutional neural networks (CNN) perform very well in this task. As a result, there is plenty of research on various architectures, for example, region-based convolutional neural networks (R-CNN) [29]. The idea of R-CNN is to start with a selective search [30]—a region-proposal procedure to pick out regions in the image that may contain objects of interest. In the next step, CNN extracts a feature vector from each region proposal, then a classification model assigns classes and scores to the extracted vectors. In the last step, a non-maximum suppression algorithm rejects image proposals with large intersection over union (IoU) overlap with higher-scored image proposals.

While R-CNN achieves satisfactory results, the drawback of this approach is the speed of training and prediction. To overcome these issues, Faster R-CNN [31] replaces the use of selective search with CNN. As a result, Faster R-CNN takes an entire image and processes it through a region-proposal network and then through a neural network that predicts classes of objects. This increases the detection speed, yet the prediction time is still not fast enough for real-time applications.

The Single-Shot Detector (SSD) is one of the fastest ways to achieve accurate object detection [32]. SSD consists of a feature-extractor backbone and SSD head. A backbone is a pre-trained classification neural network with a removed fully connected classification layer. The SSD head consists of convolutional layers added to the backbone to find the most appropriate bounding boxes. There is also a mobile variant of SSD, SSDLite [33], where regular convolutions in the SSD head are replaced with separable convolutions, which reduces both the parameter count and the computational cost compared to regular SSD. In Section 2.4, we used MobileNetV3-small as a backbone feature extractor in SSDLite, which is the same combination that the authors of MobileNetV3 used in their benchmarks Section 2.4. The MobileNet architecture is based on depth-wise separable convolutions that reduce the number of parameters [34]. In MobileNetV2, the authors introduced new inverted residual blocks [33] and in MobileNetV3 they added squeeze and excitation layers [35]. The MobileNetV3-small architecture is a variant targeted to low-resource use cases and we have chosen it for experiments because of its low number of parameters, which assures their ability to catch relationships in the data based on a small number of samples.

Another object-detection architecture is You Only look Once (YOLO) [36,37], which has become very popular in recent years. YOLOv5 is composed of three parts: backbone, neck, and detection networks. The backbone CNN aggregates image features that are processed in the neck network, creating Feature Pyramid Networks [38]. Finally, the detection network predicts each object's class, probability, and bbox position. In experiments in Section 2.4, we have used two small YOLOv5 variants, the smallest variant nano (YOLOv5n) and variant small with ghost bottleneck (YOLOv5s-ghost). The small number of parameters means that the model has a chance to perform well on the low-volume datasets that are typical for railway OD. Moreover, YOLO's good performance in a wide variety of applications implies that this architecture has great potential for railway applications as well.

We measured models' performance with mean average precision (mAP) and mean average recall (mAR) [39]. Precision measures how well a model finds true positives and recall measures the proportion of true positive predictions,

$$Precision_t = \frac{TP_t}{TP_t + FP_t}, \quad (1)$$

$$Recall_t = \frac{TP_t}{TP_t + FN_t}, \quad (2)$$

where $TP_t$ denotes the number of true positives, $FP_t$ denotes number of false positives, and $FN_t$ denotes the number of false negatives. The value of $t$ determines the IoU overlap above which bboxes are considered to be the same; thus, if the IoU value for predicted and true bboxes is greater than $t$, the predicted bbox is considered to be correct.

The average precision is the area under the precision–recall curve obtained by plotting the precision and recall values as a function of model's confidence. The mAP@t is the mean of the average precision values over all classes with a given IoU overlap threshold $t$. For example, mAP@0.5 is the mean average precision for an IoU overlap threshold $t$ equal to 0.5. In the experiments, we use also mAP@[0.5,0.95], which is an average of the mAP values for different IoU thresholds, starting from 0.5 and finishing at 0.95 with a step of 0.05.

The average recall is the area over a recall–IoU threshold for IoU $\in$ [0.5, 1] and mAR is the mean of the average recall across all classes. mAR n denotes that mAR is calculated based on the top $n$ bboxes detected in the image. In the experiments we have used an mAR of 100.

### 2.3. Background Image Extraction

In this Section, we describe our novel method, named Background Image Extraction (BIE), which cuts out areas without bboxes from the railway photo and then joins them into a new image. Adding background images to the training set is a common procedure to improve model performance—the same stands for detecting railway components where the background consists mostly of ballast. Adding background images with no objects to detect to the training phase allows neural networks to learn what they should avoid detecting, which improves the performance of their predictions. Due to the distinctive composition of the railway track, we came up with a railway-targeted method of extracting background images. Figure 6 shows the general idea behind BIE. A bbox-based mask is extracted based on a labeled image based on the position of the track component. Then, a mask is used to cut out areas in the image that are merged into a new background image.

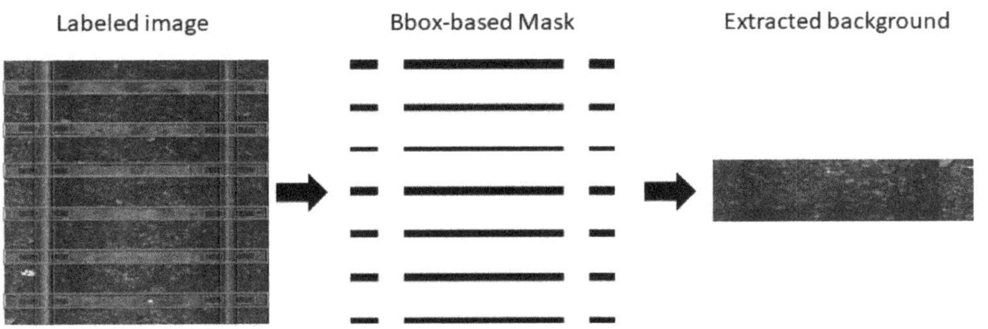


**Figure 6.** A diagram of the BIE method.

Algorithm 1 shows the procedure of mask extraction; its result is an array of the same size as an input image. In the mask, values of 0 represent the background and 225 non-background areas. Initially, each pixel within the mask has a value of 0. To identify the background, i.e., the area with ballast only, bboxes of rails are pulled to the top and bottom edges of the image, while bboxes of fasteners and sleepers are pulled to the side edges of the image. Then, the area of the pulled bboxes is treated as non-background and cropped out by setting the values of the corresponding pixels to 255. Such an approach ensures that after cropping, the union of the remaining parts will form a rectangle that contains only ballast. The newly created background image can then be used to enhance the training dataset. Figure 7 shows example backgrounds extracted with BIE.

**Algorithm 1** Mask extraction from railway image labeled with bboxes. The dot denotes a reference to the bbox property; thus, bbox.label means a class of bbox, bbox.width and bbox.height are its width and height, Additionally bbox.x_left and bbox.y_top denote the x coordinate of the left edge and y coordinate of the top edge of the bbox, respectively.

```
Require: n: image width, m: image height, x_margin: x coordinate bbox margin used for
   mask extraction, y_margin: y coordinate bbox margin used for mask extraction
   image_mask ← zeros(n, m)                    ▷ An array of size n × m filled with zeros.
   for bbox in bboxes do
       if bbox.label is "rail" then
           x_left = bbox.x_left - x_margin
           if x_left < 0 then
               x_left ← 0
           end if
           box_width ← bbox.width + x_margin * 2
           y_top ← 0                                  ▷ Extend rail to whole image height.
           box_height ← bbox.height
       else
           x_left ← 0                   ▷ Extend non-rail elements to whole image width.
           box_width ← n
           y_top ← bbox.y_top - y_margin
           if y_top < 0 then
               y_top ← 0
           end if
           box_height ← bbox.height * n + y_margin * 2
       end if
                               ▷ Set area in mask related to the adjusted bbox to 255.
       image_mask[top_y : top_y + box_height, top_x : top_x + box_width] = 255
   end for
```

**Figure 7.** Example backgrounds extracted with BIE.

All background images used in the experiments in Section 2.4 were created with BIE with x_margin 30 px and y_margin 90 px. Backgrounds with a width or height smaller than 100 px were filtered out.

### 2.4. *Experiment*

The aim of the experiment is to establish a sufficient data volume to train an efficient railway object detection model. Railway datasets usually are small and consist of images that are similar to each other. Moreover, infrastructure objects are also similar and are of a regular, rectangular shape. To find the number of images sufficient to obtain an accurate detector, we prepared training subsets of different sizes and trained the most common object-detection models. The training was carried out with and without the background-extraction method.

2.4.1. Data Acquisition and Dataset

The data were collected on 2 March 2022 and 13 April 2022 at Warszawa Grochów motive power depot in Poland. All images are grayscale and come from line-scan cameras (raL4096-24gm - Basler racer, Basler AG, Ahrensburg, Germany). placed on the draisine running on the railway tracks. The photos contain track sections with wooden and concrete sleepers and do not contain switches. The dataset consists of 348 labeled images, including 299 short images of size 2083 px × 500 px and 49 long images of size 2083 px × 2100 px. Rails, sleepers, and fasteners are labeled on the images with bboxes. Figures 8 and 9 show the annotations of short and long photos, respectively.

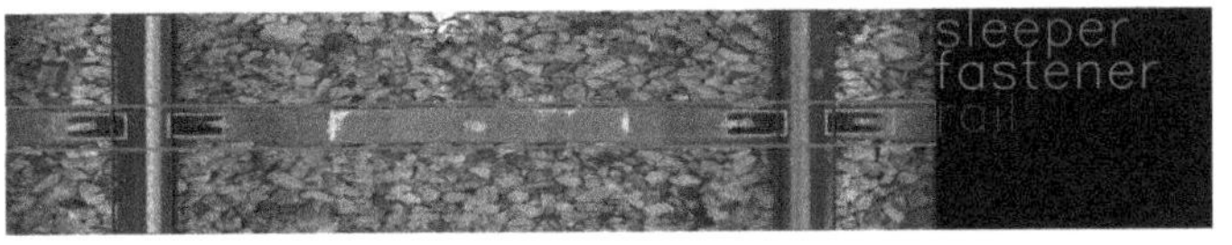


**Figure 8.** Example of a labeled short image of size 2083 px × 500 px with concrete sleepers.

**Figure 9.** Example of a labeled long image of size 2083 px × 2100 px with wooden sleepers.

2.4.2. Experiment Settings

The experiments were conducted on an AMD Ryzen 5 4600H CPU (Advanced Micro Devices, Inc., Santa Clara, California, United States) with Radeon Graphics (3.00 GHz, 32 GB RAM) and an NVIDIA GeForce RTX 2060 (CUDA version 12.0) (Nvidia Corporation, Santa Clara, California, United States) with Python 3.8.10 in the 64-bit Windows 10 business operating system.

We split the dataset into training, validation, and testing subsets of sizes 300, 24, and 24. We then took subsets of the training set of sizes 240, 180, 120, 90, 60, and 30, where each successive subset is contained in the preceding subset. The sizes of all subsets, along with the number of extracted backgrounds, are in Table 1.

We compared three neural networks: YOLOv5n, YOLOv5s-ghost, and MobileNetV3-small. Since the characteristic of the railway OD task is the small data volume, we have chosen architectures that have a small number of parameters and therefore, do not require much training data and have a fair chance to fit well. The models were trained for 100 epochs with default hyperparameters on all training subsets, both with and without extracted backgrounds, a total of 14 datasets. Detailed information about the hyperparameter values is in Table 2. For each training subset, the best model across all epochs was chosen based on its performance on the validation subset, then the final model performances were computed on the testing subset with mAP@0.5, mAP@[0.5, 0.95], and mAR 100.

**Table 1.** Dataset splits for experiment with numbers of observations.

| Split | Number of Full Railway Images | Number of Background Images Extracted with BIE | Total Number of Images |
|---|---|---|---|
| training subset 30 | 30 | 12 | 42 |
| training subset 60 | 60 | 25 | 85 |
| training subset 90 | 90 | 34 | 124 |
| training subset 120 | 120 | 49 | 169 |
| training subset 180 | 180 | 69 | 249 |
| training subset 240 | 240 | 86 | 326 |
| training subset 300 | 300 | 106 | 406 |
| validation subset | 24 | 7 | 31 |
| testing subset | 24 | - | 24 |

**Table 2.** Values of models' hyperparameters.

| Hyperparameter | YOLOv5n | YOLOv5s-Ghost | MobileNetV3-Small |
|---|---|---|---|
| Number of epochs | 100 | 100 | 100 |
| Batch size | 16 | 16 | 32 |
| Image size (in pixels) | 640 × 640 | 640 × 640 | 320 × 320 |
| Learning rate | 0.001 | 0.001 | 0.01 |

Changes in the hyperparameter values listed in Table 2 influence the model predictions and performance. For a smaller number of epochs, the models might not be able to learn relationships in the data and therefore, might achieve a poor quality on both the training and testing subsets. For a larger number of epochs, the models will learn the data better, but there is a risk of overfitting to the training data and poor generalization. Setting a higher value of learning rate causes a faster loss decrease but increases the risk of missing the optimal minimum. For the lower value of learning rate, the decrease in the loss is lower; therefore, training will take longer and there is a risk of falling into a local minimum. Resizing images to a smaller size will lead to their poor quality; some details can be missed and therefore, the model will not be able to detect objects properly. Too-large image sizes will make it harder for models to properly fit the data when there is a small number of images in the training subset.

All background images used in the experiments in Section 2.4 were created with BIE with the hyperparameter values described in Table 3. Backgrounds with a width or height smaller than 100 px were filtered out.

**Table 3.** Values of BIE hyperparameters.

| Hyperparameter | Value |
|---|---|
| x_margin | 30 px |
| y_margin | 30 px |
| Background width or height minimal size | 100 px |

## 3. Results

Tables 4–6 show the results of YOLOv5n, YOLOv5s-ghost, and MobileNetV3-small on the testing subset. The YOLOv5n architecture achieved the best performance in terms of all performance measures; its variant trained on datasets with background extraction performed the best or not far below the best result. YOLOv5s-ghost and MobileNetV3-small performed significantly worse in terms of mAP@0.5, mAP@[0.5, 0.95], and mAR. This illustrates an advantage of the YOLOv5n model as a railway object detector, which is the smallest of the neural networks taken into consideration. The task is relatively simple and the dataset so small that the large number of features in more capacious architectures caused overfitting.

**Table 4.** Values of mAP@0.5 on testing subset. Bold values are the highest ones for each size of the training subset. mAP@0.5 measures how well a model finds true objects on the image, allowing a 50% margin of error for the bbox area.

| Method | 30 obs. | 60 obs. | 90 obs. | 120 obs. | 180 obs. | 240 obs. | 300 obs. |
|---|---|---|---|---|---|---|---|
| YOLOv5n | 0.563 | 0.899 | 0.919 | **0.933** | 0.941 | 0.934 | **0.942** |
| BIE + YOLOv5n | **0.843** | **0.908** | **0.923** | 0.931 | **0.947** | **0.938** | 0.936 |
| YOLOv5s-ghost | 0.00 | 0.079 | 0.589 | 0.712 | 0.848 | 0.879 | 0.906 |
| BIE + YOLOv5s-ghost | 0.008 | 0.352 | 0.778 | 0.816 | 0.858 | 0.899 | 0.910 |
| MobileNetV3-small | 0.113 | 0.229 | 0.258 | 0.169 | 0.343 | 0.337 | 0.408 |
| BIE + MobileNetV3-small | 0.276 | 0.149 | 0.176 | 0.281 | 0.348 | 0.322 | 0.369 |

**Table 5.** Values of mAP@[0.5, 0.95] on testing subset. Bold values are the highest ones for each size of the training subset. mAP@[0.5, 0.95] measures how well a model finds true objects in the image, averaging over different margins of error for the bbox area.

| Method | 30 obs. | 60 obs. | 90 obs. | 120 obs. | 180 obs. | 240 obs. | 300 obs. |
|---|---|---|---|---|---|---|---|
| YOLOv5n | 0.208 | 0.531 | 0.582 | 0.615 | 0.641 | 0.647 | 0.663 |
| BIE + YOLOv5n | **0.427** | **0.593** | **0.636** | **0.617** | **0.651** | **0.659** | **0.667** |
| YOLOv5s-ghost | 0.00 | 0.020 | 0.218 | 0.326 | 0.480 | 0.499 | 0.555 |
| BIE + YOLOv5s-ghost | 0.002 | 0.120 | 0.395 | 0.395 | 0.479 | 0.541 | 0.559 |
| MobileNetV3-small | 0.039 | 0.067 | 0.084 | 0.060 | 0.109 | 0.116 | 0.161 |
| BIE + MobileNetV3-small | 0.099 | 0.026 | 0.042 | 0.087 | 0.093 | 0.120 | 0.108 |

**Table 6.** Values of mAR 100 on testing subset. Bold values are the highest ones for each size of the training subset. mAR 100 measures the proportion of the top 100 correctly detected objects to all objects in the image.

| Method | 30 obs. | 60 obs. | 90 obs. | 120 obs. | 180 obs. | 240 obs. | 300 obs. |
|---|---|---|---|---|---|---|---|
| YOLOv5n | 0.391 | 0.596 | 0.645 | **0.686** | **0.707** | 0.711 | 0.718 |
| BIE + YOLOv5n | **0.528** | **0.666** | **0.702** | 0.683 | **0.707** | **0.717** | **0.723** |
| YOLOv5s-ghost | 0.00 | 0.074 | 0.334 | 0.420 | 0.549 | 0.584 | 0.626 |
| BIE + YOLOv5s-ghost | 0.020 | 0.208 | 0.469 | 0.483 | 0.552 | 0.607 | 0.630 |
| MobileNetV3-small | 0.069 | 0.124 | 0.132 | 0.109 | 0.167 | 0.165 | 0.216 |
| BIE + MobileNetV3-small | 0.142 | 0.065 | 0.091 | 0.140 | 0.140 | 0.185 | 0.160 |

Figure 10 shows the relationship between the training subset size and model performance. The plot shows that for training subsets that consist of 120 or fewer observations, YOLO models trained with BIE performed better than their variants trained without BIE. For training subsets containing more than 120 observations, there was no noticeable improvement of the YOLOv5n model trained with BIE when compared to its variant trained without BIE. Nevertheless, including background images in training subsets larger than 120 observations improved YOLOv5s-ghost. MobileNetv3-small did not show a noticeable quality increase after adding background images. The plots show that for training subsets consisting of more than 120 observations, the relative improvement of YOLOv5n's performance is slight. Therefore, 120 is a sufficient number of observations to train an accurate model. In addition, when BIE is applied, even 90 observations are enough to achieve the same performance as 120 observations without BIE. YOLOv5n, compared to YOLOv5s-ghost and MobileNetv3-small, consists of a smaller number of parameters; therefore, it is expected to require fewer observations to achieve a good performance, which is consistent with the experimental results. BIE augments the training set with additional background images to teach algorithms which objects should not be detected and it can be seen that for small datasets it does indeed improve the quality of YOLO models. In contrast, it does not improve the quality of MobileNet, presumably because the model consists of the largest number of parameters and needs a larger volume of data.

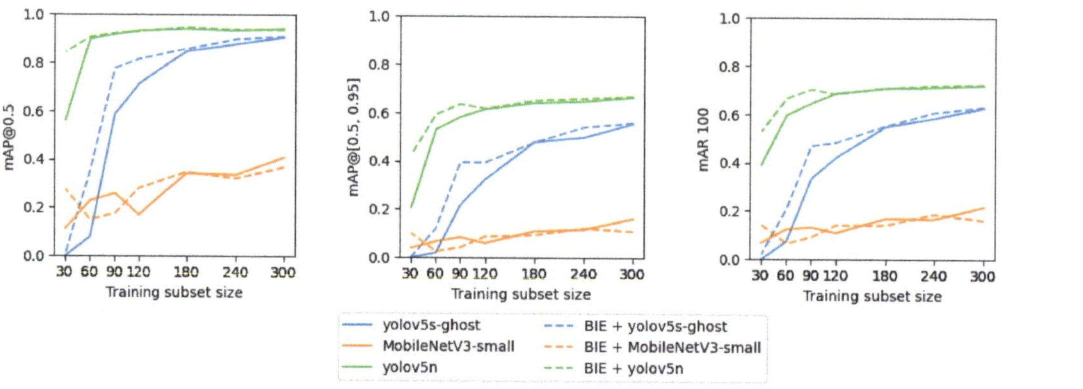


**Figure 10.** The relationship between the size of the training subset (x-axis) and model performance on the corresponding testing subset (y-axis). Each plot corresponds to a different performance measure. Colors mark neural network architectures and line types mark the presence of background images extracted with BIE in the training subset. This plot is the visualization of Tables 4–6.

To showcase the phenomena observed throughout the entire test set, we present a sample of images. Figures 11–16 show the representative visualizations of example railway component detections from the testing subset. All figures contain the results of models trained on datasets with Background Image Extraction. The red arrows and numbers on figures point to incorrectly detected bboxes. YOLOv5n returns more accurate detections when compared to YOLOv5s-ghost and MobileNetv3-small. A comparison between Figures 11 and 12 and Figures 14 and 15 shows that YOLOv5s-ghost detects sleepers and fasteners as well as YOLOv5n, but incorrectly detects rail bboxes, which are too short (Figure 12) and overlapping (Figure 15). In turn, Figure 13 shows that on short images MobileNetV3-small detects additional fasteners in inaccurate places, which may be caused by the fact that MobileNetV3 is a backbone for SSDLite, which is a variant of Single-Shot Detector (SSD) and SSD is known for worse detections on small objects [40]. Figure 16 shows that on long images MobileNetV3-small detects too-large bboxes for fasteners and sleepers, which might be caused by the small number of long images in the dataset and therefore, the model has overfitted the short images.

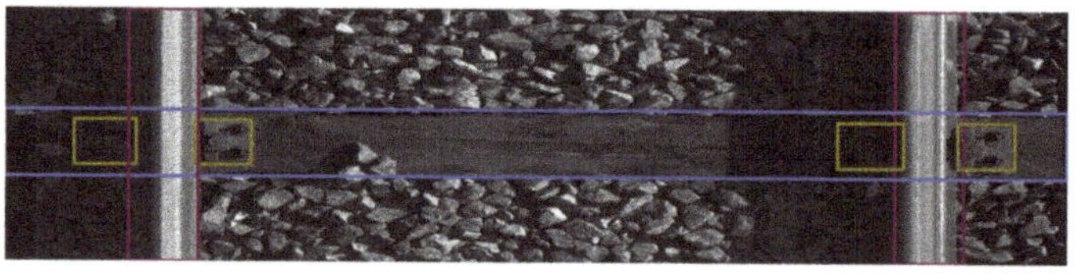

**Figure 11.** Example BIE + YOLOv5n model prediction for short image.

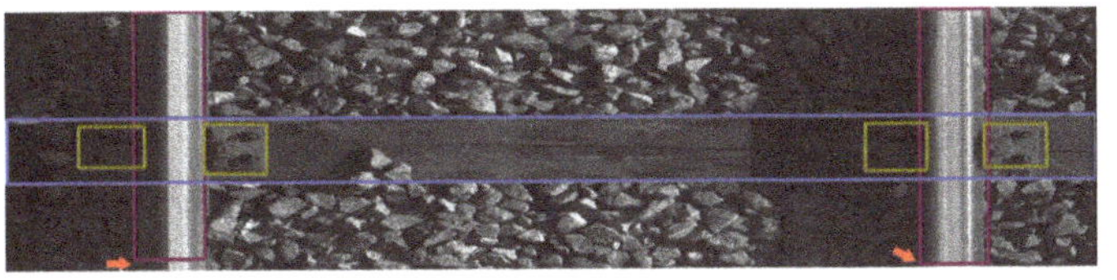

**Figure 12.** Example BIE + YOLOv5s-ghost model prediction for short image with marked incorrectly detected bboxes.

The red arrows in Figure 12 point to incorrectly detected rail bboxes—they are too short.

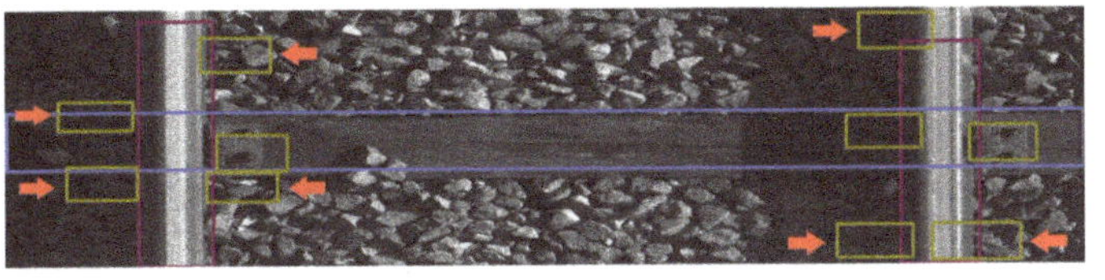

**Figure 13.** Example BIE + MobileNetV3-small model prediction for short image with marked incorrectly detected bboxes.

The red arrows in Figure 13 point to fastener bboxes that are detected in incorrect places.

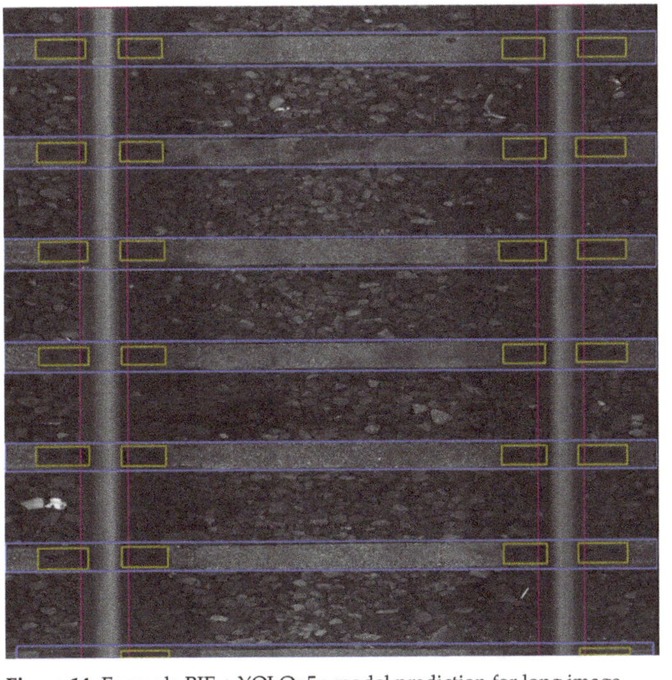

**Figure 14.** Example BIE + YOLOv5n model prediction for long image.

The red arrows in Figure 15 point to places where the detected rail bboxes are overlapping.

The red arrows in Figure 16 with the number 1 point to examples of undetected fastener bboxes, with number 2 to detected fastener bboxes that are too large, with number 3 to detected sleeper bboxes that are too large, with number 4 to a detected sleeper bbox that is in the incorrect place, and with number 5 to detected rail bboxes that are too short.

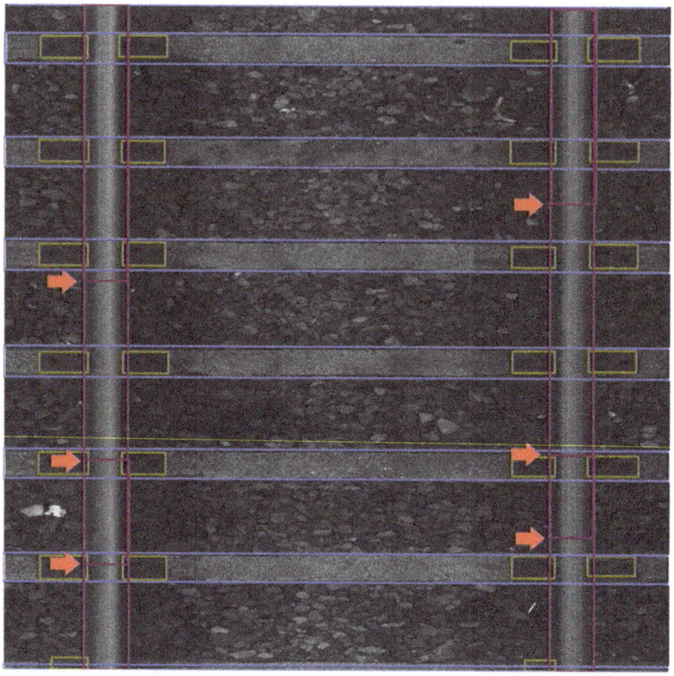

**Figure 15.** Example BIE + YOLOv5s-ghost model prediction for long image with marked incorrectly detected bboxes.

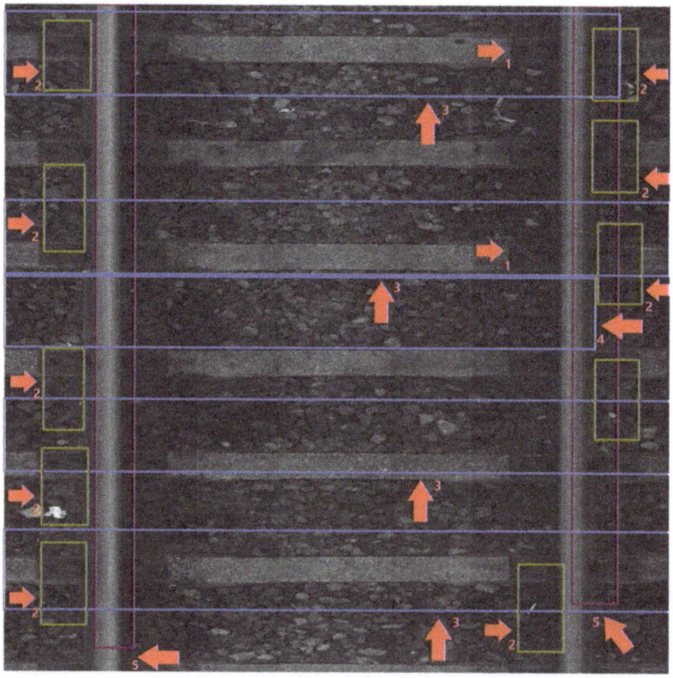

**Figure 16.** Example BIE + MobileNetV3-small model prediction for long image with marked incorrectly detected bboxes.

## 4. Discussion

In summary, the experiments show that the task of railway component detection is relatively simple, and a training set consisting of 120 labeled observations is sufficient to train an efficient model. In addition, the results show that BIE may enrich a small dataset and reduce the number of observations needed to train an accurate model from 120 to 90. The model that performed best is YOLOv5n, which is the smallest of the considered architectures, supporting the hypothesis that the task is simple and does not require much labeled data.

## 5. Conclusions

In this paper, we searched for a sufficient data volume for the detection of railway infrastructure components. As a result of this study, the following findings have been made:

- In total, 120 training observations are enough to train an efficient YOLO model. At the same time, the authors of YOLO recommend using over 1500 images per class and over 10000 labeled objects for best training results (https://github.com/ultralytics/yolov5/wiki/Tips-for-Best-Training-Results), accessed on 3 July 2023, which is approximately 100 times more than was needed in our experiment. Taking this into account, a sufficient detector of the railway objects requires a relatively small amount of data, which is desirable since labeled railway images are not easily available;
- The number of observations required to train an efficient railway OD model can be reduced to 90 observations after applying our method BIE, which allows for background extraction from the training subset. The use of background images is common in OD tasks since backgrounds are usually simple to acquire, which is different for railway backgrounds, which cannot have any images that do not contain railway components. These should be photos composed of the ballast alone, which requires additional effort to obtain them. Thus, this paper's result that BIE can be used to extract backgrounds from training images is an important finding;
- The best model for the railway object detection task is YOLOv5n, which is the smallest of the YOLO models, and therefore, is more robust for overfitting to small datasets.

In summary, this paper's results demonstrate the great potential of neural networks for detecting railway infrastructure objects. With a limited amount of data labeling, it is possible to obtain adequate models that can support people in railway track condition analysis.

**Author Contributions:** Conceptualization, A.G., M.B. and T.R.; data curation, Z.B. and M.B.; software A.G., Z.B. and M.B.; visualization, A.G.; writing—original draft, A.G. and Z.B.; writing—review and editing, T.R.; supervision, T.R. All authors have read and agreed to the published version of the manuscript.

**Funding:** This work was financially supported by the National Center for Research and Development (Poland) grant POIR.01.01.01-00-1131/20-00.

**Data Availability Statement:** Data unavailable for public sharing due to privacy reasons.

**Acknowledgments:** We would like to acknowledge Aleksander Hernik and Marcin Baran for their support with data gathering, and Korneliusz Lewczuk and Ignacy Gloza for their contributions to the code development.

**Conflicts of Interest:** The authors declare no conflict of interest.

## Abbreviations

The following abbreviations are used in this manuscript:

| | |
|---|---|
| bbox | Bounding Box |
| BIE | Background Image Extraction |
| CNN | Convolutional Neural Network |
| IoU | Intersection Over Union |
| KNN | K-Nearest Neighbors |
| ML | Machine Learning |
| mAP | Mean Average Precision |
| mAR | Mean Average Recall |
| OD | Object Detection |
| R-CNN | Region-based Convolutional Neural Networks |
| SSD | Single-Shot Detector |
| SVM | Support Vector Machines |
| YOLO | You Only Look Once |
| YOLOv5n | YOLO version 5 nano |
| YOLOv5s-ghost | YOLO version 5 small with ghost bottleneck |

## References


1. Banister, D. Cities, mobility and climate change. *J. Transp. Geogr.* **2011**, *19*, 1538–1546. [CrossRef]
2. Xia, T.; Zhang, Y.; Crabb, S.; Shah, P. Cobenefits of Replacing Car Trips with Alternative Transportation: A Review of Evidence and Methodological Issues. *J. Environ. Public Health* **2013**, *2013*, 1–14. [CrossRef]
3. Kim, N.S.; Wee, B.V. Assessment of $CO_2$ emissions for truck-only and rail-based intermodal freight systems in Europe. *Transp. Plan. Technol.* **2009**, *32*, 313–333. [CrossRef]
4. Xia, W.; Zhang, A. High-speed rail and air transport competition and cooperation: A vertical differentiation approach. *Transp. Res. Part B Methodol.* **2016**, *94*, 456–481. [CrossRef]
5. Liu, X.; Saat, M.R.; Barkan, C.P.L. Analysis of Causes of Major Train Derailment and Their Effect on Accident Rates. *Transp. Res. Rec.* **2012**, *2289*, 154–163. [CrossRef]
6. Gawlak, K. Analysis and assessment of the human factor as a cause of occurrence of selected railway accidents and incidents. *Open Eng.* **2023**, *13*, 1–3. [CrossRef]
7. Nakhaee, M.C.; Hiemstra, D.; Stoelinga, M.; van Noort, M. The Recent Applications of Machine Learning in Rail Track Maintenance: A Survey. In *Reliability, Safety, and Security of Railway Systems. Modelling, Analysis, Verification, and Certification*; Springer International Publishing: Berlin/Heidelberg, Germany, 2019; pp. 91–105. [CrossRef]
8. Li, Y.; Trinh, H.; Haas, N.; Otto, C.; Pankanti, S. Rail Component Detection, Optimization, and Assessment for Automatic Rail Track Inspection. *IEEE Trans. Intell. Transp. Syst.* **2014**, *15*, 760–770. [CrossRef]
9. Manikandan, R.; Balasubramanian, M.; Palanivel, S. Machine Vision Based Missing Fastener Detection in Rail Track Images Using SVM Classifier. *Int. J. Smart Sens. Intell. Syst.* **2017**, *10*, 574–589. [CrossRef]
10. Ghiasi, A.; Ng, C.T.; Sheikh, A.H. Damage detection of in-service steel railway bridges using a fine k-nearest neighbor machine learning classifier. *Structures* **2022**, *45*, 1920–1935. [CrossRef]
11. Santur, Y.; Karaköse, M.; Akin, E. Random forest based diagnosis approach for rail fault inspection in railways. In Proceedings of the 2016 National Conference on Electrical, Electronics and Biomedical Engineering (ELECO), Bursa, Turkey, 1–3 December 2016; pp. 745–750.
12. Hsieh, C.C.; Hsu, T.Y.; Huang, W.H. An Online Rail Track Fastener Classification System Based on YOLO Models. *Sensors* **2022**, *22*, 9970. [CrossRef]
13. Gibert, X.; Patel, V.M.; Chellappa, R. Deep Multitask Learning for Railway Track Inspection. *IEEE Trans. Intell. Transp. Syst.* **2017**, *18*, 153–164. [CrossRef]
14. Zhu, Y.; Sekiya, H.; Okatani, T.; Yoshida, I.; Hirano, S. Acceleration-based deep learning method for vehicle monitoring. *IEEE Sensors J.* **2021**, *21*, 17154–17161. [CrossRef]
15. Lorenzen, S.R.; Riedel, H.; Rupp, M.M.; Schmeiser, L.; Berthold, H.; Firus, A.; Schneider, J. Virtual Axle Detector Based on Analysis of Bridge Acceleration Measurements by Fully Convolutional Network. *Sensors* **2022**, *22*, 8963. [CrossRef] [PubMed]
16. Cha, Y.J.; Choi, W.; Suh, G.; Mahmoudkhani, S.; Büyüköztürk, O. Autonomous Structural Visual Inspection Using Region-Based Deep Learning for Detecting Multiple Damage Types. *Comput.-Aided Civ. Infrastruct. Eng.* **2018**, *33*, 731–747. [CrossRef]
17. Guedes, A.; Silva, R.; Ribeiro, D.; Vale, C.; Mosleh, A.; Montenegro, P.; Meixedo, A. Detection of Wheel Polygonization Based on Wayside Monitoring and Artificial Intelligence. *Sensors* **2023**, *23*, 2188. [CrossRef]
18. Ni, Y.Q.; Zhang, Q.H. A Bayesian machine learning approach for online detection of railway wheel defects using track-side monitoring. *Struct. Health Monit.* **2021**, *20*, 1536–1550. [CrossRef]
19. Ta, Q.B.; Huynh, T.C.; Pham, Q.Q.; Kim, J.T. Corroded Bolt Identification Using Mask Region-Based Deep Learning Trained on Synthesized Data. *Sensors* **2022**, *22*, 3340. [CrossRef]

20. Tan, L.; Tang, T.; Yuan, D. An Ensemble Learning Aided Computer Vision Method with Advanced Color Enhancement for Corroded Bolt Detection in Tunnels. *Sensors* **2022**, *22*, 9715. [CrossRef]
21. Datta, D.; Hosseinzadeh, A.Z.; Cui, R.; Lanza di Scalea, F. Railroad Sleeper Condition Monitoring Using Non-Contact in Motion Ultrasonic Ranging and Machine Learning-Based Image Processing. *Sensors* **2023**, *23*, 3105. [CrossRef]
22. Kaewunruen, S.; Sresakoolchai, J.; Huang, J.; Zhu, Y.; Ngamkhanong, C.; Remennikov, A.M. Machine Learning Based Design of Railway Prestressed Concrete Sleepers. *Appl. Sci.* **2022**, *12*, 311. [CrossRef]
23. Zhuang, L.; Wang, L.; Zhang, Z.; Tsui, K.L. Automated vision inspection of rail surface cracks: A double-layer data-driven framework. *Transp. Res. Part C: Emerg. Technol.* **2018**, *92*, 258–277. [CrossRef]
24. Chen, J.; Liu, Z.; Wang, H.; Núñez, A.; Han, Z. Automatic Defect Detection of Fasteners on the Catenary Support Device Using Deep Convolutional Neural Network. *IEEE Trans. Instrum. Meas.* **2018**, *67*, 257–269. [CrossRef]
25. Meixedo, A.; Ribeiro, D.; Santos, J.; Calçada, R.; Todd, M.D. *Real-Time Unsupervised Detection of Early Damage in Railway Bridges Using Traffic-Induced Responses*; Springer: Berlin/Heidelberg, Germany, 2022.
26. Suh, G.; Cha, Y.J. Deep faster R-CNN-based automated detection and localization of multiple types of damage. In Proceedings of the Sensors and Smart Structures Technologies for Civil, Mechanical, and Aerospace Systems 2018 (SPIE), Denver, CO, USA, 4–8 March 2018; Volume 10598, pp. 197–204.
27. Giben, X.; Patel, V.M.; Chellappa, R. Material classification and semantic segmentation of railway track images with deep convolutional neural networks. In Proceedings of the 2015 IEEE International Conference on Image Processing (ICIP), Quebec City, QC, Canada, 27–30 September 2015; pp. 621–625. [CrossRef]
28. Wang, T.; Yang, F.; Tsui, K.L. Real-Time Detection of Railway Track Component via One-Stage Deep Learning Networks. *Sensors* **2020**, *20*, 4325. [CrossRef] [PubMed]
29. Girshick, R.; Donahue, J.; Darrell, T.; Malik, J. Rich Feature Hierarchies for Accurate Object Detection and Semantic Segmentation. In Proceedings of the Proceedings of the IEEE Conference on Computer Vision and Pattern Recognition (CVPR), Columbus, OH, USA, 23–28 June 2014 .
30. Uijlings, J.R.R.; van de Sande, K.E.A.; Gevers, T.; Smeulders, A.W.M. Selective Search for Object Recognition. *Int. J. Comput. Vis.* **2013**, *104*, 154–171. [CrossRef]
31. Ren, S.; He, K.; Girshick, R.; Sun, J. Faster R-CNN: Towards Real-Time Object Detection with Region Proposal Networks. In *Proceedings of the Advances in Neural Information Processing Systems, Montreal, QC, Canada, 7–12 December 2015*; Cortes, C., Lawrence, N., Lee, D., Sugiyama, M., Garnett, R., Eds.; Curran Associates, Inc.: Red Hook, NY, USA, 2015; Volume 28.
32. Liu, W.; Anguelov, D.; Erhan, D.; Szegedy, C.; Reed, S.; Fu, C.Y.; Berg, A.C. SSD: Single Shot MultiBox Detector. In *Proceedings of the Computer Vision—ECCV 2016, Amsterdam, The Netherlands, 11–14 October 2016*; Leibe, B., Matas, J., Sebe, N., Welling, M., Eds.; Springer: Cham, Switzerland, 2016; pp. 21–37.
33. Sandler, M.; Howard, A.; Zhu, M.; Zhmoginov, A.; Chen, L.C. MobileNetV2: Inverted Residuals and Linear Bottlenecks. In Proceedings of the 2018 IEEE/CVF Conference on Computer Vision and Pattern Recognition, Salt Lake City, UT, USA, 18–23 June 2018; pp. 4510–4520. [CrossRef]
34. Howard, A.G.; Zhu, M.; Chen, B.; Kalenichenko, D.; Wang, W.; Weyand, T.; Andreetto, M.; Adam, H. MobileNets: Efficient Convolutional Neural Networks for Mobile Vision Applications. *arXiv* **2017**, arXiv:1704.04861.
35. Howard, A.; Sandler, M.; Chen, B.; Wang, W.; Chen, L.C.; Tan, M.; Chu, G.; Vasudevan, V.; Zhu, Y.; Pang, R.; et al. Searching for MobileNetV3. In Proceedings of the 2019 IEEE/CVF International Conference on Computer Vision (ICCV), Seoul, Republic of Korea, 27 October–2 November 2019; pp. 1314–1324. [CrossRef]
36. Redmon, J.; Divvala, S.K.; Girshick, R.B.; Farhadi, A. You Only Look Once: Unified, Real-Time Object Detection. *arXiv* **2015**, arXiv:1506.02640.
37. Guo, K.; He, C.; Yang, M.; Wang, S. A pavement distresses identification method optimized for YOLOv5s. *Sci. Rep.* **2022**, *12*, 3542. [CrossRef]
38. Lin, T.Y.; Dollar, P.; Girshick, R.; He, K.; Hariharan, B.; Belongie, S. Feature Pyramid Networks for Object Detection. In Proceedings of the Proceedings of the IEEE Conference on Computer Vision and Pattern Recognition (CVPR), Honolulu, HI, USA, 21–26 July 2017.
39. Padilla, R.; Passos, W.L.; Dias, T.L.B.; Netto, S.L.; da Silva, E.A.B. A Comparative Analysis of Object Detection Metrics with a Companion Open-Source Toolkit. *Electronics* **2021**, *10*, 279. [CrossRef]
40. Fang, L.; Zhao, X.; Zhang, S. Small-objectness sensitive detection based on shifted single shot detector. *Multimed. Tools Appl.* **2019**, *78*, 13227–13245. [CrossRef]

*sensors*

MDPI

*Article*

# A Machine-Learning-Based Approach for Railway Track Monitoring Using Acceleration Measured on an In-Service Train

**Abdollah Malekjafarian [1,*], Chalres-Antoine Sarrabezolles [2], Muhammad Arslan Khan [1] and Fatemeh Golpayegani [3]**

1 Structural Dynamics and Assessment Laboratory, School of Civil Engineering, University College Dublin, D04V1W8 Dublin, Ireland
2 The École Nationale des Travaux Publics de l'État (ENTPE), 69518 Lyon, France
3 School of Computer Science, University College Dublin, D04V1W8 Dublin, Ireland
* Correspondence: abdollah.malekjafarian@ucd.ie

**Abstract:** In this paper, a novel railway track monitoring approach is proposed that employs acceleration responses measured on an in-service train to detect the loss of stiffness in the track sub-layers. An Artificial Neural Network (ANN) algorithm is developed that works with the energies of the train acceleration responses. A numerical model of a half-car train coupled with a track profile is employed to simulate the train vertical acceleration. The energy of acceleration signals measured from 100 traversing trains is used to train the ANN for healthy track conditions. The energy is calculated every 15 m along the track, each of which is called a slice. In the monitoring phase, the trained ANN is used to predict the energies of a set of train crossings. The predicted energies are compared with the simulated ones and represented as the prediction error. The damage is modeled by reducing the soil stiffness at the sub-ballast layer that represents hanging sleepers. A damage indicator (DI) based on the prediction error is proposed to visualize the differences in the predicted energies for different damage cases. In addition, a sensitivity analysis is performed where the impact of signal noise, slice sizes, and the presence of multiple damaged locations on the performance of the DI is assessed.

**Keywords:** machine learning; railway infrastructure monitoring; track damage detection; SHM; acceleration; in-service train measurements; drive-by monitoring; ANN

**Citation:** Malekjafarian, A.; Sarrabezolles, C.-A.; Khan, M.A.; Golpayegani, F. A Machine-Learning-Based Approach for Railway Track Monitoring Using Acceleration Measured on an In-Service Train. *Sensors* **2023**, *23*, 7568. https://doi.org/10.3390/s23177568

Academic Editor: Jiawei Xiang

Received: 20 July 2023
Revised: 28 August 2023
Accepted: 30 August 2023
Published: 31 August 2023



## 1. Introduction

In recent years, with the increasing global demand for mass transportation and freight, the maintenance of existing transport infrastructure has become important. Railway systems are vital components of transportation systems, which provide a reliable, cost-efficient, and sustainable transportation mode. Railway services are commonly seen as a safe mode of travel (or freight moving) with low tariffs, reliable speeding, and low environmental impact [1]. Most of the existing railway infrastructure is aged and requires continuous monitoring to keep it in service, which requires enormous cost [2]. Moreover, these structures are subjected to heavier axle loads, faster train speeds, and greater frequencies of trains, which have resulted in rapid deterioration over time [3]. Hence, efficient and reliable infrastructure monitoring systems are needed to ensure these systems run smoothly at a reasonable cost.

Railways tracks are the first contact between the heavy, traversing trains and the railway structure below. For that reason, the track receives significant amounts of stresses (both longitudinal and transverse), resulting in severe deterioration over time [3–8]. The structural condition of the railway tracks needs to be monitored regularly and essential repairs need to be planned to provide an effective first train–rail contact. Conventionally, railway tracks are monitored through regular visual inspection by walking along the track with the help of hand-held surveying equipment, e.g., tachometer or leveler. This method is expensive to perform frequently and intensive when the whole network is considered [9,10].

Moreover, this method detects only major faults that are visually detectable, and it is not efficient on detecting inner damages, which are only apparent when a train passes over. Another commonly implemented method for monitoring tracks is the use of a specialized vehicle, i.e., Track Recording Vehicle (TRV), which is equipped with optical and inertial sensors to collect track geometric information while traversing [11]. These vehicles can measure the level and alignment of the rail, track gauge, cross level, rail twist, curvature/curve radius, gradient, and track position, using GPS and displacement sensors [11,12]. According to the European regulations, the geometric characterization of the railway track has to be measured at a regular frequency in accordance with the EN13848 standard [12], which can be implemented using TRVs. If any characteristic is above the tolerance limit, temporary speed restrictions or a suspension of train operations may be applied. Although TRVs provide reliable inspection data, they are extremely expensive to deploy, and they create traffic disruptions during inspection in the form of occupied tracks and restricted traffic speeds.

Recent research on the use of sensing technology and computational power has opened new opportunities to improve railway track monitoring systems and to provide accurate and cost-effective solutions [9,13]. A common practice is the installation of sensors on the track and monitoring track accelerations to detect any faults. For example, Liu et al. [14] proposed a method that uses a distributed network of sensors installed on the track, which measures vertical accelerations and strains of the track. Similarly, many researchers have proposed the use of fiber Bragg grating (FBG) strain sensors to monitor the behavior of the track [15,16]. Although direct track monitoring systems yield effective results, these are not feasible for the complex and widely spread network of railway infrastructure [3].

Indirect methods of monitoring have received much attention in recent years, which are based on the installation of sensors on a passing train for track health monitoring and railway bridge monitoring, also known as "drive-by" monitoring [17–21]. In these methods, train components closer to the track, e.g., suspensions, are instrumented to measure their dynamic responses, which can also be real-time data of a passing train. In this way, several passenger trains can be instrumented and used for continuous railway track monitoring, which can detect damage at its early stage. Several types of sensors, like laser technology [22], cameras [23,24], and inertial sensors [25], have been tested to develop drive-by track monitoring systems. Lederman et al. [20] proposed an energy-based method to inspect track changes using acceleration from the train body. They used the energy of the signals measured from several passes of an in-service train and employed a feature detection indicator to identify changes in the track over time. In order to find the track stiffness profile, Nafari et al. [22] used the relative vertical distance between the rail surface at two points measured from the train, which can be acquired using a laser-based rolling deflection measurement system. Although these approaches have shown efficacy in the results, they have targeted one specific damage case for analysis. Malekjafarian et al. [26] employed the Hilbert–Huang transformation in order to obtain the instantaneous amplitudes of the acceleration signals measured on a train. They proposed a representation of the energy of the signal as a function of train localization to detect track irregularities. Malekjafarian et al. [3] showed that the bogie-filtered displacement (BFD) can be numerically obtained using the measured drive-by acceleration and the system was validated using an in-service Irish rail train. It is also seen that the BFD is sensitive to train speed and signal noise [3]. However, with the use of statistics of train passes with different forward speeds, BFD can effectively detect the loss of stiffness on-track.

Several approaches have been made to model the train–track interaction [27,28]. OBrien et al. [28] proposed a method for inferring a track longitudinal profile from drive-by measurement using a numerical model. The train was modeled as a 10-degree-of-freedom (DOF) half-train and the track was modeled as a beam resting on a three-layer sprung mass system representing the ballast, pads, and sleepers. OBrien et al. [25] compared the results between the numerical simulations and field test measurements to validate the modeling scheme. They used an uneven longitudinal track profile for realistic representation of the track–train interaction. Using this method, dynamic responses of the train as well as railway structures can be measured with the help of numerical methods [29].

Machine learning techniques have recently been shown to be a feasible approach for drive-by monitoring of railway tracks. These algorithms create a neural network model that needs to be trained using the known data and parameters in time from a benchmark condition of a structure. Then, the model is tested using the data with unknown condition information with the help of hidden layers and neurons [30,31]. There have been significant developments in the application of machine learning techniques to SHM and damage detection. Avci et al. [30] described a comprehensive overview of the use of machine learning for vibration-based damage detection methods that can be implemented for any type of structure. Bridge structures have commonly been studied for developing SHM systems using machine learning algorithms [32]. Neves et al. [33] and Gonzalez and Karoumi [34] applied an Artificial Neural Network (ANN) that was trained using railway bridge accelerations to identify structure behavior and to develop damage detection systems using long-term data. Santos et al. [35] used a machine learning algorithm on bridge inspection data and identified inaccuracies caused by inspection issues. The effect of temperature changes on bridges was also studied using the Kalman-filter-based ANN, which has shown to eliminate the temperature effects for bridge health monitoring [36]. There are many researchers who have used cluster-based and data-driven approaches for bridge health monitoring [37–40] and have shown promising results for effective monitoring systems. Malekjafarian et al. [31,41] recently applied the concepts of machine learning on drive-by monitoring of the structures. They proposed the use of an ANN model using vehicle data, which can detect bridge frequencies and cracks on the deck. However, the road profile roughness and temperature effects have shown sensitivity to the drive-by approaches. Similarly, the learning-based approaches have been implemented for railway track monitoring, e.g., to detect and classify the severity of rail corrugation [42], wheelflat [43], and track profile [6,25,26,44], and have produced reasonable results for detecting track damages.

In this paper, a novel two-stage railway track condition monitoring approach is developed using the responses measured on a passing train and a machine learning algorithm. In the first stage, an ANN is created and trained using the energy of the accelerations collected from a fleet of trains passing over a healthy track at different forward speeds. The trained ANN is then used to predict the energy of the signals in the monitoring phase. The prediction error, which is the difference between the predicted and the simulated accelerations from each train pass, is used as the damage indicator. In the second stage, a Gaussian fitting method is applied to the prediction errors under healthy and damaged conditions. This process improves the damage detection capabilities of the system and has been shown to effectively detect damaged track conditions by reasonably increasing the fitted prediction error.

## 2. Numerical Modeling

In this paper, a coupled Train–Track Interaction (TTI) model developed by Cantero et al. [45,46] is employed. The TTI model consists of two sub-models—the train and the track (see Figure 1)—which interact and couple at the contact points. These sub-models and the coupling process are explained and illustrated in the following subsections.

### 2.1. Track model

Train tracks generally have five components—rails, pads, sleepers, ballast, and subgrade—that can be represented with various levels of sprung layers [45]. In this paper, the track model (see Figure 2) consists of a beam at the surface and three sub-levels modeled by a sprung mass. The beam represents the rail, and the sub-levels represent the sleepers, the ballast, and the subgrade [3,28,47–50]. The rail beam is created as an Euler–Bernoulli Finite Element (FE) beam, each of which has four degrees of freedom: two translations and two rotations. Two beam elements are modeled between each pair of sleepers with constant properties such as mass per unit length ($m$), modulus of elasticity ($E$), and the second moment of area ($J$). The springs and masses, representing the sub-levels, are located under each sleeper joint spaced at regular intervals of $L_S$. Geometric and mechanical properties of the track are adapted from Zhai et al. [50] and presented in Table 1.

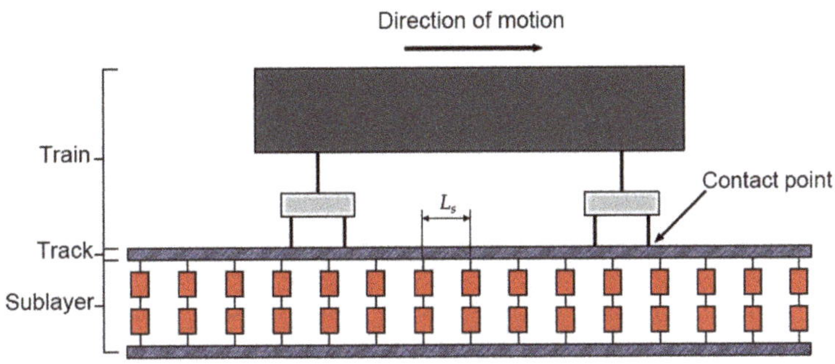


**Figure 1.** Schematic of the coupled system.

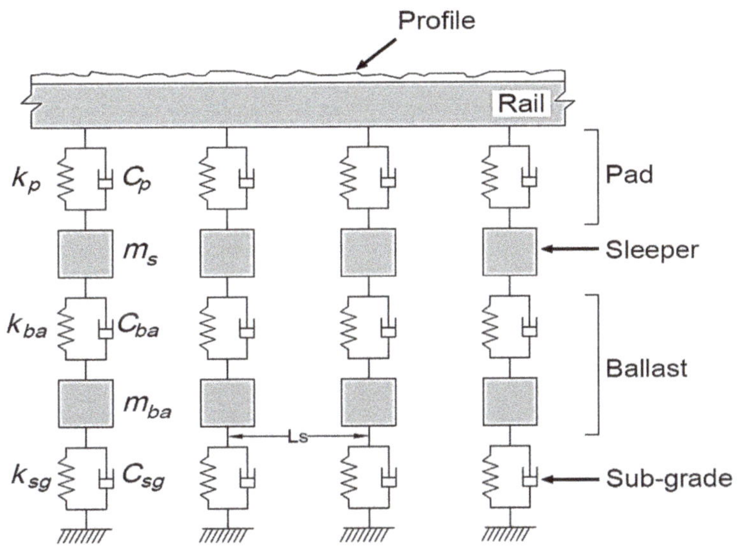


**Figure 2.** Track numerical model.

**Table 1.** Properties of the track.

| Property | Unit | Value |
|---|---|---|
| Elastic modulus of rail | N/m$^2$ | $2.059 \times 10^{11}$ |
| Rail cross-sectional area | m$^2$ | 1 |
| Rail second moment of area | m$^4$ | $3.217 \times 10^{-5}$ |
| Rail mass per unit length | kg/m | 60.64 |
| Rail pad stiffness | N/m | $6.5 \times 10^7$ |
| Rail pad damping | Ns/m | $7.5 \times 10^4$ |
| Sleeper mass (half) | kg | 125.5 |
| Sleeper spacing | m | 0.545 |
| Ballast stiffness | N/m | $137.75 \times 10^6$ |
| Ballast damping | Ns/m | $5.88 \times 10^4$ |
| Ballast mass | kg | 531.4 |
| Subgrade stiffness mean | N/m | $77.5 \times 10^6$ |
| Subgrade damping | Ns/m | $3.115 \times 10^4$ |

A total length of 1000 m is used for the track, which appropriately allows for continuous welded rails (commonly more than 180 m long [3]). In this paper, a long length of track is considered to maintain stable distance between each end of the boundary conditions. To add realistic parameters to the model, the stiffness of the ballast is considered in a non-uniform way with an average of $7.7495 \times 10^7$ N/m and a standard deviation of $4.5824 \times 10^6$ N/m. The ballast stiffness, in the space domain, is shown in Figure 3. In addition, a rail profile of class 4 irregularities is generated randomly, according to the US Federal Rail Administration guide. This profile is assumed to be constant in all the simulations in this study. However, in some cases, the profile might change over time, which might not be because of a defect on the track. Therefore, the authors acknowledge that this assumption might not truly reflect real life applications, but it is considered here for simplification. This represents a condition of the surface of the conventional rail, which comes from its power spectral density function [51,52].

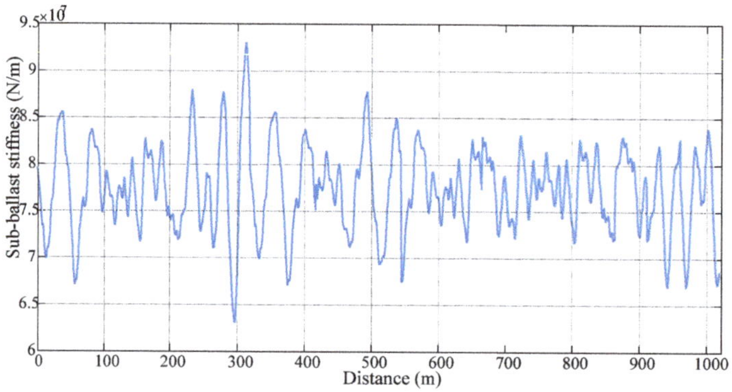


**Figure 3.** Distribution of the sub-ballast stiffness over distance.

The dynamic responses at any location change with time depending on the train position. The FE modeling vectors containing the location of each DOF and its interaction with the corresponding DOFs are created in one matrix, each for stiffness, mass, and damping parameters, using MATLAB software R2021b [53]. Dynamic responses of the modeled track to a time varying force are given by the system of equations at each time-step:

$$M_t\ddot{y}_t + C_t\dot{y}_t + K_t y_t = f_{int} \tag{1}$$

where $M_t$, $C_t$, and $K_t$ are the mass, damping, and stiffness matrices of the model, respectively; and $\ddot{y}_t$, $\dot{y}_t$, and $y_t$ are the respective vectors of acceleration, velocity, and displacement. $f_{int}$ represents the time-dependent dynamic interaction forces between the train and the track.

*2.2. Train Model*

A half-train system, as shown in Figure 4, is modeled using MATLAB software R2021b [53] to represent the train. This type of system is adapted from the literature [45,47,50,54], which has been used in multiple TTI-related studies. The train car model has 10 degrees of freedom: 7 vertical translations (four for the wheels, two for the bogies, and one for the main body) and 3 rotations in the plane (two for the bogies and one for the main body). $m_w$, $m_b$, and $J_b$ are the mass of the wheelsets, bogie mass, and moment of inertia, respectively; and $m_v$ and $J_v$ represent body mass and its moment of inertia, respectively. Viscous dampers with $c_{pa}$ damping and a spring with $k_{pa}$ stiffness are used to connect the wheels with the bogies to form a primary suspension. Likewise, a viscous damper with $C_s$ damping and a spring with $k_s$ stiffness are used for connecting the bogie with the main body. Table 2 shows the mechanical properties of the half-train system, which are adapted from Cantero et al. [45]. It should be noted that in this study, the wheelsets are considered as lumped masses and are assumed to be fixed to the track with no separation being allowed. This means that the system will be modeled as 6 degrees of freedom in the global equations of motion.

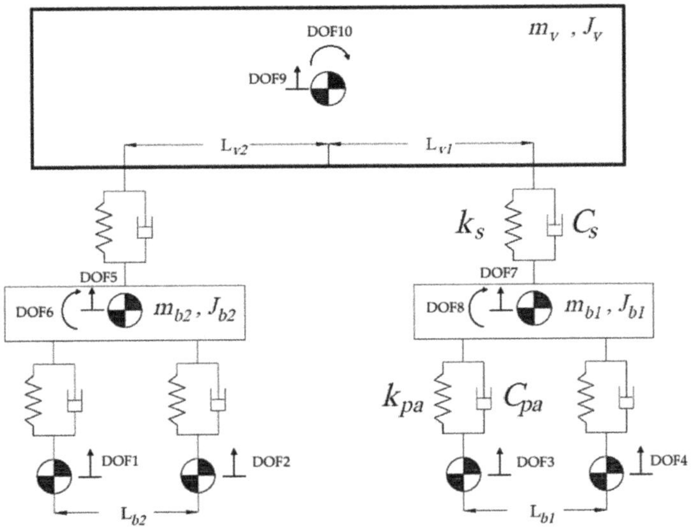


**Figure 4.** Train half-car numerical model.

**Table 2.** Properties of the train.

| Property | Symbol | Unit | Value |
|---|---|---|---|
| Wheelset mass | $m_w$ | kg | 1843.5 |
| Bogie mass | $m_b$ | kg | 59,364.2 |
| Car body mass | $m_v$ | kg | 5630.8 |
| Moment of inertia of bogie | $J_b$ | kg·m$^2$ | 9487 |
| Moment of inertia of main body | $J_v$ | kg·m$^2$ | $1.723 \times 10^6$ |
| Primary suspension stiffness | $k_{pa}$ | N/m | $2.399 \times 10^6$ |
| Secondary suspension stiffness | $k_s$ | N/m | $0.8858 \times 10^6$ |
| Primary suspension damping | $c_{pa}$ | Ns/m | $30 \times 10^3$ |
| Secondary suspension damping | $C_s$ | Ns/m | $45 \times 10^3$ |
| Distance between car body center of mass and bogie pivot | $L_{v1}$, $L_{v2}$ | m | 5.73 |
| Distance between axles | $L_{b1}$, $L_{b2}$ | m | 3 |

The dynamic responses of the vehicle can be measured using the equations of motion represented by the second-order differential equation:

$$M_v\ddot{y}_v + C_v\dot{y}_v + K_v y_v = f_v \tag{2}$$

where $M_v$, $C_v$, and $K_v$ are the mass, damping, and stiffness matrices of the train model, respectively; and $\ddot{y}_v$, $\dot{y}_v$, and $y_v$ represent vectors of train acceleration, velocity, and displacement, respectively. The dynamic interaction forces applied to the vehicle DOFs through the track profile and rail displacements are contained in the vector $f_v$.

### 2.3. Coupling of Train–Track Models

The train and the track models are combined at the wheels, which represent contact points, to form a coupled TTI system. Equations (1) and (2) are combined in a way that the corresponding DOFs from each model couple together, resulting in global matrices of the system:

$$\begin{bmatrix} M_v & 0 \\ 0 & M_t \end{bmatrix} \begin{bmatrix} \ddot{y}_v \\ \ddot{y}_t \end{bmatrix} + \begin{bmatrix} C_v & C_{v,t} \\ C_{t,v} & C_t \end{bmatrix} \begin{bmatrix} \dot{y}_v \\ \dot{y}_t \end{bmatrix} + \begin{bmatrix} K_v & K_{v,t} \\ K_{t,v} & K_t \end{bmatrix} \begin{bmatrix} y_v \\ y_t \end{bmatrix} = F \tag{3}$$

where $M$, $C$, and $K$ are global mass, damping, and stiffness matrices, respectively; and $F$ is the time-varying vector of interactive force applied by the train to the track. These matrices are calculated at each time-step according to the changing location of the traversing train. Static forces, caused by gravity and the track profile, are also included in the force vector. Equation (3) is solved using the numerical Wilson-Theta integration technique [29]. A value of Wilson-Theta ($\theta$) of 1.420815 is used to ensure unconditional stability in the integration process.

### 2.4. Modeling of Damage

In this paper, the percentage loss of track ballast stiffness is used as the damaged condition. A traversing train transfers the moving force through the tracks to the ground, and a change in the stiffness in any component of the system impacts the vehicle's accelerations [45,47]. Track variability may create a local stress concentration under the sleepers. This could lead to a loss of stiffness under the sleeper and a soft soil formation in the subgrade layer. In the long-term, it may create a loss of contact between the ballast and the sleepers, which are also known as hanging sleepers [55]. If hanging sleepers are not detected at the right time, they may accelerate the track deterioration and cause more damage. In the numerical model, the damage is simulated by reducing the ballast stiffness by 10%, 30%, 50%, and 70% of the healthy state. These damage conditions represent partial or nearly complete loss of the ballast underneath the track, which is one of the major issues commonly seen by the track inspectors [3,24]. The acceleration response from the traversing train is measured using health and damaged track cases and compared using the machine learning and data-driven approach.

## 3. The Proposed Algorithm for Track Monitoring

A novel track monitoring algorithm is proposed in this paper that consists of two phases in correlation to detect track damages. The first phase utilizes an ANN model to estimate the energy of train acceleration responses while transversing the irregular track with healthy conditions. This involves training a neural network to predict data from simulated acceleration measurements. In the second phase, a test data point is predicted using the neural network and compared with the simulated data. Also, the healthy and the damaged data cases are separated and compared to show the changes in the predicted data from the trained ANN. These two phases are illustrated as a flowchart in Figure 5, which shows signal processing steps involved in each phase.

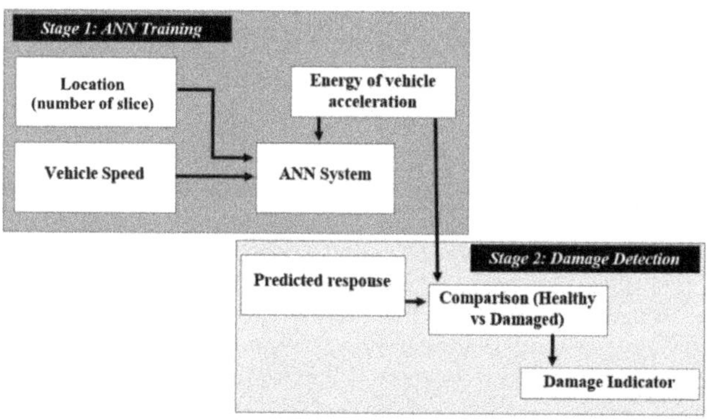


**Figure 5.** The proposed machine learning approach.

The ANN is a commonly used algorithm for monitoring structures and is used to predict the energy content of an in-service train response, considering various energy bands and train speeds. During the monitoring stage (phase 2), the calculated energy spectrum from each vehicle passage is compared to that predicted by the ANN and a damage indicator (DI) is evaluated for each traversing train.

### 3.1. ANN Background

A neural network is a very useful and evolving tool for predicting one or more desired outcomes in complex systems, using a history of data collection and previous outcomes. This tool has been used several times in research studies to solve predictions to nonlinear problems, pattern recognition, or optimization [30,39,56]. Neural networks consist of incoming data, hidden data, outgoing data, weights and bias, an activation function, and a summing node [36,57]. Each level incorporates several units of calculation called a neuron [30,41], which takes its input data from the previous level and provides output data for the next level. The input level provides the input data of the network, which are fed to the hidden level. Each hidden level has a number of neurons that will calculate an output using all the inputs of the input level and a predefined set of weights and bias. This output can be either fed to the next level or directly to the output layer. The output level analyzes all the input produced in the last hidden level and produces the last output of the whole algorithm.

Using the MATLAB deep learning toolbox, the ANN is implemented in this paper to a set of vehicle acceleration data. The ANN consists of an input layer with 3 input neurons, two hidden layers (containing 20 neurons each), and an output layer. The output layer provides the predicted results that, in this case, is the predicted energy response. Figure 6 shows the schematic of the ANN where each neuron calculates a single output data point from the inputs at the previous levels. In Figure 6, $S_i$ is the output from the neuron $n_i$ of the previous level, $w_i$ is the weight associated with $n_i$, $b$ is the neuron bias, and $S$ is some transform function, typically sigmoid. Activation and transform functions are given in Equations (4) and (5), respectively. The main role of the activation function is to transform the summed weighted input into an output value, which will be used to feed into the next hidden layer.

$$\text{Activation}: \sum_{i=1}^{N} w_i S_i \tag{4}$$

$$\text{Transform}: S\left(b + \sum_{i=1}^{N} w_i S_i\right) \tag{5}$$

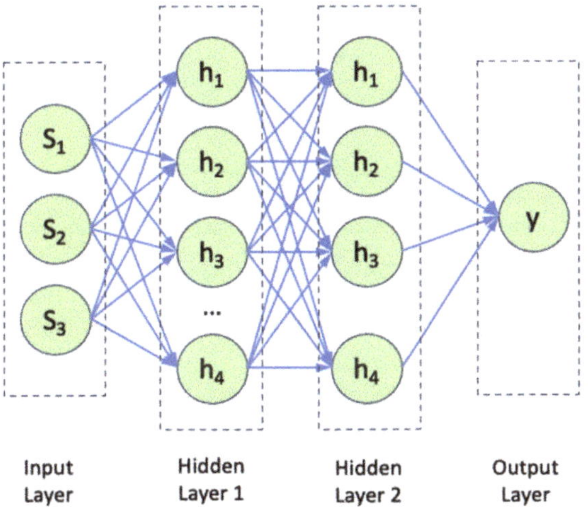


**Figure 6.** Flowchart of output calculations using neurons.

In this paper, the ANN is implemented using the supervised learning approach, which is based on the difference between the predicted and the simulated outputs. To train the ANN model, a Levenberg–Marquardt backpropagation (LMBP) algorithm [58] is applied, which operates in a closed loop to minimize the differences between the predicted and measured signals. The hidden layers of the ANN contain two hyperbolic tangent (TanH) activation functions, and the output layer contains a linear activation function. In this process, the ANN must reach a state of $\tau^* = \tau$, where $\tau^*$ is the vector of optimized parameters and $\tau$ is the vector of neural network parameters. The number of parameters depends on the number of neurons, at each sub-level of learning. As described in Equation (6), for a couple $(s, y)$, the *arg* operator minimizes the value predicted by the neural network using the input data and the actually expected data to calculate the $\tau^*$ value:

$$\tau^* = arg\{|\mathrm{ANN}(\tau, s_i) - y_i|\} \tag{6}$$

Hidden layers perform as neuron nodes in-between inputs and outputs, which allow the neural network to learn complicated features. This is performed by the multiple neurons in each hidden layer that carry an assigned function and weight, depending on the errors coming from each iteration. The correct number of hidden levels and neurons is necessary to achieve accurate results with the least computational effort required. The LMBP process is applied to the ANN, in which, during training, random weights are assigned to the neurons initially, the inputs are passed through the hidden level to give a predicted output value, and an error value is calculated between the predicted and actual output value. An iterative process is created that adjusts the weight at each cycle to minimize the error. LMBP facilitates convergence to a stable system with low computational effort and time, and it combines the aspects of the steepest-descent method and the Gauss–Newton method to improve the accuracy of the output with fewer iterations required [58,59]. Therefore, the LMBP algorithm is used in this paper to utilize the ANN architecture more efficiently.

### 3.2. The Proposed ANN Model

A novel ANN model is developed that predicts the energy level of an in-service train acceleration, given that the speed of the train is known. For the train response, the energy of the vertical acceleration signal of the first bogie is chosen, as it weighs uniformly and consistently for most of the trains due to engine weight. The acceleration signals are transformed from the time domain to the space domain using the location and speed of the train. The space domain is then divided into several segments of the same size where the energy of the signals is calculated for each segment. The number of segments is used as the second input in the ANN. The energy of the vertical acceleration signal will be summed up for each segment and will used as the output.

$$e_{k,j} = \ddot{u}_{k,j}{}^2 \tag{7}$$

where $e_{k,j}$ is the energy of spatial point $k$ for run $j$ and $\ddot{u}_{k,j}$ is the vertical acceleration of spatial point $k$ for run $j$.

The independent variable $s$ is formed of the vector containing

$$s_{i,j} = \left(i, v_j\right) \tag{8}$$

where $i$ is the segment number (so the spatial location) and $v$ is the speed, constant over the whole passage, for run $j$.

The dependent variable can be translated to

$$E_{i,j} = \sum_{I} e_{k,j} \tag{9}$$

where $I$ represents the set of points included in segment $i$, $e$ is the energy at special point $k$, and $j$ represents the number of the run.

### 3.3. Damage Indicator

The damage indicator is formed using the prediction error, which is the square root of the difference between the predicted and simulated data for each segment and each pass.

$$Pe_{i,j} = \{\text{ANN}(\tau^*, s_{i,j}) - E_{i,j}\}^2 \tag{10}$$

where $Pe$ is the prediction error of segment $i$ and run $j$, and ANN is the operator that predicts the data using the ANN trained in the training phase.

The prediction error might vary significantly depending on the speed of the train passage. This results from a stochastic distribution that is characterized by a normal distribution to recognize the healthy structure from the damaged one. In this article, prediction errors are considered to follow a Gaussian process with a mean of $\mu$ and standard deviation of $\sigma$.

$$Pe_{i,j} \sim (\mu, \sigma) \tag{11}$$

The prediction errors for a healthy structure will be quite low, and they will remain low as long as a healthy structure is predicted. This means that the energy predicted by the ANN is at the same level as the energy measured by the train during the testing phase. A damage indicator is then introduced by comparing the prediction error with the mean of the prediction errors of the healthy structure, divided by the standard deviation of the prediction errors of the healthy structure.

$$DI_{i,j} = \frac{Pe_{i,j} - \mu_{training}}{\sigma_{training}} \tag{12}$$

where $\mu_{training}$ and $\sigma_{training}$ are, respectively, the mean and the standard deviation of the prediction errors of the healthy structure.

## 4. Result of the Machine Learning

In this section, the proposed ANN model is analyzed using the numerical TTI model shown in Section 2. A 1100 m long track is modeled and a track section in between 700 m and 1000 m is used to test the proposed algorithm to avoid the boundary conditions and to stabilize the DOFs. The damage is modeled as a loss of stiffness of the ballast, and four damaged cases are simulated. These include reductions in stiffness down to 10%, 30%, 50%, and 70% of the healthy stiffness. The damage is simulated as a local loss and a 5 m section (from 800 m to 805 m length) is chosen to assess the efficacy of the approach. Figure 7 illustrates the stiffness of the ballast with healthy and damaged case scenarios.

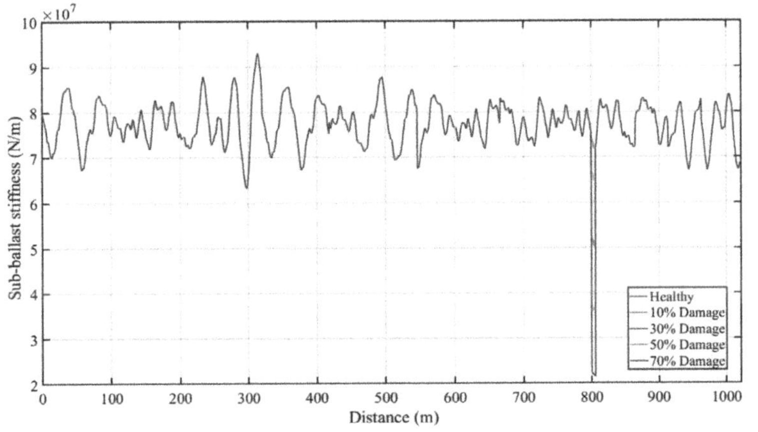


**Figure 7.** Sub-ballast stiffness profile with different damaged conditions at 800–805 m section.

A fleet of trains traversing the track with the roughness profile is simulated, and the train bogie acceleration energies are calculated using a sampling rate of 500 Hz. For a healthy case, a fleet of 100 trains is selected, which is used to train the proposed ANN model. For each run, the train velocity is chosen randomly following a normal distribution with a mean of 110 km/h and a standard deviation of 12 km/h. Similarly, a fleet of the same size each is used for the four damaged cases. The distributions of randomly chosen velocities for the healthy and damaged cases are shown in Figure 8a,b. In this section, no signal noise is added in the simulations and the energy magnitudes are calculated for each time step.

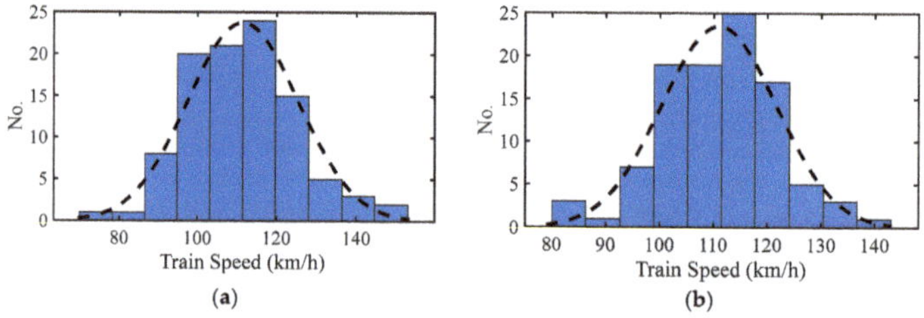


**Figure 8.** Distribution of train velocities for (**a**) healthy case and (**b**) damaged case.

The healthy track case is used to train the ANN model, which adjusts and stabilizes the weights of the neurons in each layer. The trained ANN model is used to predict the output data for the next four fleets of the damaged cases. These predicted data are compared with the simulated data and the error and damage indicators are calculated. In this section, 300 m of the track around the damage location is chosen and sliced into segments of 15 m (resulting in 20 slices), and the energy of the vertical acceleration of the train is predicted and compared with the interval of each segment.

The change in the damage indicator with the change in damage percentage is illustrated as a contour plot in Figure 9 on a logarithmic scale that allows us to make a better observation of the energy behavior with damage. In this figure, the number of runs is indicative of one healthy and four damaged cases, meaning that the first 100 runs represent the healthy condition, 101–200 runs represent the 10% damaged condition, 201–300 runs represent the 30% damaged condition, 301–400 runs represent the 50% condition, and 401–500 runs correspond to the 70% damaged condition. The runs are not sorted according to the velocities and are arranged randomly as chosen. It is also noted that a lower damage percentage, e.g., 10% loss of stiffness, is difficult to detect using the simple ANN analysis, and more research would be required for detecting a low magnitude of damages. However, the proposed ANN model can be a suitable system to identify the damage before any critical event and help prevent serious consequences.

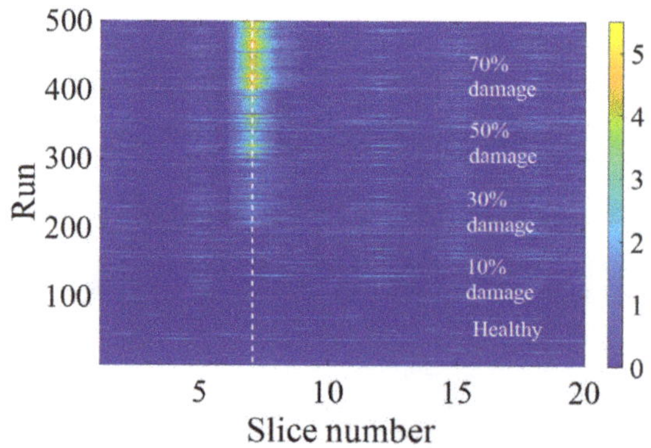


**Figure 9.** Damaged indicator by slice and run-on logarithm scale (contour plot).

## 5. Sensitivity Analysis

In this section, the sensitivity analysis is also carried out to ensure robustness of the ANN model. Three new scenarios are simulated, and the analysis is repeated to assess the efficacy of the proposed system: (i) change in energy band, (ii) adding signal noise, and (iii) multiple damage locations. In the previous analysis, the healthy and damaged tracks were sliced with a size of 15 m.

### *5.1. Size of Segments*

In this analysis, the impact of the size of the slices on the effectiveness of the proposed approach is studied. The analysis in the previous section is repeated with three different lengths of segments—(i) 30 slices of 10 m, (ii) 60 slices of 5 m, and (iii) 150 slices of 2 m. The damage location and the train properties are kept similar to the previous analysis, and the performance of the proposed model is assessed using finer sizes of segments. This is important to see the ability of the ANN model to identify the location of the damage and its magnitude with a changing level of computational effort. Figure 10a–c illustrate the results of the first sensitivity analysis where three different lengths of segments are used. It can be seen in Figure 10b,c that with a smaller segment size (finer segment), there is a loss of precision and clarity in the results. It should be noted that there is expected to be a trade-off between the size of the area being affected on the track and the size of slices. In this case, the damaged segment is on a 5 m section of the track; however, it can be seen that a slice size of 10 m shows better results compared to the 2 m. This can be considered as an important finding when it comes to real-life applications of the method.

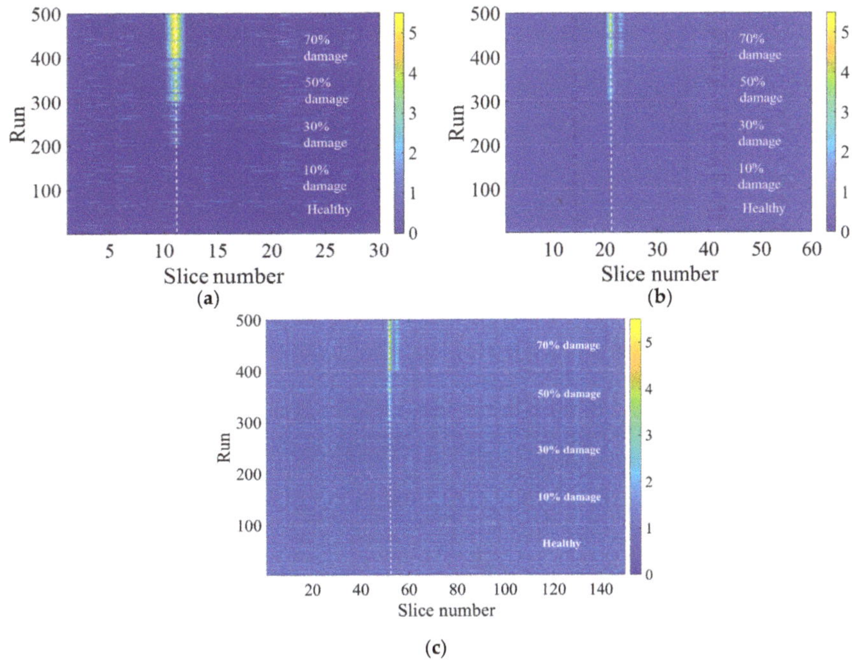


**Figure 10.** Damaged indicator by slice and run-on logarithm scale (contour plot): (**a**) 10 m, (**b**) 5 m, (**c**) 2 m.

This may be caused by the comparison between the damaged area length and the size of the segment, e.g., as shown in Figure 10c where the size of the segments is less than the actual length of the damaged area.

### 5.2. Noise Assessment

A second sensitivity analysis is carried out by repeating the analysis from Section 4 with the addition of random noise in the acceleration signal. In field testing, sensors tend to have a certain magnitude of imperfection, which results in random perturbations in the signals. In addition, other sources of error such as changes in rail roughness profile and the temperature effects can be considered as sources of noise. In order to evaluate the impact of these parameters on the accuracy of the results, the noisy signal is generated using a commonly used equation (Equation (13)) that creates a white noise vector randomly using a normal distribution with a mean of 0 and a standard deviation of a percentage of the signal amplitude [60,61]. The white noise vector is then added to the calculated responses to generate noisy/polluted responses. It can be mathematically represented as

$$\ddot{u}^{polluted} = \ddot{u} + E_p \times N_{noise} \times \sigma(\ddot{u}) \tag{13}$$

where $\ddot{u}$ is the calculated response and $\sigma(\ddot{u})$ is its standard deviation, $N_{noise}$ is a standard normal distribution vector with zero mean value and unit standard deviation, $E_p$ is the noise level, and $\ddot{u}^{polluted}$ is the polluted response of the model. For the analysis in this section, two percentages of the standard deviation 5% and 10% are used to add the noise to the train bogie accelerations.

For each noise level, a new set of healthy track data using noisy accelerations is measured to train the ANN model. Also, a set of 400 runs are simulated for the damaged cases to assess the damage indicator with added signal noise. The contour plots of the damage indicators from the analysis, using the noisy signals, are shown in Figure 11, where Figure 11a,b represent the results with 5% and 10% noise in the signals, respectively. It can be seen from Figure 11 that the proposed ANN model for monitoring the sub-ballast stiffness of the tracks works with reasonable results and is reliable under realistic conditions.

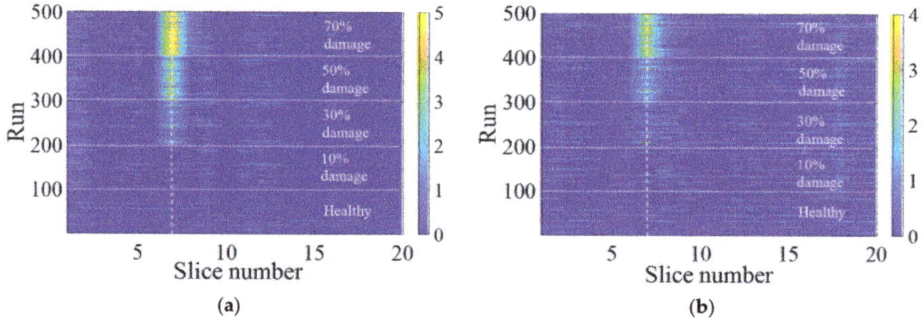


**Figure 11.** Influence of the noise on the DI (plotted on logarithmic scale): (**a**) with 5% signal noise and (**b**) with 10% signal noise.

### 5.3. Multiple Damage Locations

In this section, the proposed approach is also tested with damages at multiple locations. In this analysis, two sections of the track are considered with a 50% loss of ballast stiffness. The sections are 800 m to 805 m and 950 m to 955 mm, and a slice length of 5 m is chosen for the analysis. Figure 12 illustrates the results of the proposed ANN analysis damage indicator with two damaged locations. It can be seen in the figure that there is a significant increase in the signal energy differences at the two damaged locations (slice 20 and 50, respectively). Although there are some other visible slices that have shown a change in magnitude of damage indicator, these differences are not as significant as the differences at the damaged slices. This analysis proves that the proposed approach is effective even if there are multiple locations of damage.

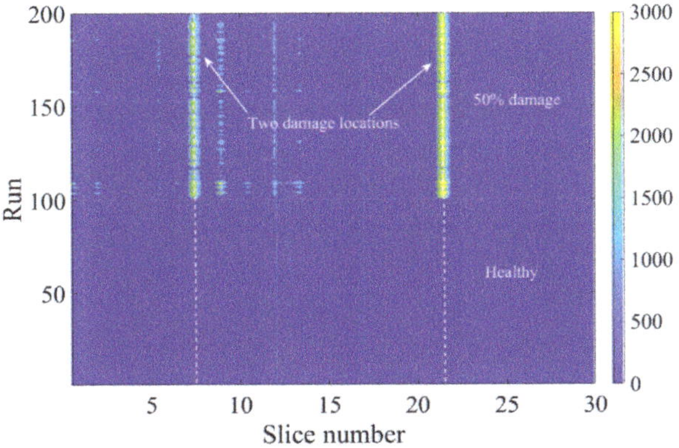


**Figure 12.** The results of the damage indicator with two damaged sections: 800–805 m (slice 8) and 950–955 m (slice 22).

## 6. Conclusions

A novel railway track damage detection approach is proposed in this paper using a machine learning technique that combines an Artificial Neural Network model (ANN) and a Gaussian process to detect the loss of track sub-ballast stiffness. The ANN is trained using energy responses of 100 simulated vertical train accelerations traversing over a healthy track. Using the trained ANN, the energy responses are predicted and the prediction error for each passage of trains is calculated using the square of the difference between the simulated and the predicted responses. The prediction error is assessed using different track sub-ballast stiffnesses and a Damage Indicator (DI) based on the prediction error is proposed. In order to interpret the prediction errors and to minimize the error in the machine learning process, the DI is defined using a Gaussian process and is used to normalize the distribution of the prediction errors. The numerical study demonstrates that this novel approach is effective in detecting changes in sub-ballast stiffness and is able to locate the area of damage. Although the approach is tested for the sub-ballast stiffness loss, other types of rail damages may also be monitored (by training the algorithm with different damage cases) and therefore will be part of our future studies. This paper provides a theoretical concept and numerical validation for track damage detection using the ANN. However, a full-scale real-life demonstration of the approach is recommended as part of future work to test the resilience of the approach on real-life tracks where environmental variations and other physical phenomena might limit the effectiveness. A high-accuracy positioning system, to record the train location in time and to calculate the average speed, is an essential element in such installations. In addition, the rail and track profile are assumed to be constant during the training and testing phase. However, it should be noted this will not be necessary in real-life applications. Therefore, further studies need to be carried out to address this drawback.

**Author Contributions:** A.M.: Conceptualization, writing, review; C.-A.S.: software, analysis, original draft preparation; M.A.K.: validation, writing; F.G.: conceptualization, methodology, review, editing. All authors have read and agreed to the published version of the manuscript.

**Funding:** This publication has emerged from research conducted with the financial support of Science Foundation Ireland under Grant number 20/FFP-P/8706.

**Institutional Review Board Statement:** Not applicable.

**Informed Consent Statement:** Not applicable.

**Data Availability Statement:** Not applicable.

**Conflicts of Interest:** The authors declare no conflict of interest.

## References


1. Schneider, A. *Railway Safety Research–A Cross-Disciplinary Literature Review*; Universität Koblenz, Universitätsbibliothek: Koblenz, Germany, 2020.
2. Gao, N.; Touran, A. Cost overruns and formal risk assessment program in US rail transit projects. *J. Constr. Eng. Manag.* **2020**, *146*, 05020004. [CrossRef]
3. Malekjafarian, A.; Obrien, E.J.; Quirke, P.; Cantero, D.; Golpayegani, F. Railway Track Loss-of-Stiffness Detection Using Bogie Filtered Displacement Data Measured on a Passing Train. *Infrastructures* **2021**, *6*, 93. [CrossRef]
4. Kaewunruen, S.; Remennikov, A.M. Field trials for dynamic characteristics of railway track and its components using impact excitation technique. *NDT E Int.* **2007**, *40*, 510–519. [CrossRef]
5. Koks, E.E.; Rozenberg, J.; Zorn, C.; Tariverdi, M.; Vousdoukas, M.; Fraser, S.A.; Hall, J.W.; Hallegatte, S. A global multi-hazard risk analysis of road and railway infrastructure assets. *Nat. Commun.* **2019**, *10*, 2677. [CrossRef]
6. Quirke, P.; Obrien, E.J.; Bowe, C.; Cantero, D.; Malekjafarian, A. The calibration challenge when inferring longitudinal track profile from the inertial response of an in-service train. *Can. J. Civ. Eng.* **2022**, *49*, 274–288. [CrossRef]
7. Azim, M.R.; Gül, M. Damage detection of steel girder railway bridges utilizing operational vibration response. *Struct. Control Health Monit.* **2019**, *26*, e2447. [CrossRef]
8. Rageh, A.; Azam, S.E.; Linzell, D.G. Steel railway bridge fatigue damage detection using numerical models and machine learning: Mitigating influence of modeling uncertainty. *Int. J. Fatigue* **2020**, *134*, 105458. [CrossRef]
9. Barke, D.; Chiu, W.K. Structural Health Monitoring in the Railway Industry: A Review. *Struct. Health Monit.* **2005**, *4*, 81–93. [CrossRef]
10. Malekjafarian, A.; OBrien, E.J.; Golpayegani, F. Indirect monitoring of critical transport infrastructure: Data analytics and signal processing. In *Data Analytics for Smart Cities*; CRC Press: Boca Raton, FL, USA, 2018; pp. 157–176.
11. Berggren, E.G.; Nissen, A.; Paulsson, B.S. Track deflection and stiffness measurements from a track recording car. *Proc. Inst. Mech. Eng. Part. F J. Rail Rapid Transit.* **2014**, *228*, 570–580. [CrossRef]
12. Andrade, A.R.; Teixeira, P.F. Unplanned-maintenance needs related to rail track geometry. In *Proceedings of the Institution of Civil Engineers-Transport*; Thomas Telford Ltd.: London, UK, 2014.
13. Wei, Z.; Núñez, A.; Li, Z.; Dollevoet, R. Evaluating Degradation at Railway Crossings Using Axle Box Acceleration Measurements. *Sensors* **2017**, *17*, 2236. [CrossRef]
14. Chong, L.; Jiahong, W.; Zhixin, Z.; Junsheng, L.; Tongqun, R.; Hongquan, X. Design and evaluation of a remote measurement system for the online monitoring of rail vibration signals. *Proc. Inst. Mech. Eng. Part. F J. Rail Rapid Transit.* **2016**, *230*, 724–733. [CrossRef]
15. Mennella, F.; Laudati, A.; Esposito, M.; Cusano, A.; Cutolo, A.; Giordano, M.; Campopiano, S.; Breglio, G. Railway monitoring and train tracking by fiber Bragg grating sensors. In *Third European Workshop on Optical Fibre Sensors*; International Society for Optics and Photonics: Bellingham, WA, USA, 2007.
16. Kerrouche, A.; Boyle, W.; Gebremichael, Y.; Sun, T.; Grattan, K.; Täljsten, B.; Bennitz, A. Field tests of fibre Bragg grating sensors incorporated into CFRP for railway bridge strengthening condition monitoring. *Sens. Actuators A Phys.* **2008**, *148*, 68–74. [CrossRef]
17. Fitzgerald, P.C.; Malekjafarian, A.; Cantero, D.; Obrien, E.J.; Prendergast, L.J. Drive-by scour monitoring of railway bridges using a wavelet-based approach. *Eng. Struct.* **2019**, *191*, 1–11. [CrossRef]
18. Bocciolone, M.; Caprioli, A.; Cigada, A.; Collina, A. A measurement system for quick rail inspection and effective track maintenance strategy. *Mech. Syst. Signal Process.* **2007**, *21*, 1242–1254. [CrossRef]
19. Molodova, M.; Li, Z.; Dollevoet, R. Axle box acceleration: Measurement and simulation for detection of short track defects. *Wear* **2011**, *271*, 349–356. [CrossRef]
20. Lederman, G.; Chen, S.; Garrett, J.; Kovačević, J.; Noh, H.Y.; Bielak, J. Track-monitoring from the dynamic response of an operational train. *Mech. Syst. Signal Process.* **2017**, *87*, 1–16. [CrossRef]
21. Bowe, C.; Quirke, P.; Cantero, D.; OBrien, E.J. Drive-by structural health monitoring of railway bridges using train mounted accelerometers. In Proceedings of the 5th ECCOMAS Thematic Conference on Computational Methods in Structural Dynamics and Earthquake Engineering, Crete Island, Greece, 25–27 May 2015.
22. Nafari, S.F.; Gül, M.; Roghani, A.; Hendry, M.T.; Cheng, J.R. Evaluating the potential of a rolling deflection measurement system to estimate track modulus. *Proc. Inst. Mech. Eng. Part. F J. Rail Rapid Transit.* **2018**, *232*, 14–24. [CrossRef]
23. Bar-Am, M. On-Train Rail Track Monitoring System. U.S. Patent 8942426B2, 27 January 2015.
24. Fosburgh, B.A.; Nichols, M.E.; Holmgren, P.M.; Larsson, N.T. Railway Track Monitoring. U.S. Patent 9810533B2, 7 November 2017.
25. Obrien, E.J.; Quirke, P.; Bowe, C.; Cantero, D. Determination of railway track longitudinal profile using measured inertial response of an in-service railway vehicle. *Struct. Health Monit.* **2018**, *17*, 1425–1440. [CrossRef]
26. Malekjafarian, A.; Obrien, E.; Quirke, P.; Bowe, C. Railway Track Monitoring Using Train Measurements: An Experimental Case Study. *Appl. Sci.* **2019**, *9*, 4859. [CrossRef]
27. Zhai, W.; Han, Z.; Chen, Z.; Ling, L.; Zhu, S. Train–track–bridge dynamic interaction: A state-of-the-art review. *Veh. Syst. Dyn.* **2019**, *57*, 984–1027. [CrossRef]
28. Obrien, E.J.; Bowe, C.; Quirke, P.; Cantero, D. Determination of longitudinal profile of railway track using vehicle-based inertial readings. *Proc. Inst. Mech. Eng. Part. F J. Rail Rapid Transit.* **2017**, *231*, 518–534. [CrossRef]

29. Wahlbin, L.B.; Bathe, K.-J.; Wilson, E.L. *Numerical Methods in Finite Element Analysis*; Prentice-Hall: Hoboken, NJ, USA, 1976.
30. Avci, O.; Abdeljaber, O.; Kiranyaz, S.; Hussein, M.; Gabbouj, M.; Inman, D.J. A review of vibration-based damage detection in civil structures: From traditional methods to Machine Learning and Deep Learning applications. *Mech. Syst. Signal Process.* **2021**, *147*, 107077. [CrossRef]
31. Malekjafarian, A.; Golpayegani, F.; Moloney, C.; Clarke, S. A Machine Learning Approach to Bridge-Damage Detection Using Responses Measured on a Passing Vehicle. *Sensors* **2019**, *19*, 4035. [CrossRef] [PubMed]
32. Peng, J.; Zhang, S.; Peng, D.; Liang, K. Application of machine learning method in bridge health monitoring. In Proceedings of the 2017 Second International Conference on Reliability Systems Engineering (ICRSE), Beijing, China, 10–12 July 2017.
33. Neves, A.C.; González, I.; Leander, J.; Karoumi, R. Structural health monitoring of bridges: A model-free ANN-based approach to damage detection. *J. Civ. Struct. Health Monit.* **2017**, *7*, 689–702. [CrossRef]
34. Gonzalez, I.; Karoumi, R. BWIM aided damage detection in bridges using machine learning. *J. Civ. Struct. Health Monit.* **2015**, *5*, 715–725. [CrossRef]
35. Santos, A.; Figueiredo, E.; Silva, M.; Santos, R.; Sales, C.; Costa, J.C.W.A. Genetic-based EM algorithm to improve the robustness of Gaussian mixture models for damage detection in bridges. *Struct. Control Health Monit.* **2017**, *24*, e1886. [CrossRef]
36. Corbally, R.; Malekjafarian, A. A data-driven approach for drive-by damage detection in bridges considering the influence of temperature change. *Eng. Struct.* **2022**, *253*, 113783. [CrossRef]
37. Diez, A.; Khoa, N.L.D.; Alamdari, M.M.; Wang, Y.; Chen, F.; Runcie, P. A clustering approach for structural health monitoring on bridges. *J. Civ. Struct. Health Monit.* **2016**, *6*, 429–445. [CrossRef]
38. Goi, Y.; Kim, C.-W. Damage detection of a truss bridge utilizing a damage indicator from multivariate autoregressive model. *J. Civ. Struct. Health Monit.* **2017**, *7*, 153–162. [CrossRef]
39. Khodabandehlou, H.; Pekcan, G.; Fadali, M.S. Vibration-based structural condition assessment using convolution neural networks. *Struct. Control Health Monit.* **2019**, *26*, e2308. [CrossRef]
40. Kordestani, H.; Zhang, C.; Shadabfar, M. Beam Damage Detection Under a Moving Load Using Random Decrement Technique and Savitzky–Golay Filter. *Sensors* **2019**, *20*, 243. [CrossRef]
41. Malekjafarian, A.; Moloney, C.; Golpayegani, F. Drive-by bridge health monitoring using multiple passes and machine learning. In *European Workshop on Structural Health Monitoring*; Springer: Berlin/Heidelberg, Germany, 2020.
42. Kaewunruen, S.; Sresakoolchai, J.; Zhu, G. Machine learning aided rail corrugation monitoring for railway track maintenance. *Struct. Monit. Maint.* **2021**, *8*, 151–166.
43. Sresakoolchai, J.; Kaewunruen, S. Wheel flat detection and severity classification using deep learning techniques. *Insight—Non-Destr. Test. Cond. Monit.* **2021**, *63*, 393–402. [CrossRef]
44. Quirke, P.; Obrien, E.J.; Bowe, C.; Malekjafarian, A.; Cantero, D. Estimation of railway track longitudinal profile using vehicle-based inertial measurements. In *Sustainable Solutions for Railways and Transportation Engineering, Proceedings of the International Congress and Exhibition "Sustainable Civil Infrastructures: Innovative Infrastructure Geotechnology", Cairo, Egypt, 24–28 November 2018*; Springer: Berlin/Heidelberg, Germany, 2018.
45. Cantero, D.; O'Brien, E.J.; González, A. Modelling the vehicle in vehicle—Infrastructure dynamic interaction studies. *Proc. Inst. Mech. Eng. Part. K J. Multi-Body Dyn.* **2010**, *224*, 243–248. [CrossRef]
46. Cantero, D. VEqMon2D—Equations of motion generation tool of 2D vehicles with Matlab. *SoftwareX* **2022**, *19*, 101103. [CrossRef]
47. Cantero, D.; Arvidsson, T.; OBrien, E.; Karoumi, R. Train–track–bridge modelling and review of parameters. *Struct. Infrastruct. Eng.* **2016**, *12*, 1051–1064. [CrossRef]
48. Zhai, W.; Wang, K.; Cai, C. Fundamentals of vehicle–track coupled dynamics. *Veh. Syst. Dyn.* **2009**, *47*, 1349–1376. [CrossRef]
49. Nguyen, K.; Goicolea, J.; Galbadón, F. Comparison of dynamic effects of high-speed traffic load on ballasted track using a simplified two-dimensional and full three-dimensional model. *Proc. Inst. Mech. Eng. Part. F J. Rail Rapid Transit.* **2014**, *228*, 128–142. [CrossRef]
50. Zhai, W.; Wang, K.; Lin, J. Modelling and experiment of railway ballast vibrations. *J. Sound. Vib.* **2004**, *270*, 673–683. [CrossRef]
51. Quirke, P.; Cantero, D.; Obrien, E.J.; Bowe, C. Drive-by detection of railway track stiffness variation using in-service vehicles. *Proc. Inst. Mech. Eng. Part. F J. Rail Rapid Transit.* **2017**, *231*, 498–514. [CrossRef]
52. Lei, X.; Noda, N.-A. Analyses of dynamic response of vehicle and track coupling system with random irregularity of track vertical profile. *J. Sound. Vib.* **2002**, *258*, 147–165. [CrossRef]
53. Kwon, Y.W.; Bang, H. *The Finite Element Method Using MATLAB*; CRC Press: Boca Raton, FL, USA, 2018.
54. González, A. Vehicle-bridge dynamic interaction using finite element modelling. In *Finite Element Analysis*; Moratal, D., Ed.; InTechOpen: Rijeka, Croatia, 2010; pp. 637–662.
55. Kaynia, A.M.; Park, J.; Norén-Cosgriff, K. Effect of track defects on vibration from high speed train. *Procedia Eng.* **2017**, *199*, 2681–2686. [CrossRef]
56. Jin, C.; Jang, S.; Sun, X.; Li, J.; Christenson, R. Damage detection of a highway bridge under severe temperature changes using extended Kalman filter trained neural network. *J. Civ. Struct. Health Monit.* **2016**, *6*, 545–560. [CrossRef]
57. Cavadas, F.; Smith, I.F.; Figueiras, J. Damage detection using data-driven methods applied to moving-load responses. *Mech. Syst. Signal Process.* **2013**, *39*, 409–425. [CrossRef]
58. Sapna, S.; Tamilarasi, A.; Kumar, M.P. Backpropagation learning algorithm based on Levenberg Marquardt Algorithm. *Comp. Sci. Inform. Technol. (CS IT)* **2012**, *2*, 393–398.

59. Yu, H.; Wilamowski, B.M. Levenberg-marquardt training. In *Industrial Electronics Handbook*; CRC Press: Boca Raton, FL, USA, 2011; Volume 5, p. 1.
60. Keenahan, J.; Obrien, E.J.; McGetrick, P.J.; Gonzalez, A. The use of a dynamic truck–trailer drive-by system to monitor bridge damping. *Struct. Health Monit.* **2014**, *13*, 143–157. [CrossRef]
61. Khan, M.A.; McCrum, D.P.; Prendergast, L.J.; Obrien, E.J.; Fitzgerald, P.C.; Kim, C.-W. Laboratory investigation of a bridge scour monitoring method using decentralized modal analysis. *Struct. Health Monit.* **2021**, *20*, 3327–3341. [CrossRef]

*sensors*

MDPI

*Article*

# Freight Wagon Digitalization for Condition Monitoring and Advanced Operation

**Iker Moya [1,2,*], Alejandro Perez [1,2], Paul Zabalegui [1,2], Gorka de Miguel [1,2], Markos Losada [1,2], Jon Amengual [1,2], Iñigo Adin [1,2] and Jaizki Mendizabal [1,2]**

[1] CEIT-Basque Research and Technology Alliance (BRTA), Manuel Lardizabal 15, 20018 Donostia/San Sebastián, Spain; aperez@ceit.es (A.P.); pzabalegui@ceit.es (P.Z.); gdemiguel@ceit.es (G.d.M.); mlosada@ceit.es (M.L.); jamengualm@ceit.es (J.A.); iadin@ceit.es (I.A.); jmendizabal@ceit.es (J.M.)

[2] Universidad de Navarra, Tecnun, Manuel Lardizabal 13, 20018 Donostia/San Sebastián, Spain

* Correspondence: imoya@ceit.es

**Abstract:** Traditionally, freight wagon technology has lacked digitalization and advanced monitoring capabilities. This article presents recent advancements in freight wagon digitalization, covering the system's definition, development, and field tests on a commercial line in Sweden. A number of components and systems were installed on board on the freight wagon, leading to the intelligent freight wagon. The digitalization includes the integration of sensors for different functions such as train composition, train integrity, asset monitoring and continuous wagon positioning. Communication capabilities enable data exchange between components, securely stored and transferred to a remote server for access and visualization. Three digitalized freight wagons operated on the Nässjo–Falköping line, equipped with strategically placed monitoring sensors to collect valuable data on wagon performance and railway infrastructure. The field tests showcase the system's potential for detecting faults and anomalies, signifying a significant advancement in freight wagon technology, and contributing to an improvement in freight wagon digitalization and monitoring. The gathered insights demonstrate the system's effectiveness, setting the stage for a comprehensive monitoring solution for railway infrastructures. These advancements promise real-time analysis, anomaly detection, and proactive maintenance, fostering improved efficiency and safety in the domain of freight transportation, while contributing to the enhancement of freight wagon digitalization and supervision.

**Keywords:** railway; digitalization; freight; monitoring; wagon; infrastructure

**Citation:** Moya, I.; Perez, A.; Zabalegui, P.; de Miguel, G.; Losada, M.; Amengual, J.; Adin, I.; Mendizabal, J. Freight Wagon Digitalization for Condition Monitoring and Advanced Operation. *Sensors* **2023**, *23*, 7448. https://doi.org/10.3390/s23177448

Academic Editors: Araliya Mosleh, Diogo Ribeiro, Abdollah Malekjafarian and Maria D. Martínez-Rodrigo

Received: 20 July 2023
Revised: 16 August 2023
Accepted: 19 August 2023
Published: 27 August 2023



## 1. Introduction

In the freight train sector, there is a significant lack of knowledge about the state of deterioration of the railway infrastructure and the trains themselves, mainly due to the absence of digitization and advanced monitoring. Reliant solely on manual inspections and visual assessments, the industry has faced significant challenges in optimizing maintenance and ensuring operational efficiency. However, recent advancements in freight wagon digitalization have brought about a paradigm shift in this sector.

The digitalization of freight trains is a crucial advancement aimed at creating modern functionalities that provide a cost-effective and appealing service, while also offering improved operational opportunities to operators and infrastructure managers. These modern functionalities encompass intelligence, detection, actuation, and communication capabilities.

Moreover, the digitalization of freight trains aligns closely with the principles of Industry 4.0, ushering in a new era of interconnected and intelligent systems. Embracing this transformative approach, the freight train sector can harness the power of well-established technologies such as sensor deployment and element virtualization. These technologies, already matured and successfully applied in domains like Industry 4.0, offer great potential for revolutionizing the railway industry [1]. By strategically integrating sensors, the digitalization process enables real-time monitoring of vital aspects such as train composition, train integrity, wagon asset condition, and continuous wagon positioning. The seamless communication facilitated by these advanced technologies fosters a data-driven ecosystem that empowers operators and infrastructure managers with valuable insights for enhanced decision-making and proactive maintenance strategies. Thus, the utilization of these mature technologies becomes a cornerstone in advancing the efficiency, reliability, and safety of freight services in the contemporary railway landscape.

According to its principles, the transport industry must significantly enhance the cost competitiveness and dependability of freight services to fulfil the ambitious goals outlined in the Transport White Paper [2] for the advancement of rail freight. These goals include nearly doubling rail freight usage compared to 2005, achieving a 30% shift of road freight over distances exceeding 300 km to modes such as rail or waterborne transport by 2030, and surpassing a 50% shift by 2050. Consequently, it is crucial to improve the cost-effectiveness and reliability of freight services to meet these objectives successfully.

Rail freight must adopt a cost-effective and appealing approach to entice shippers and divert freight from the congested road network. The challenge at hand entails two key aspects:

- Establishing a new service-oriented profile for rail freight services that prioritizes punctual deliveries at competitive prices. This entails integrating operations with other modes of transportation, incorporating innovative value-added services to cater to customer needs, and striving for operational excellence.
- Enhancing productivity by addressing existing operational and systemic weaknesses, including interoperability issues. This can be achieved by seeking cost-effective solutions, optimizing the utilization of current infrastructure, and embracing technology transfer from other sectors to enhance rail freight operations.

By addressing these challenges, rail freight can position itself as a reliable and efficient alternative, contributing to the shift of freight from the congested road network while providing a cost-effective and attractive service to shippers.

The freight railway environment presents a set of formidable challenges characterized by its extensive geographical distribution, harsh environmental conditions, and stringent energy considerations.

This article presents a comprehensive overview of the recent developments in freight wagon digitalization, focusing on the definition, development, and field tests conducted on a commercial line in Sweden. With the integration of a wide range of components and systems, the concept of the intelligent freight wagon has emerged. This digitalization process involves the strategic installation of sensors that enable various functionalities, including train composition, train integrity, wagon asset monitoring, and continuous wagon positioning. Furthermore, advanced communication capabilities facilitate seamless data exchange between these components.

To validate the effectiveness of this digitalization approach, field tests were carried out on three freight wagons operating on the operational line between Nässjo and Falköping in Sweden. These wagons were equipped to monitor the behavior of the train, enabling the detection of faults or anomalies in both the wagons and the railway infrastructure. This integrated approach not only enhances safety but also lays the foundation for a comprehensive monitoring solution for railway infrastructures, enabling real-time analysis, anomaly detection, and proactive maintenance.

The remaining sections of this article are organized as follows. First, the existing work on the digitalization of freight wagons is described (Section 2). Subsequently, a comprehensive overview of the developed system, including its services and functionalities, is provided (Section 3). Following that, the test campaign is presented, outlining the methods and procedures employed (Section 4). The results and discussions are then presented in Section 5, offering valuable insights and highlighting the significance of data acquisition. In Section 6, the main results and their implications are summarized, together with suggestions for future research and potential areas for improvement based on the findings.

## 2. Related Work

This section describes the related work for on-board monitoring for rolling stock and infrastructure condition determination found in the literature and in EU research projects.

Shift2Rail [3] is the first European rail initiative to seek focused research and innovation (R&I) and market-driven solutions by accelerating the integration of new and advanced technologies into innovative rail product solutions. Shift2Rail promotes the competitiveness of the European rail industry and meets changing EU transport needs. R&I carried out under this Horizon 2020 initiative develops the necessary technology to complete the Single European Railway Area (SERA).

One of the main objectives of TD3.8 Intelligent Asset Management Strategies (IAMS) is to shift towards a tailor-made maintenance approach by using the necessary tools for information management and decision support. This enhances the need to digitalize railway assets. Information is derived from the data obtained on board and on field. One of the most needed digitalizations is in the freight railway subsector, focused on the IP5 pillar for Shift2Rail. These activities are mostly based on the successful progress of TD5.1 fleet digitization and automation and mostly TD5.3 smart freight wagon concepts. For condition monitoring on the freight subsector, TD5.3.3 extended market wagons and TD5.3.4 telematics and electrification have made the greater efforts and they have been delivered on the documentation, demonstrations and results presented.

As for EU Rail, FP3 [4] and FP5 [5] are the pillars concerned and they have just started their activities, so there is no published information nor are there any conclusions related to railway onboard and infrastructure condition monitoring.

From the academic and scientific point of view, Figure 1 presents the time evolution of the research papers related to on-board monitoring for rolling stock and infrastructure condition determination. The increasing number of papers since 2016 proves the growing interest in this field in the past few years. It also shows that the technology and techniques are in the right place to serve the needs of the railway industry. Research works such as the one discussed in [6] show that the deployment of sensors on freight wagons allows, indeed, the detection and transmission of multiple status information regarding the maintenance and safety of these rolling elements.

The most cross-cited papers from the comprehensive list of references [7–46] represent the current state of the art in the field of condition monitoring for railway infrastructure. These articles primarily focus on advanced monitoring techniques and track quality assessment, including the findings of supervised experiments conducted on Polish railway lines using the electric multiple unit (EMU-ED74) equipped with a prototype track quality monitoring system [17]. The system incorporates a track quality indicator (TQI) algorithm, which utilizes a given transformation to preprocess the acceleration signals. This preprocessing is employed to extract the fundamental dynamics from the measured data, enabling a more comprehensive evaluation of the geometrical track quality. A comparative analysis is conducted to assess the performance of the proposed approach against other existing methods. This solution is the core of a track inspection system on board an electrified unit, which is a first step but it is not directly employable on freight wagons mainly due to power and location constraints. In addition, this paper presents further advanced features for freight train operation.

The second most cited [14], from 2021, is a survey which presents a comprehensive examination of the current literature conducted to provide an updated and content-driven analysis. This theoretical analysis is also of great interest to the research presented in this paper as it identifies the key contributors who have significantly influenced the progress of research in the specific area of interest. Using a coupled methodology that combines bibliometric performance analysis and a systematic literature review, the authors are able to identify the influential researchers, journals, and papers in the field. The findings of this study not only highlight the research trends pertaining to the analyzed area but also shed light on future research directions, particularly from an engineering standpoint. The main trends have also been considered in the research presented in this paper.

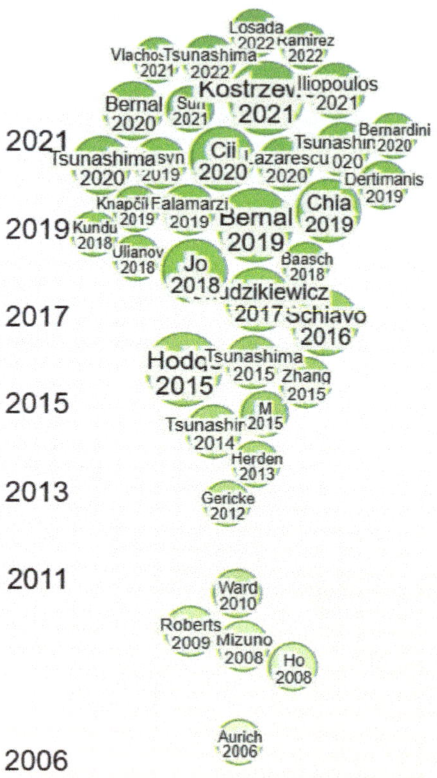


**Figure 1.** Time evolution of the research papers related to on-board monitoring of rolling stock and infrastructure condition determination [7–46].

The following list discloses the different referrals and nuances collected by this research for the definition of "condition monitoring":

- Direct measurement of relevant signals with time and/or frequency domain signal processing. Collection and real-time recording of digital and analogue signals using distributed transducers.
- Detecting and identifying deterioration in component structures and infrastructure performance in operating conditions. Continuous or periodic monitoring options.
- Alarm tool for maintenance. Distinguishing between normal and abnormal conditions and thresholding techniques for alarm systems.
- Implementing proactive condition monitoring technology. Tracking technical degradation and implementing preventive activities.
- Fault detection and diagnosis systems with intelligent algorithms. Condition-based monitoring for prognosis and diagnosis of component degradation.
- Ensuring safe and cost-effective train operation.
- Gathering and processing data for design, availability, reliability, and maintenance support.

Enhanced infrastructure monitoring of various elements such as bridges, viaducts, tunnels, crosses, rail gaps, frozen soil, and leaky feeders can yield significant benefits in terms of efficiency and safety. Neglecting safety and security monitoring of railway infrastructure poses risks such as train collisions, derailments, terrorism, and wagon failures. Notably, infrastructure or rolling stock failures still account for 35% of train delays, indicating the potential for substantial performance enhancements through intelligent systems in railway freight management [28].

From the alternatives listed above, refs. [40] and [23] categorize them into three levels here introduced and expanded in the picture below:

- Level 1 Data Logging and Event Recording Systems. When major incidents occur, they are used primarily to provide conclusive evidence. Equipment and operations are generally recorded digitally. This type of system can be used to detect faults in certain assets whose operation time or logic changes under fault conditions. Such systems are generally devoid of any data analysis. Typically, remote access is available to the systems, and data are logged locally.
- Level 2 Event Recording and Data Analysis Equipment. In addition to Level 1, this offers basic data analysis options, including statistical or sequence analysis. It is generally equipped with additional communication modules for remote access to data and analysis. In general, these systems are used for fault detection or the investigation of allegations but are unable to predict future failures.
- Level 3 Online Health Monitoring Systems. These systems are defined as the highest level of condition monitoring. These devices gathers digital and analogue (digitized) signals from monitored equipment, analyze them into characteristic signatures, compare them with an internal database of healthy and simulated faulty operation modes, and signal alarms and fault diagnosis information to operators. Expert systems, knowledge bases, and look-up tables are standard analysis techniques.

As a complement, Figure 2 illustrates an example of an intelligent infrastructure framework for railways [47]. It completes the level categorization with examples of uses and services that could be served with the equipment and strategy put in place for the monitoring.

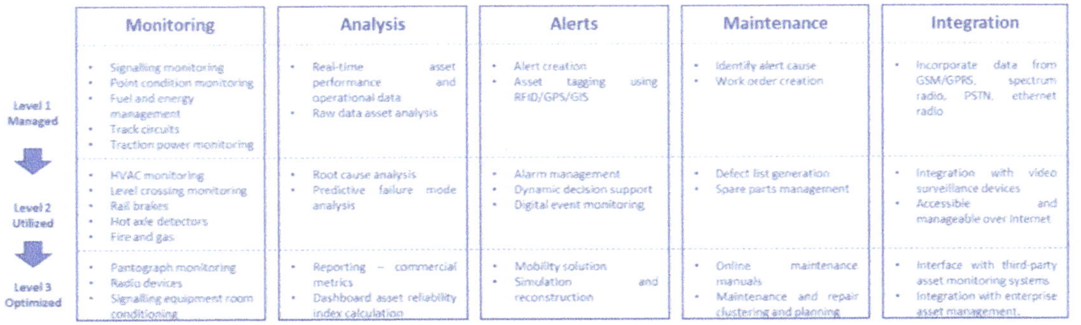


**Figure 2.** Intelligent infrastructure framework for railways [47].

The main conclusion from the analysis shown in this section is that a connected, distributed, and integrated system, with more layers and distributed acquisition and processing subsystems, is able to provide more useful information. The work presented in this paper is the result of the work performed in the TD5.3 smart freight wagon concepts topic and the result is part of the final demonstration performed as the conclusive activity for condition monitoring.

## 3. Perspective of the System: A High-Level Overview

The digitalization framework for freight wagons presented in this section is applicable to wagon assets and infrastructure monitoring and is designed to acquire and monitor data from various sensors installed on the wagons, enabling efficient and reliable operation. The monitoring with several sensors on each bogie is complemented with train composition, train integrity and positioning, which data are combined and converged for more accurate processing of the raw data. The visual representation, displayed in Figure 3, provides a clear and concise overview of the system's architecture. It showcases the various components and their interconnections, offering a comprehensive understanding of how the system is designed. By referring to Figure 3, one can easily grasp the hierarchical structure, the flow of data, and the relationships between different modules within the system.

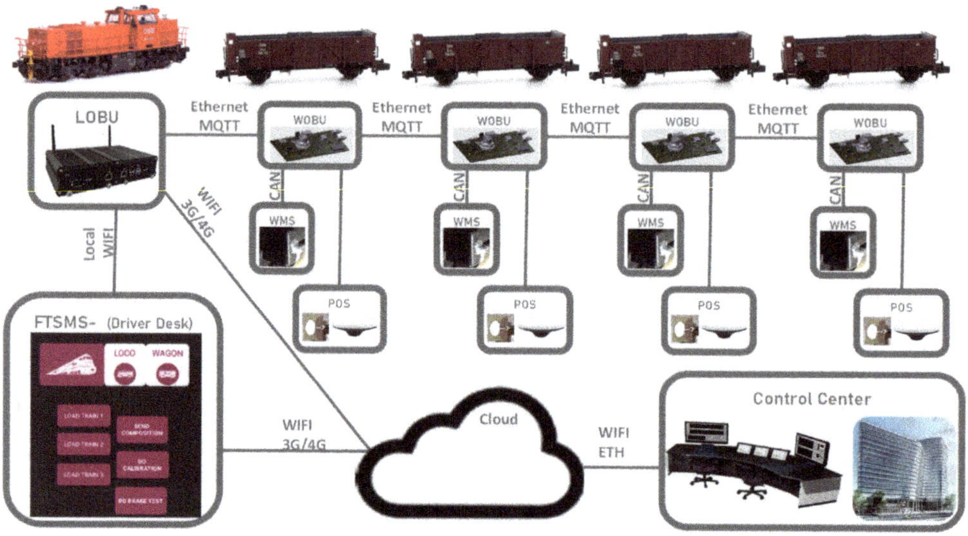


**Figure 3.** Architecture of the digitalization framework for freight wagons.

The locomotive on-board unit (LOBU) is responsible for controlling and enabling all communications. The LOBU serves as a central hub, storing and processing data received from all the connected wagons. It plays a crucial role in coordinating data exchange and ensuring seamless integration of information.

The system architecture comprises several hardware components that work together to enable data acquisition, storage, and analysis. Each freight wagon is equipped with a wagon on-board unit (WOBU), which serves as a local data storage and communication device. The WOBU collects and stores persistent information about the wagon, such as its identification, type, and available functionalities. It also acts as a gateway for sensor data acquisition.

The HW definition of the wagon on-board unit system deployed in each wagon is presented below; this is a connected multiprocessing platform. This architecture consists of two controllers for processing, which are a mainstream microcontroller (STM32F105RC) and a system on module (SOM) (iMX8-based), a series of devices for sensorization, communications and the power supply system [20].

A customized card was employed, featuring an SODIMM type connector, to interface with a VAR-SOM-MX8M-MINI [20]. It incorporates a certified railway connector, designed for railway applications, to enable the wiring of a CAN bus in addition to the Ethernet and USB interfaces. Figure 4 depicts the block diagram of the designed hardware (HW). The mechanical dimensions measure 150 mm $\times$ 95 mm $\times$ 45 mm.

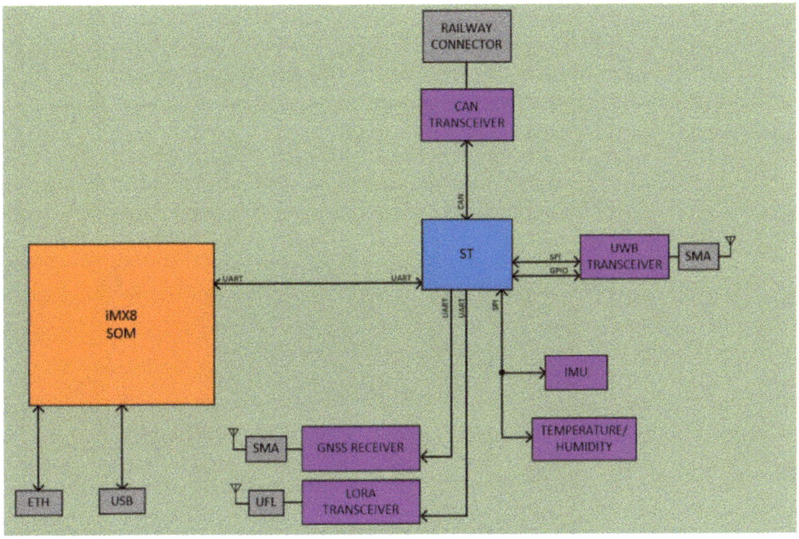


**Figure 4.** Block diagram of the WOBU proposed HW [10].

Connectivity between the LOBU and WOBU, as well as between multiple WOBUs, is established through a scalable communication infrastructure. This ensures efficient data exchange and synchronization, enabling real-time monitoring and analysis capabilities.

In addition to the LOBU and WOBUs, the system includes a driver desk, which provides a user interface for direct connectivity to the on-board system. The driver desk allows for efficient interaction and communication with the system, facilitating control and monitoring of various functionalities.

Furthermore, the system incorporates the control center, a centralized platform for control and monitoring. The control center retrieves data from the cloud storage and enables remote monitoring and analysis of the acquired information. It serves as a comprehensive management tool, providing insights into train performance, wagon behavior, and infrastructure evaluation. Advanced algorithms can be applied within the control center to derive valuable conclusions and optimize decision-making processes.

The system encompasses essential functionalities such as train composition, train integrity, continuous wagon positioning, and spring monitoring. These functionalities play a crucial role in organizing wagons, ensuring connectivity and safety, tracking wagon location, and detecting spring faults.

In the following sub-sections, the specific functionalities implemented within the system are explored, providing detailed explanations of their capabilities and the benefits they offer for comprehensive freight wagon digitalization and condition monitoring.

### *3.1. Train Composition*

This functionality is a pivotal aspect of freight train operations. It involves strategically organizing and assembling wagons to create an efficient transport unit. Its importance lies in optimizing various aspects of freight operations, such as weight distribution, load balancing, and overall train performance. By carefully arranging wagons and ensuring seamless connectivity between them, logistics managers can achieve optimal resource allocation, streamlined logistics processes, and enhanced operational efficiency. Additionally, efficient train composition reduces stress on rail infrastructure, minimizing wear and tear. Accurate identification and tracking of wagons within the train formation enable real-time monitoring, cargo identification, and efficient resource utilization.

In the developed system, each wagon is equipped with a wagon on-board unit (WOBU) that stores essential information such as wagon identification, type, number of bogies, and available functionalities. This WOBU functionality is triggered upon request from the driver desk, as mentioned earlier in this subsection. The driver desk initiates a discovery process through the LOBU (locomotive on-board unit), which communicates with the connected wagons. In response, each wagon provides persistent information along with the real-time status of the connected sensors. The LOBU processes these data to establish the current composition of the train and subsequently notifies the driver desk to display the updated train information as shown in Figure 5. This streamlined process ensures effective communication and seamless coordination between the different components of the system.

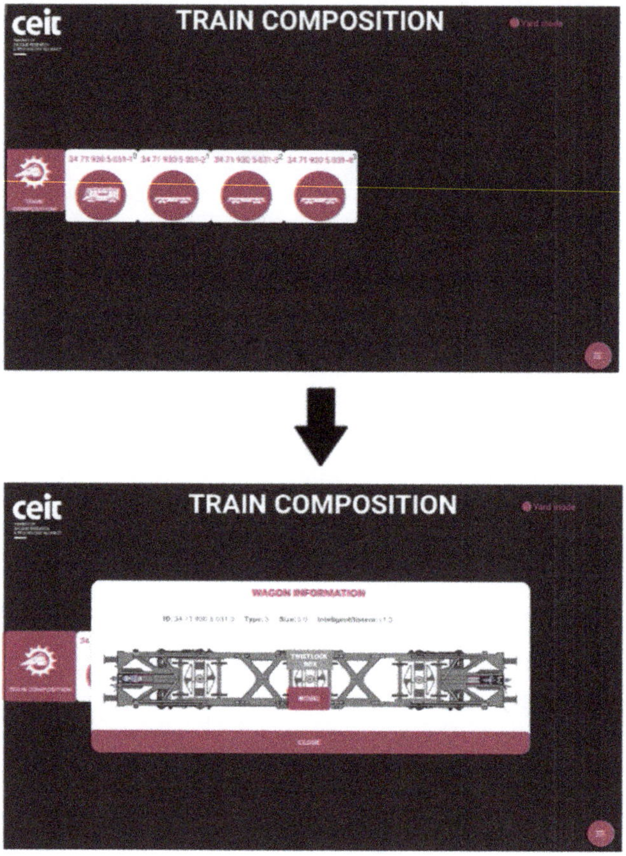


**Figure 5.** Driver desk visualization of the train composition.

This advanced train composition functionality offers a comprehensive solution for managing the composition of freight trains. Leveraging digitalization technologies, our system provides real-time insights into train formation, enabling logistics operators to make informed decisions regarding load distribution, coupling order, and overall train configuration. This not only optimizes train performance but also enhances safety, reduces operational costs, and improves the overall efficiency of freight transportation.

### 3.2. *Wagon Positioning*

The integration of a position stamp in freight wagon monitoring services is crucial for accurate evaluation and efficient tracking. It provides timestamped records of wagon locations throughout their journey, enhancing safety, optimizing operations, and enabling digitalization in freight transportation.

Accurate and real-time location tracking is a primary reason for implementing a position stamp. It allows stakeholders to precisely track wagon locations, ensuring safety and enabling proactive measures in response to deviations or incidents. Real-time tracking also facilitates efficient resource allocation, optimized loading/unloading operations, and informed decision-making.

The use of a position stamp optimizes maintenance schedules and resource allocation. Continuous monitoring helps identify maintenance requirements, minimizing breakdown risks and maximizing operational efficiency. The data from position stamps provide insights into wagon utilization patterns, informing resource allocation decisions, routing optimization, and fleet management practices.

The integration of position stamps supports comprehensive digitalization. Time-stamped position data enable efficient documentation, data-driven decision-making, and advanced analytics. Leveraging this data, including machine learning algorithms, helps identify optimization opportunities, improve route planning, and enhance supply chain visibility.

To provide time-stamped position data, the proprietary hardware of WOBUs (wagon on-board units) employs single-frequency multi-constellation GNSS receivers. These receivers translate satellite signals into messages and estimated satellite receiver distances.

The algorithm utilizes GPS and Galileo observables to estimate WOBU positions along the train's route. Due to suboptimal satellite visibility caused by the GNSS antennas' lateral location between freight containers, a least squares estimation algorithm is employed. This algorithm allows recalculation of positions based on the required information, without considering past measurements or results. It provides positions despite harsh railway environments. Multiple WOBUs offer position redundancy for post-processing and error analysis. Figure 6 shows the driver desk visualization of the wagon positioning functionality.

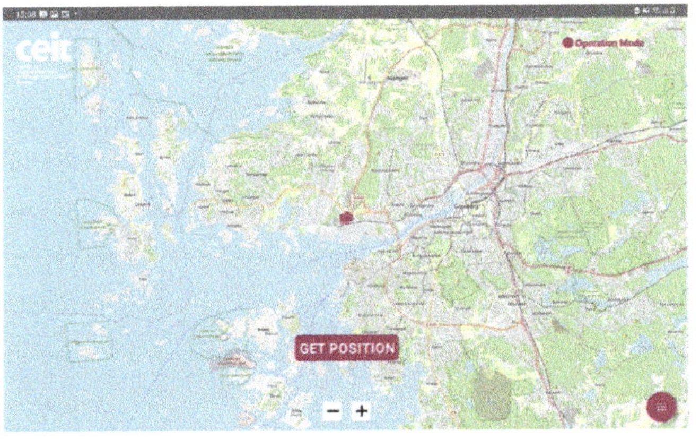


**Figure 6.** Driver desk visualization of the wagon positioning.

### 3.3. *Train Integrity*

The monitoring of the integrity of a freight train has turned out to be an essential requirement to operate the train in a safe way. To have the confirmation of the integrity of the train, the operator ensures that the full train is travelling towards its destination and no goods have been left in the way. Moreover, if this system is used as part of the safety critical signaling system used for the operation of the railway, the occupancy of the lane can be increased due to knowledge of the position and completeness of the train and its wagons. This subsection introduces the different train integrity classes defined in the X2RAIL-4 project [48] and how they could be used alone, or in a combined way to ensure the integrity of the freight train.

X2RAIL-4 project defined three train integrity classes depending on the technology used to measure it:

- Train integrity class 1: This relies on wired net connectivity. All the wagons are wired, forming a net that goes from the locomotive to the tail of the train. The LOBU, placed in the locomotive, is continuously monitoring the wired composition functionality to verify that all the WOBUs connected at the beginning of the operation are still connected to the network. Any fault detected in the aforementioned network generates an alarm message in the train integrity class 1 function.
- Train integrity class 2: This relies on the coherence between the velocities measured at the head and tail of the train. The velocity of the train is continuously measured both at the head and tail of the train. These velocities are then compared. If there is a difference bigger than a threshold programmed for the lane in which the freight train is operating, a train integrity class 2 alarm is raised in the system.
- Train integrity class 3: This relies on the distance measured between wagons. The head and the tail of each of the wagons are equipped with an ultra-wideband (UWB) anchor. These anchors are used to calculate the distance between the tail of a wagon and the head of the next wagon. If the distance measured is less than the maximum distance between coupled wagons plus a security margin calculated depending on the maximum gap between wagons, a train integrity class 3 alarm is raised in the system.

The freight train can have deployed one or more of the introduced train integrity monitoring classes, as shown in Figure 7. In the case that only one of the classes has been installed, the whole train integrity function will be performed according to that class. If there is more than one class installed, the status of all of them will be taken into account, and the joint train integrity will be calculated taking into account the outputs of the existing classes and their probabilities of false alarms.

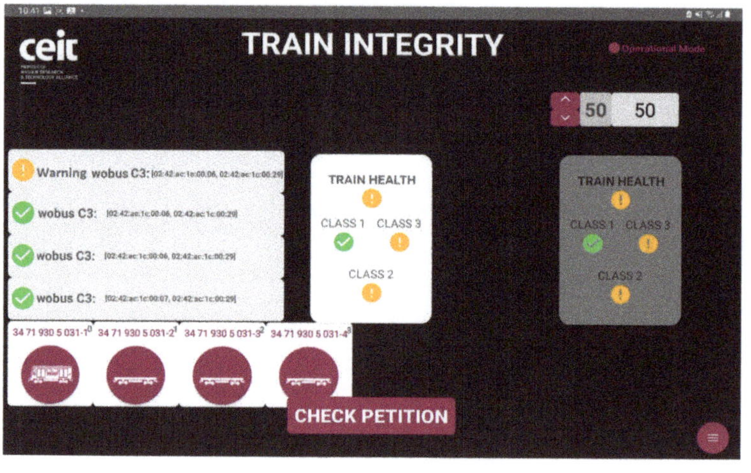


**Figure 7.** Driver desk visualization of the train integrity.

### 3.4. Wagon Monitoring System

Ensuring safe and efficient freight wagon operation relies heavily on monitoring springs. Springs play a vital role in absorbing shocks and vibrations, enabling a smooth ride while safeguarding cargo and wagon integrity. However, the springs endure extreme loads and adverse conditions throughout their service life. Factors like wear, fatigue, and severe impacts can lead to deterioration and loss of functionality over time, negatively impacting wagon performance and potentially leading to accidents.

Continuous monitoring of springs on freight wagons offers multiple benefits. First, it allows early detection of any deterioration or damage to the springs, which helps to prevent catastrophic failures and accidents. In addition, regular monitoring facilitates predictive maintenance, which means that springs can be replaced or repaired before serious problems occur. This not only improves safety, but also reduces operating costs by avoiding unplanned outages and optimizing maintenance schedules, minimizing disruptions, and ensuring a constant flow of goods. In addition, this spring monitoring functionality, working together with the wagon positioning function described above, facilitates the detection of possible faults in the track structure.

The spring monitoring functionality is performed by measuring the accelerations in the buffers of the bogies of the wagons, both at the top and at the bottom of the springs, as shown in Figure 8.

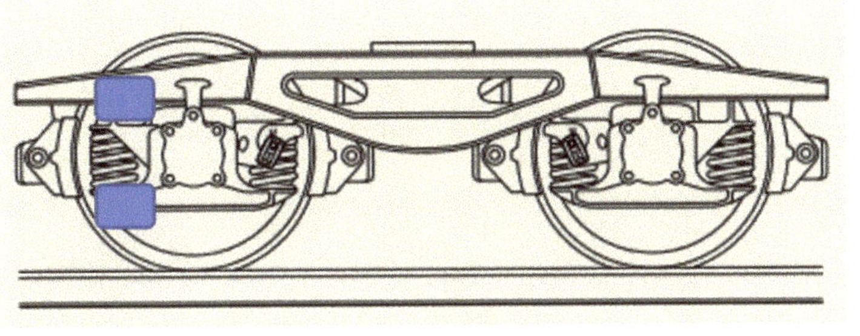

**Figure 8.** Vibration monitoring points on springs.

Measuring the accelerations that occur in this part of the wagon allows us to check whether the weight of the load in the wagon is balanced or not, to know the state of wear of the dampers themselves, as well as possible defects in the track infrastructure due to the vibration produced by the transit of the wheels of the wagon over possible imperfections in the infrastructure.

3.4.1. Sensor Node HW Implementation

To collect data on the vibrations occurring in the springs, the development of the HW shown in Figure 9 was proposed.

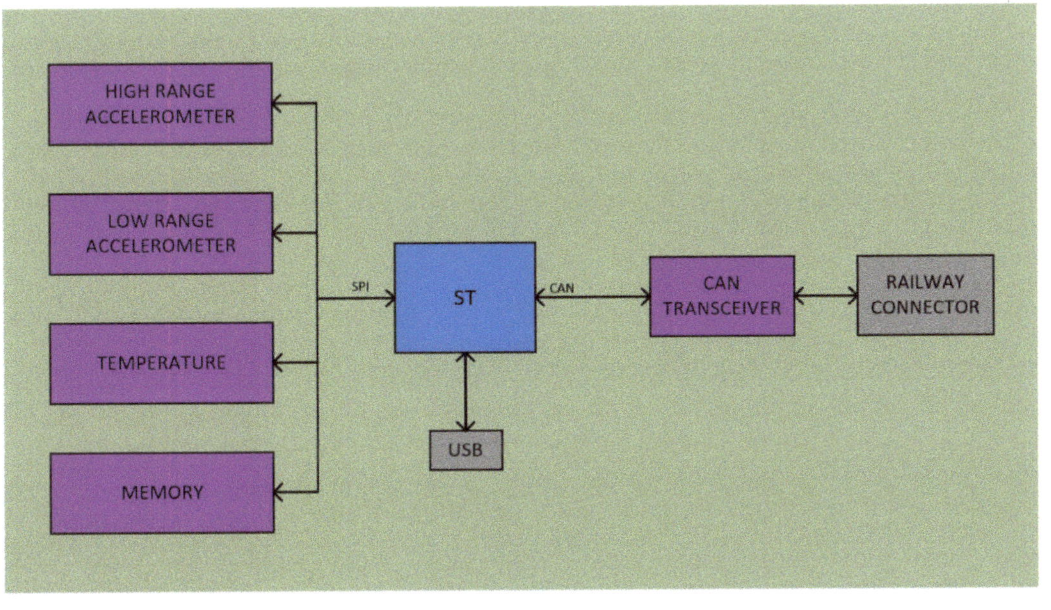


**Figure 9.** Block diagram of the sensor node proposed HW.

This hardware consists mainly of an STM32F105RC microcontroller, which is responsible for managing the data measured by the two accelerometers proposed in this design. This controller incorporates an ARM Cortex-M3 32-bit RISC core operating at 72 MHz frequency. The two accelerometers proposed are the ADXL345 and the ADXL357, which allow one to select one or the other depending on the accuracy or sensitivity to be obtained in the measurements.

#### 3.4.2. Data Collection and Wagon Monitoring System Functionality Flow

To obtain the acceleration data at the springs, a network consisting of four sensor nodes connected through the CAN interface to the WOBU was deployed in every wagon. The arrangement of these sensor nodes in the wagon can be seen in Figure 10.

**Figure 10.** The arrangement of sensor nodes in the wagon.

The data flow for the spring monitoring functionality is represented in Figure 11. Data from the three axes (x, y and z) is collected by the sensor nodes in 10-s time windows and sent to the WOBU through the CAN network. When the data is received at the WOBU, it stores it in its internal memory and sends it through the ETH network to the LOBU, which is in charge of storing all the information and processing the data coming from the wagons.

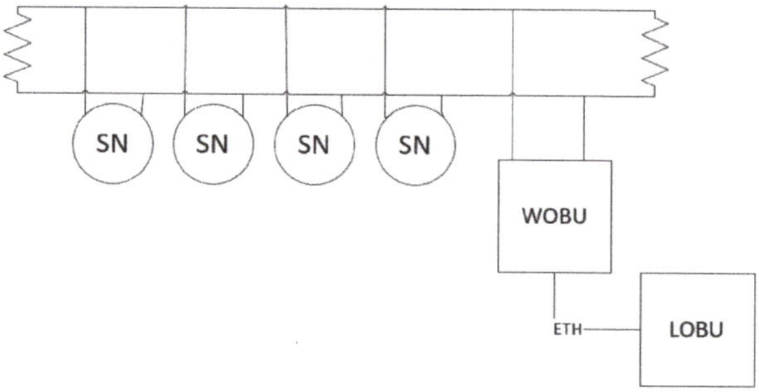


**Figure 11.** Schematic of the hardware involved in collecting acceleration data at the springs.

From the tablet, through a request to the LOBU, we can visualize the data as shown in Figure 12. The driver desk application on the tablet allows us to select which spring we want to monitor as well as to make a comparison between different springs and coordinate axes and different measurements within the same spring.

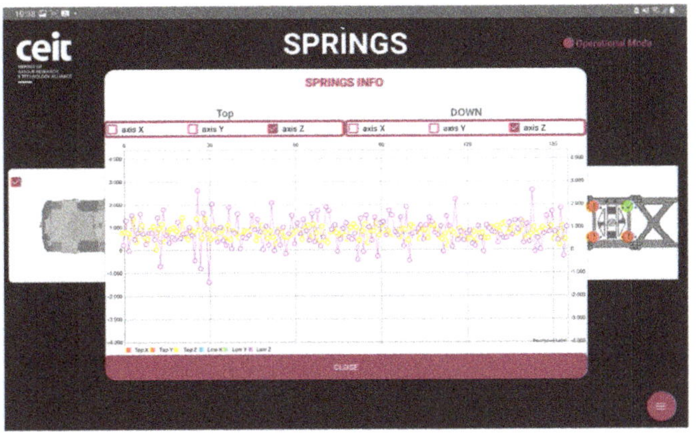


**Figure 12.** Driver desk visualization of the wagon monitoring system measurements.

## 4. Test Campaign

This section provides a detailed description of the test campaign conducted to rigorously test and validate the functionalities outlined in the previous section. The test campaign was an integral part of the FR8RAIL-IV European project [49], aimed at evaluating the performance and effectiveness of the developed system. The campaign took place in Sweden from 22 May to 26 May 2023, and involved a planned route that encompassed various aspects of freight train operations. In this section, we will delve into the duration of the test campaign, the specific types of wagons utilized, the characteristics of the track, and the strategic placement of sensors, electronics, and antennas. These insights and findings from the test campaign are instrumental in assessing the reliability and efficiency of the digitalization and monitoring solution for freight wagons.

The journey, as shown in Figure 13, commenced at Nässjo station at 10 a.m., with the train arriving at Göteborg at 3 p.m. This allowed for the loading and unloading of containers in the wagons. At 6 p.m., the train departed from Göteborg, reaching Falköping at 10 p.m.

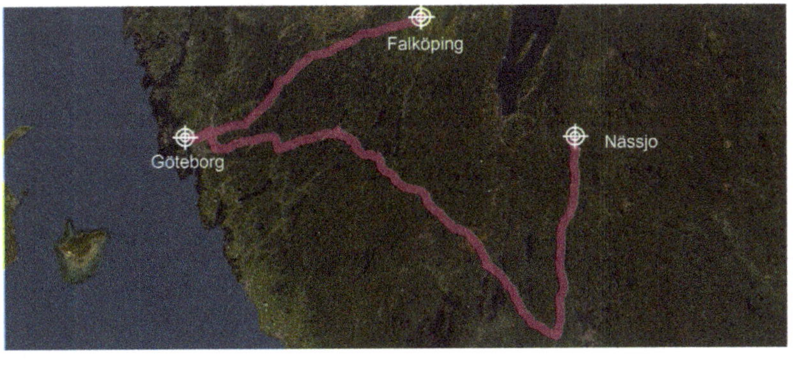


**Figure 13.** Journey undertaken by the freight train during the test campaign between Nässjo and Falköping.

The average speed throughout the journey was maintained at 35 km/h, and the train made several stops at intermediate stations. The route encompassed various track sections, including track changes and a mid-journey train orientation reversal.

Figure 14 presents the freight wagons employed in the test campaign "Sggrss 80' | 6-axle articulated intermodal wagon". The train consisted of 21 wagons, with wagons 15, 16, and 17 being selected for monitoring, each carrying two containers. This strategic selection enabled comprehensive data collection for analysis.

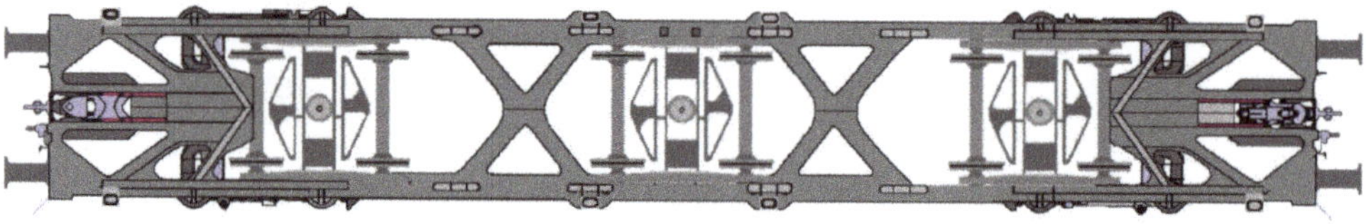

**Figure 14.** Schematic of the Sggrss 80′ I 6-axle articulated intermodal wagon [50].

In terms of hardware placement, as depicted in Figure 15, the electronic components and antennas were carefully installed in the middle section of the wagons, specifically in the stairwell area. This location ensured easy accessibility for maintenance purposes for the validation phase of the functionalities.

**Figure 15.** Installation of electronic components and antennas.

The test campaign provided valuable insights into the performance and functionality of the developed system under realistic operational conditions. The collected data serve as a crucial foundation for further analysis, validation, and enhancement of the digitalization and monitoring solution for freight wagons.

## 5. Results and Discussion

The primary objective, as mentioned earlier, was to acquire a substantial amount of data from various functionalities and establish an effective monitoring system, conditions included, with the intention of conducting comprehensive analysis in the future. The following are the results obtained for each functionality:

- Train composition: The system successfully obtained and displayed real-time information about the train, its wagons, and the connected sensors. The data acquisition and visualization were performed accurately, enabling efficient monitoring of the train's composition, which is a key feature for train operation but also for the processing of the data incoming from the sensors.
- Train integrity: Continuous checks were carried out to ensure the train's integrity while in motion. While the system effectively detected integrity breaches, there were occasional false positives at very low speeds. This aspect is being addressed for further improvement.
- Train positioning: Real-time visualization of the train's current location and its individual wagons was achieved. The system provided accurate positioning information,

allowing for effective monitoring and tracking of the train's movement and its derivative on components and infrastructure monitoring.

- Wagon monitoring system: The continuous monitoring and real-time visualization of the accelerometers (both upper and lower) for each spring shown in this paper, provide valuable insights into the dynamic behavior of the wagon throughout its operation. By analyzing the combined data from these accelerometers, it becomes possible to assess the wagon's response to the condition of the railway infrastructure. Figures 16 and 17 illustrate the recorded data, offering a comprehensive understanding of how the wagon interacts with the track, thereby facilitating effective maintenance planning and optimizing the overall performance of the system.

Every operational data acquisition was logged with its corresponding timestamp and associated position. This detailed recording ensures that any faults, defects, or alarms can be precisely located and identified, facilitating prompt action and maintenance interventions.

During the validation tests, the train was constantly monitored through various means, including direct and remote connections from the driver desk, as well as from a centralized control center. With this feature deployed, the comprehensive monitoring approach ensures continuous oversight of the train's operations, allowing for a quick response to any anomalies or emergencies.

All the collected information is also stored in a remote server, ready for future processing. This long-term storage enables thorough analysis and processing of the data to derive meaningful insights, contributing to enhanced operational efficiency and informed decision-making.

The obtained results not only validate the successful achievement of the objectives but also lay a robust foundation for conducting future in-depth analysis and deriving valuable insights from the accumulated data. Depending on the specific focus of the future analysis, this rich dataset can be instrumental in detecting anomalies, failures, and wear and tear, both within the wagons and across the track infrastructure. By leveraging this comprehensive monitoring approach, we can see significant potential for enhancing maintenance strategies, identifying potential issues proactively, and optimizing the overall performance and safety of both the rolling stock and the track system.

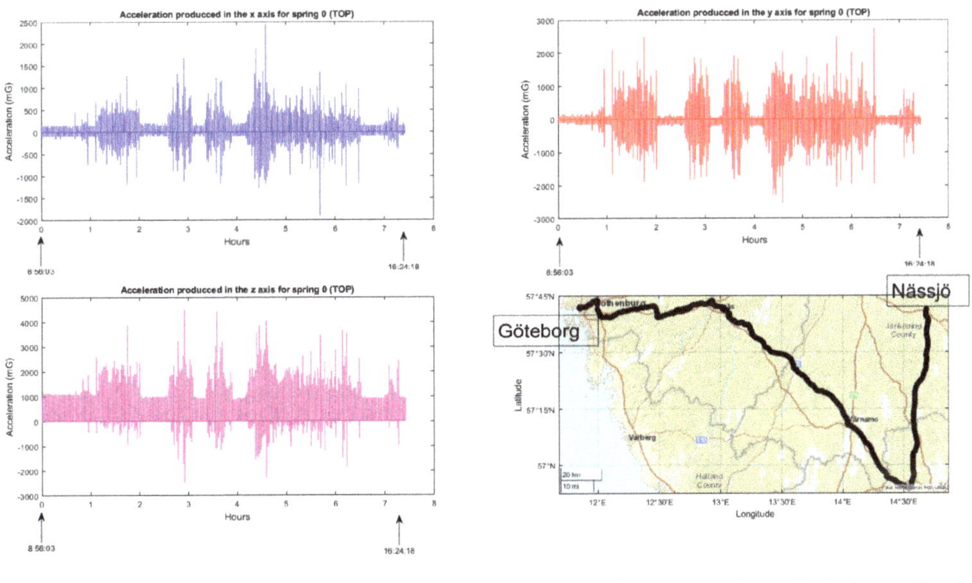


**Figure 16.** Accelerometer measurements above the wagon springs, displaying acceleration values in the x, y, and z axes together with travel route information.

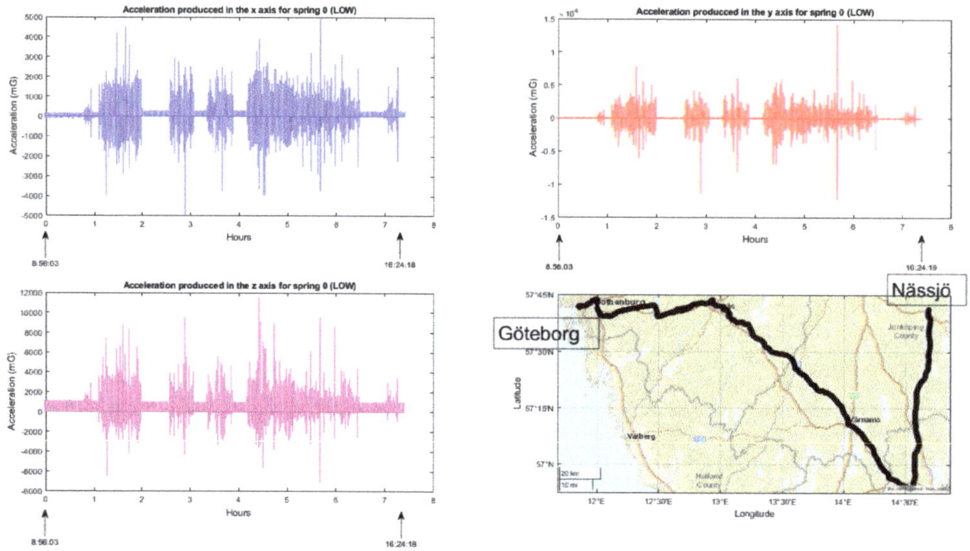


**Figure 17.** Accelerometer measurements under the wagon springs, displaying acceleration values in the x, y, and z axes together with travel route information.

## 6. Conclusions

Based on the comprehensive analysis and findings presented in this study, the following key conclusions can be drawn:

First and foremost, the successful development and implementation of the digitalization system for freight wagons have not only addressed the limitations of traditional operational and manual inspections but have also showcased the immense potential for enhancing the monitoring and management of components and railway infrastructures. By integrating advanced sensors and monitoring technologies, the system has enabled accurate and real-time monitoring of train composition, train integrity, wagon asset monitoring, and continuous wagon positioning.

The primary objective of the system is comprehensive data gathering and monitoring. Collaborating with stakeholders, research institutions, and the railway industry is crucial for successful digitalization implementation. This approach brings diverse perspectives, enhances understanding of industry needs, optimizes resource utilization, and accelerates innovation deployment. By tailoring the system to specific demands, safety standards, and regulations, its overall effectiveness and acceptance in the industry are enhanced. Collaborative efforts foster knowledge exchange, driving further advancements in digitalization strategies to meet evolving freight transportation needs, ensuring informed decisions, improved efficiency, and enhanced safety.

Furthermore, the validation and testing campaign conducted on the operating line in Sweden provided crucial information on the real-world performance and functionality of the system. The data collected during the campaign served as the basis for the subsequent analysis, validation and improvement of the freight car digitization and monitoring solution. This data-driven approach allows for continuous improvement and optimization of the system, resulting in increased efficiency and reliability. The success of freight car monitoring in harsh and inaccessible environments demonstrates the system's adaptability and its potential to provide comprehensive monitoring capabilities in a variety of operating conditions.

In addition, it is important to recognize the growing significance of effective data management in the context of large-scale digitalization efforts. As the volume of data generated by the digitalization system increases, adopting advanced data management techniques, such as big data analytics and machine learning algorithms, becomes essential. These cutting-edge technologies offer invaluable insights and opportunities for predictive maintenance, optimized resource allocation, and enhanced overall system performance. Therefore, the future direction of this research should focus on exploring these areas further to leverage the full potential of digitalization in freight wagon monitoring. By integrating big data analytics and machine learning algorithms, the system's capabilities can be greatly enhanced, enabling proactive maintenance practices, and ultimately leading to improved operational efficiency and cost-effectiveness.

The monitoring of multiple wagons enables a thorough assessment of individual wagon behavior. Moreover, the combination of data from these wagons offers the opportunity to extract valuable insights regarding the overall condition of the railway infrastructure. Analyzing patterns and trends derived from the collective data can help identify potential defects or issues in the track infrastructure, enhancing maintenance planning and ensuring optimal system performance.

In conclusion, the digitalization of freight wagons and the integration of advanced monitoring capabilities offer transformative potential for the railway industry. By harnessing the power of real-time data, stakeholders can optimize operational efficiency, enhance safety measures, and improve the overall performance of railway infrastructures. As technology continues to advance, the successful implementation of digitalization in the freight wagon industry requires addressing emerging challenges. Ensuring interoperability among different systems, prioritizing data security and privacy, investing in research and innovation, and providing comprehensive training for stakeholders are essential steps. By proactively tackling these challenges, the industry can unlock the full potential of digitalization, leading to improved safety, efficiency, and overall performance of the railway infrastructure.

**Author Contributions:** Conceptualization, I.M. and J.M.; methodology, I.M., A.P., P.Z., G.d.M., M.L. and I.A.; software, I.M., A.P., P.Z., G.d.M., M.L. and J.A.; validation, I.M., A.P., P.Z., G.d.M., M.L. and J.A..; formal analysis, I.M., P.Z., G.d.M., M.L. and I.A.; investigation, I.M., A.P., P.Z., G.d.M., M.L. and I.A.; writing—original draft preparation, I.M. and J.M.; writing—review and editing, all authors; project administration, I.A. and J.M.; funding acquisition, J.M. All authors have read and agreed to the published version of the manuscript.

**Funding:** The project FR8RAIL IV received funding from the Shift2Rail Joint Undertaking (JU) under grant agreement No. 101004051. The JU receives support from the European Union's Horizon 2020 research and innovation program and the Shift2Rail JU members other than the Union.

Europe's Rail

Horizon 2020
European Union Funding
for Research & Innovation

**Acknowledgments:** We would like to express our gratitude to Jonas Eriksson (Lindholmen Science Park), Anders Ekmark (Lindholmen Science Park), and Hans Arvidsson (SMP SvenskMaskinprovning AB), for their valuable support throughout the field test full process. Their contribution was key in shaping our research and helping us to overcome challenges. This joint effort significantly contributed to the success and impact of this scientific article, and we are deeply appreciative of their commitment to advancing knowledge in the field of freight railways.

**Conflicts of Interest:** This dissemination of results reflects only the authors' view, and the Shift2Rail Joint Undertaking is not responsible for any use that may be made of the information it contains.

## References


1. Gerhátová, Z.; Zitrický, V.; Klapita, V. Industry 4.0 Implementation Options in Railway Transport. *Transp. Res. Procedia* **2021**, *53*, 23–30. [CrossRef]
2. Transforming Europe's Rail System. Available online: https://rail-research.europa.eu/wp-content/uploads/2020/07/20200705_Partnership_High-Level-Paper.pdf (accessed on 18 July 2023).
3. Shift2Rail Projects. Available online: https://projects.shift2rail.org (accessed on 18 July 2023).
4. Flagship Project 3. Available online: https://projects.rail-research.europa.eu/eurail-fp3 (accessed on 18 July 2023).
5. Flagship Project 5. Available online: https://projects.rail-research.europa.eu/eurail-fp5 (accessed on 18 July 2023).
6. Ußler, H.; Michler, O.; Löffler, G. Validation of multiple sensor systems based on a telematics platform for intelligent freight wagons. *Transp. Res. Procedia* **2019**, *37*, 187–194. [CrossRef]
7. Bernardini, L.; Carnevale, M.; Somaschini, C.; Matsuoka, K.; Collina, A. A numerical investigation of new algorithms for the drive-by method in railway bridge monitoring. In Proceedings of the XI International Conference on Structural Dynamics, Athens, Greece, 23–26 November 2020; Volume 1, pp. 1033–1043. [CrossRef]
8. Kundu, P.; Darpe, A.K.; Singh, S.P.; Gupta, K. A review on condition monitoring technologies for railway rolling stock. In Proceedings of the PHM Society European Conference, Philadelphia, PA, USA, 24–27 September 2018.
9. Falamarzi, A.; Moridpour, S.; Nazem, M. A Review on Existing Sensors and Devices for Inspecting Railway Infrastructure. *J. Kejuruter.* **2019**, *31*, 1–10. [CrossRef]
10. Lazarescu, M.T.; Poolad, P. Asynchronous Resilient Wireless Sensor Network for Train Integrity Monitoring. *IEEE Internet Things J.* **2020**, *8*, 3939–3954. [CrossRef]
11. Gericke, C.; Hecht, M. CargoCBM—Feature Generation and Classification for a Condition Monitoring System for Freight Wagons. *J. Phys. Conf. Ser.* **2012**, *364*, 012003. [CrossRef]
12. Sysyn, M.; Nabochenko, O.; Gerber, U.; Kovalchuk, V.; Petrenko, O. Common Crossing Condition Monitoring with on Board Inertial Measurements. *Acta Polytech.* **2019**, *59*, 423–434. [CrossRef]
13. Herden, M.O.; Friesen, U. COMORAN—Condition Monitoring for Railway Applications. *Rail Technol. Rev.* **2013**, *53*, 42–46.
14. Kostrzewski, M.; Melnik, R. Condition Monitoring of Rail Transport Systems: A Bibliometric Performance Analysis and Systematic Literature Review. *Sensors* **2021**, *21*, 4710. [CrossRef] [PubMed]
15. Ward, C.; Dixon, R.; Charles, G.; Goodall, R. Condition Monitoring of Rail Vehicle Bogies. In Proceedings of the UKACC International Conference on CONTROL 2010, Coventry, UK, 7–10 September 2010. [CrossRef]
16. Tsunashima, H.; Hirose, R. Condition monitoring of railway track from car-body vibration using time–frequency analysis. *Veh. Syst. Dyn.* **2020**, *60*, 1170–1187. [CrossRef]
17. Chudzikiewicz, A.; Bogacz, R.; Kostrzewski, M.; Konowrocki, R. Condition monitoring of railway track systems by using acceleration signals on wheelset axle-boxes. *Transport* **2017**, *33*, 555–566. [CrossRef]
18. Tsunashima, H.; Asano, A.; Ogino, M.; Yanagisawa, K.; Mori, H. Condition Monitoring of Railway Tracks Using Compact Size On-board Monitoring Device. In Proceedings of the 6th IET Conference on Railway Condition Monitoring (RCM 2014), Birmingham, UK, 17–18 September 2014. [CrossRef]
19. Dušan; Banić, M.; Milosevic, M. Condition Monitoring Technologies in Railway Maintenance. In Proceedings of the XVIII Scientific-Expert conference on Railways (RAILCON'18), Niš, Serbia, 11–12 October 2018.
20. Losada, M.; Adin, I.; Perez, A.; Ramírez, R.C.; Mendizabal, J. Connected Heterogenous Multi-Processing Architecture for Digitalization of Freight Railway Transport Applications. *Electronics* **2022**, *11*, 943. [CrossRef]
21. Mizuno, Y.; Fujino, Y.; Kataoka, K.; Matsumoto, Y. Development of a mobile sensing unit and its prototype implementation. *Tsinghua Sci. Technol.* **2008**, *13*, 223–227. [CrossRef]
22. Tsunashima, H.; Mori, H.; Ogino, M.; Asano, A. Development of Track Condition Monitoring System Using Onboard Sensing Device. In *Railway Research: Selected Topics on Development, Safety and Technology*; Intech: Rijeka, Croatia, 2015. [CrossRef]
23. Ramirez, R.C.; Adin, N.; Goya, J.; Alvarado, U.; Brazalez, A.; Mendizabal, J. Freight Train in the Age of Self-Driving Vehicles. A Taxonomy Review. *IEEE Access* **2022**, *10*, 9750–9762. [CrossRef]
24. Schiavo, A.L. Fully Autonomous Wireless Sensor Network for Freight Wagon Monitoring. *IEEE Sens. J.* **2016**, *16*, 9053–9063. [CrossRef]
25. Hoelzl, C.; Arcieri, G.; Ancu, L.; Banaszak, S.; Kollros, A.; Dertimanis, V.; Chatzi, E. Fusing Expert Knowledge with Monitoring Data for Condition Assessment of Railway Welds. *Sensors* **2023**, *23*, 2672. [CrossRef]
26. Ho, S. Futuristic railway condition monitoring system. In Proceedings of the IET International Conference on Railway Engineering 2008 (ICRE 2008), Hong Kong, China, 25–28 March 2008; pp. 1–9. [CrossRef]
27. Baasch, B.; Roth, M.H.; Groos, J.C. In-service condition monitoring of rail tracks: On an on-board low-cost multi-sensor system for condition based maintenance of railway tracks. *Int. Verkehrswesen* **2018**, *70*, 76–79.
28. Balog, M.; Sokhatska, H.; Iakovets, A. Intelligent Systems in the Railway Freight Management. In *Advances in Manufacturing II: Volume 1-Solutions for Industry 4.0*; Springer: Cham, Switzerland, 2019. [CrossRef]
29. Jo, O.; Kim, Y.-K.; Kim, J. Internet of Things for Smart Railway: Feasibility and Applications. *IEEE Internet Things J.* **2017**, *5*, 482–490. [CrossRef]
30. Aurich, S.; Rasel, T. Method and Device for Condition Monitoring of Wheelsets or Bogies of a Rail Vehicle. 2006. Available online: https://patents.google.com/patent/WO2007082657A1/en (accessed on 16 August 2023).

31. Tsunashima, H.; Takikawa, M. Monitoring the Condition of Railway Tracks Using a Convolutional Neural Network. In *Recent Advances in Wavelet Transforms and Their Applications*; IntechOpen: Rijeka, Croatia, 2022. [CrossRef]
32. Sun, X.; Yang, F.; Shi, J.; Ke, Z.; Zhou, Y. On-Board Detection of Longitudinal Track Irregularity Via Axle Box Acceleration in HSR. *IEEE Access* **2021**, *9*, 14025–14037. [CrossRef]
33. Dertimanis, V.K.; Zimmermann, M.; Corman, F.; Chatzi, E.N. On-Board Monitoring of Rail Roughness via Axle Box Accelerations of Revenue Trains with Uncertain Dynamics. In *Model Validation and Uncertainty Quantification Proceedings of the37th IMAC, a Conference and Exposition on Structural Dynamics, Orlando, FL, USA, 28–31 January 2019*; Springer: Berlin/Heidelberg, Germany, 2020; Volume 3. [CrossRef]
34. Bernal, E.; Spiryagin, M.; Cole, C. Onboard Condition Monitoring Sensors, Systems and Techniques for Freight Railway Vehicles: A Review. *IEEE Sens. J.* **2018**, *19*, 4–24. [CrossRef]
35. Chia, L.; Bhardwaj, B.; Lu, P.; Bridgelall, R. Railroad Track Condition Monitoring Using Inertial Sensors and Digital Signal Processing: A Review. *IEEE Sens. J.* **2018**, *19*, 25–33. [CrossRef]
36. Ulianov, C.; Hyde, P.; Shaltout, R. Railway Applications for Monitoring and Tracking Systems. In *Sustainable Rail Transport. Lecture Notes in Mobility Proceedings of the RailNewcastle Talks 2016*; Springer: Berlin/Heidelberg, Germany, 2017. [CrossRef]
37. Zhang, X.; Jia, L.; Wei, X.; Ru, N. Railway track condition monitoring based on acceleration measurements. In Proceedings of the 27th Chinese Control and Decision Conference (2015 CCDC), Qingdao, China, 23–25 May 2015. [CrossRef]
38. Cii, S.; Tomasini, G.; Bacci, M.L.; Tarsitano, D. Solar Wireless Sensor Nodes for Condition Monitoring of Freight Trains. *IEEE Trans. Intell. Transp. Syst.* **2020**, *23*, 3995–4007. [CrossRef]
39. Shaikh, M.Z.; Ahmed, Z.; Chowdhry, B.S.; Baro, E.N.; Hussain, T.; Uqaili, M.A.; Mehran, S.; Kumar, D.; Shah, A.A. State-of-the-Art Wayside Condition Monitoring Systems for Railway Wheels: A Comprehensive Review. *IEEE Access* **2023**, *11*, 13257–13279. [CrossRef]
40. Roberts, C.; Goodall, R.M. Strategies and techniques for safety and performance monitoring on railways. *IFAC Proc. Vol.* **2009**, *42*, 746–755. [CrossRef]
41. Vlachospyros, G.; Iliopoulos, I.A.; Kritikakos, K.; Kaliorakis, N.; Fassois, S.D.; Sakellariou, J.S.; Deloukas, A.; Leoutsakos, G.; Giannakis, C.; Chronopoulos, E.; et al. The MAIANDROS System for Random-Vibration-Based On-Board Railway Vehicle and Track Monitoring. In Proceedings of the ASME 2021 International Design Engineering Technical Conferences and Computers and Information in Engineering Conference, Virtual, 17–19 August 2021. [CrossRef]
42. Tsunashima, H. Track Condition Monitoring Based on Car-Body Acceleration Using Time-Frequency Analysis. In *Advances in Dynamics of Vehicles on Roads and Tracks. Proceedings of the IAVSD 2019. Lecture Notes in Mechanical Engineering, Gothenburg, Sweden, 12–16 August 2019*; Springer: Berlin/Heidelberg, Germany, 12 August 2020. [CrossRef]
43. Jones, M.; Southcombe, J. Track Condition Monitoring Using Service Trains. In Proceedings of the AusRAIL PLUS 2015, Doing it Smarter. People, Power, Performance, Melbourne, VIC, Australia, 24–26 November 2015.
44. Iliopoulos, I.A.; Sakellariou, J.S.; Fassois, S.D. Track segment automated characterisation via railway–vehicle–based random vibration signals and statistical time series methods. *Veh. Syst. Dyn.* **2021**, *60*, 3336–3357. [CrossRef]
45. Bernal, E.; Spiryagin, M.; Cole, C. Ultra-Low Power Sensor Node for On-Board Railway Wagon Monitoring. *IEEE Sens. J.* **2020**, *20*, 15185–15192. [CrossRef]
46. Hodge, V.J.; O'Keefe, S.; Weeks, M.; Moulds, A. Wireless Sensor Networks for Condition Monitoring in the Railway Industry: A Survey. *IEEE Trans. Intell. Transp. Syst.* **2015**, *16*, 1088–1106. [CrossRef]
47. Josey, J. Intelligent infrastructure for next-generation rail system. *Cogniz. 20-20 Insights* **2013**, 1–8.
48. Deliverable D7.2 OTI Technology Migration, X2RAIL-4. 17 March 2023. Available online: https://projects.shift2rail.org/s2r_ip2_n.aspx?p=X2RAIL-4 (accessed on 19 July 2023).
49. FR8RAIL IV Project. Available online: https://projects.shift2rail.org/s2r_ip5_n.aspx?p=FR8RAIL%20iv (accessed on 18 July 2023).
50. Sggrss 80 | 6 Axle Wagon. Available online: https://www.greenbrier-europe.com/2022/10/08/sggrss-80-6-axle-articulated-intermodal-wagon (accessed on 18 July 2023).

sensors

MDPI

*Article*

# An Unsupervised Learning Approach for Wayside Train Wheel Flat Detection

**Mohammadreza Mohammadi [1,*], Araliya Mosleh [1], Cecilia Vale [1], Diogo Ribeiro [2], Pedro Montenegro [1] and Andreia Meixedo [1]**

1 CONSTRUCT—LESE, Faculty of Engineering, University of Porto, 4200-465 Porto, Portugal
2 CONSTRUCT, School of Engineering, Polytechnic of Porto, 4249-015 Porto, Portugal
* Correspondence: up202202392@fe.up.pt

**Abstract:** One of the most common types of wheel damage is flats which can cause high maintenance costs and enhance the probability of failure and damage to the track components. This study aims to compare the performance of four feature extraction methods, namely, auto-regressive (AR), auto-regressive exogenous (ARX), principal component analysis (PCA), and continuous wavelet transform (CWT) capable of automatically distinguishing a defective wheel from a healthy one. The rail acceleration for the passage of freight vehicles is used as a reference measurement to perform this study which comprises four steps: (i) feature extraction from acquired responses using the specific feature extraction methods; (ii) feature normalization based on a latent variable method; (iii) data fusion to enhance the sensitivity to recognize defective wheels; and (iv) damage detection by performing an outlier analysis. The results of this research show that AR and ARX extraction methods are more efficient techniques than CWT and PCA for wheel flat damage detection. Furthermore, in almost every feature, a single sensor on the rail is sufficient to identify a defective wheel. Additionally, AR and ARX methods demonstrated the potential to distinguish a defective wheel on the left and right sides. Lastly, the ARX method demonstrated robustness to detect the wheel flat with accelerometers placed only in the sleepers.

**Keywords:** wheel flat detection; wayside condition monitoring; train-track interaction; unsupervised learning

**Citation:** Mohammadi, M.; Mosleh, A.; Vale, C.; Ribeiro, D.; Montenegro, P.; Meixedo, A. An Unsupervised Learning Approach for Wayside Train Wheel Flat Detection. *Sensors* **2023**, *23*, 1910. https://doi.org/10.3390/s23041910

Academic Editor: Gilbert-Rainer Gillich

Received: 30 December 2022
Revised: 2 February 2023
Accepted: 6 February 2023
Published: 8 February 2023



## 1. Introduction

Nowadays, due to the increasing importance of railway transportation infrastructures, many studies have been conducted on their cost-effectiveness, particularly in terms of operation and maintenance costs [1–3]. One of the main responsible for the structural degradation of the railway infrastructure, particularly the track, is the operating rolling stock [4,5]. Therefore, an efficient and reliable condition assessment of the rolling stock is crucial for any infrastructure manager.

Many types of damage can affect a train's operational performance and one of the most important is defective wheels, which include two categories of defects, localized defects in the wheel tread (e.g., wheel flat, spalling and shelling), and defects around the complete wheel perimeter (e.g., wheel corrugation and polygonal wheel).

Wheel flats are the most common type of defect in train wheels and remarkably affects running safety and causes significant damage to the infrastructure, namely the rails and sleepers, due to the higher impact forces induced in the track [6]. The initial cause for the wheel flat is the friction between the wheel and rail due to braking forces, as friction can change the shape of the exterior perimeter of the wheel from round to flat. The wheel flat length is the standard for wheelset maintenance, as stated in the General Contract for the Use of Wagons [7]. For a wheel diameter larger than 840 mm, and in the presence of flat lengths greater than 60 mm, the wheelset should be immediately replaced.

Detecting defective wheels at an early stage is recommended to maintain safety, stability, and minimize maintenance costs.

To do this, an automated approach must be developed that can clearly distinguish between a healthy and damaged wheel. Therefore, finding effective methods for the early detection and identification of wheel flats is of great interest to railway administrations and rolling stock operators.

In the last few decades, researchers have proposed several onboard and wayside systems for detecting wheel defects in operation conditions, most based on the concept that the interaction force between the train and the track increases in a defective wheel [8,9]. Many onboard techniques are based on vibration, acoustic, image detection, and ultrasonic technologies [10,11]. Nevertheless, all wheels must be equipped with sensors for comprehensive diagnosis and effective wheel condition management. The high cost and maintenance problems of this method make it rarely used. Moreover, onboard detection methods are commonly used to monitor track conditions.

Alternatively, wayside measurement systems are currently the preferred solution to identify wheel flats since all wheels are evaluated during the train passage at the specific system location [12–14]. Previous research has been focused on advanced signal processing methods to eliminate signal interference and spotlight the faulty signal patterns of wheel flats. Jiang et al. [15] used the empirical mode decomposition (EMD) method to divide the signal into several intrinsic mode functions (IMF) which separates the faulty signal mode from interferences. Amini et al. [13] proposed a method based on time–spectral kurtosis (TSK) to reduce the effect of noise and highlight the faulty signal patterns of wheel flats. Mosleh et al. [16] proposed a method to distinguish a defective wheel from a healthy one based on the envelope spectrum method. Krummenacher et al. [17], by measuring the vertical wheels' force and using a sensor system permanently installed on the railway track, proposed two machine-learning methods to automatically detect a defective wheel during operation. These methods learn different types of wheel defects and predict whether a wheel has a defect. Yi-Qing et al. [18] developed a probabilistic Bayesian method using trackside strain sensors for the online condition monitoring of the wheels. They found that only using monitoring data from a single sensor may produce false-negative results, but with the data from all the deployed sensors could provide more accurate diagnostic results.

Typically, the phases for damage identification methods are related to data acquisition, feature extraction, feature normalization, data fusion, and feature classification [1,19]. The process of transforming time series data into alternative information, where the correlation with damage is easily visible, is known as feature extraction [20,21]. Typically, the auto-regressive model (AR) [5], auto-regressive model with exogenous input (ARX) [19], principal component analysis (PCA) [22], and continuous wavelet transform (CWT) [8] are employed to extract the damage-sensitive features using the dynamic responses.

One of the main challenges to detect a damaged wheel is to remove the environmental and operational effects from the dynamic responses to obtain features that are mainly sensitive to damage but insensitive to environmental and operational changes (EOVs). Therefore, to reduce the variation caused by EOVs and enhance the sensitivity to damage, feature normalization is performed by using various linear and non-linear correction models, such as, PCA [23], kernel principal component analysis (KPCA) [24], non-linear principal component analysis (NLPCA) [25], and factor analysis (FA) [26].

For feature fusion and dimension reduction, several algorithms, including neighborhood-preserving embedding (NPE) [27], neural networks [28], Mahalanobis distance [29], manifold-learning methods [30], and kernel-based methods [31], have recently been employed. The capability of the Mahalanobis distance to capture the variability in multivariate datasets has led to the widespread use of this technique [23]. This method has been used in multiple research studies with excellent results as it increases the sensitivity to the damage and can integrate data from various sensors [32].

In recent years, machine-learning (ML) approaches in combination with advanced signal processing methods have been applied for feature classification to differentiate a healthy wheel from a defective one [18,33]. Unsupervised and supervised learning are two different types of ML techniques. Unsupervised learning involves finding hidden structures in unlabeled data to classify them into meaningful categories. On the other hand, supervised learning assumes that a database's categories or hierarchy of the database are known in advance. Researchers have recently investigated supervised and unsupervised approaches for classifying data based on dedicated features, including unsupervised methods, such as, k-mean [1], self-organizing maps (SOM) [34], and cluster analysis, as well as supervised methods, such as, naive Bayes classifiers [35] and k-nearest neighbor (kNN) classifiers [36].

Most of the previous research on wheel flat detection is based on engineering field tests. However, numerical analysis is very useful for understanding the mechanism and physical consequences based on dedicated models. Additionally, models can be used for deeper comprehension and prediction in situations that cannot be reproduced in experimental tests. For example, external elements, such as noise, environmental disturbances, measurement errors, and electromagnetic interferences, easily influence the measurement process and may affect the results, causing a decrease in measurement accuracy. Additionally, numerical simulation makes it possible to define each unknown variable separately to check how it affects the results.

It should be highlighted that the initial research on this topic was developed by Mosleh et al. [5,8], who proposed an automatic wheel flat identification method based on shear and accelerometer time series evaluated on the rails. It should be noted that the CWT [8] and AR [5] methods have been used separately in each research to extract features. However, none of these studies compared the accuracy of different features. Therefore, one of the novelties of this research is the comparison of the accuracy of four different feature extraction techniques using an unsupervised learning methodology to automatically detect a defective wheel, which is a clear step forward in terms of the effectiveness of the proposed method and allows full implementation for real-world application. Therefore, a 3D numerical dynamic model of a vehicle–track coupling system was used for this purpose. The features were extracted by applying the AR, ARX, PCA, and CWT models to the measurement records. Moreover, PCA, as well as Mahalanobis distance, were used for feature modeling and data fusion, respectively. Finally, outlier and cluster analyses were applied for feature classification. The following significant contributions can be highlighted from this research work:

(1) Development of an unsupervised data-driven methodology using acceleration responses on the rail for detecting defective wheels from healthy ones;
(2) Implementation of AR, ARX, CWT, and PCA for feature extraction from multiple sensors to transform the time series measurements into damage-sensitive features, where the correlation with the damage can be more easily observed;
(3) Analysis of the performance of the four feature extraction methods considering the different number and locations of the sensors on the rails;
(4) Comparison of the sensitivity of the proposed methodologies to the side (left vs. right) of the defective wheel in a train axle;
(5) Evaluation of the effectiveness of the proposed method with respect to the minimalist layout of sensors;
(6) Improvement in wheel flat detection by applying a two-stage fusion process: in the first step, the features from each sensor are merged and, in the second stage, the multi-sensor information is fused to enhance the sensibility to the damage.

## 2. Numerical Simulation

### 2.1. Train–Track Dynamic Interaction

In this study, by using in-house software vehicle–structure interaction (VSI), the simulations for numerical train–track dynamic interaction were carried out. The vehicle–structure interaction analysis is explained and validated in detail in the work of Montenegro et al. [37] and has been used in several applications [5,16]. A 3D wheel–rail contact model couples the train to the track using Hertzian theory [38], to compute normal contact forces, and USETAB routine [39], to compute the tangential forces caused by rolling friction creep. The structural matrices from the structure (in this case, the track) and the vehicle, previously modeled in a finite element program (FE), were imported into this numerical tool, which was developed in MATLAB [40]. Although these subsystem models were initially created individually, the VSI program connects them using a fully linked technique [37]. Figure 1 represents the graphical representation of this procedure.

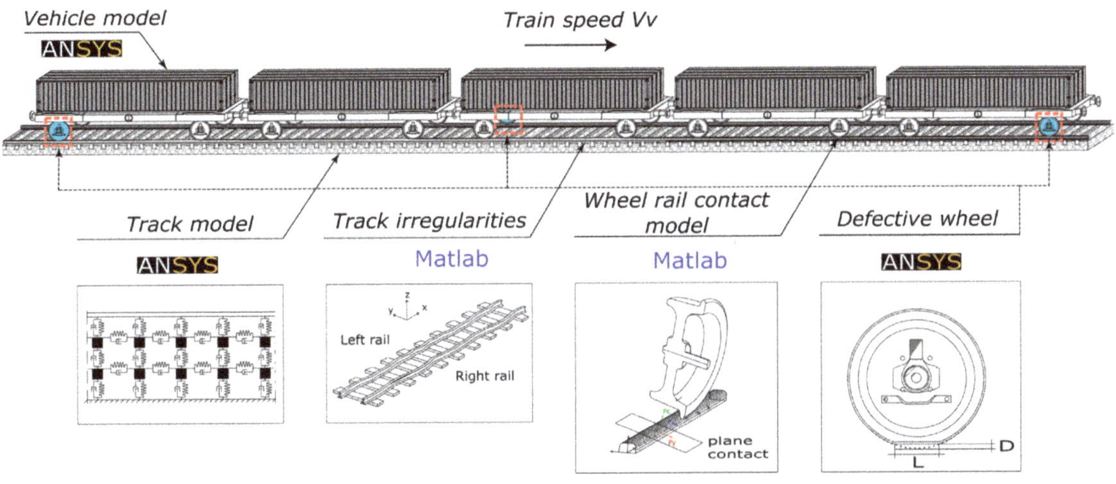


**Figure 1.** Numerical modeling of the train–track system.

The software ANSYS [41] was used to simulate the track. Beam elements were used to model the rails and sleepers, while spring–dashpot components were used to simulate the behavior of the flexible layers, i.e., the ballast, fasteners/pad, and mass point components to account for the ballast's mass as shown in Figure 1. The train was composed of five wagons of Laagrss type, each one with two axles, had also been modeled in ANSYS [41] through a multibody formulation, using mass point elements located at the center of gravity of each body, specifically the car body, and wheelsets, to simulate their mass and inertial effects. Rigid beams were also used to consider the rigid body movements of the vehicle. The characteristics of both the track and train models are fully described in the work of Mosleh et al. [16,42].

### 2.2. Virtual Wayside System

A set of eight accelerometers were considered along the track as part of the wheel flat-detecting system. Figure 2 depicts the position of the sensors in the proposed virtual wayside monitoring system. Measurement points 1 to 4 simulate the position of the accelerometers located on the right side of the track, particularly on the rail and on the sleepers; otherwise, measurement points 5 to 8 represent the sensors located on the left side of the track. In Section 4, accelerometers 1–4 were selected to depict the results. One of the main advantages of the proposed method compared to previous approaches [16,42,43] is that there is no need to install a series of sensors on the rail to monitor the whole perimeter of the wheel. Only a minimalist set of sensors are sufficient to detect a defective wheel.

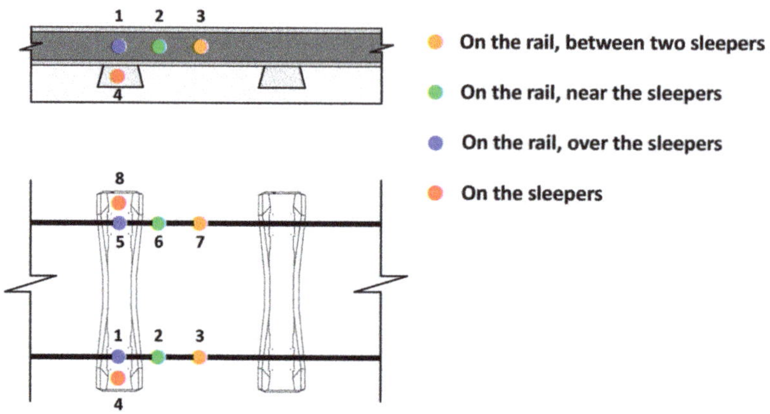


**Figure 2.** Virtual wayside monitoring system.

### 2.3. Baseline and Damaged Scenarios

For testing and validating the automatic wheel flat-identification method proposed in this work, baseline (undamaged) and damaged wheel scenarios were considered. After validation, this method can reproduce real experimental data, from different types of trains with various wheel defects, running at different speeds on the rail track with distinct rail irregularities profiles.

As shown in Figure 1, for damaged scenarios, three defective cases are considered, particularly ones located on: (i) the right wheel on the front wheelset of the first wagon (Damage 1), (ii) the left wheel of the rear wheelset of the third wagon (Damage 2); (iii) right wheel of the rear wheelset of the fifth (last) wagon (Damage 3). The lower and upper bounds of the flat length are defined by uniform distributions U (50, 100) for the three defective wheels. The wheel flat depth (D) is defined by the following expression [41]:

$$D = \frac{L^2}{16R_w}$$

where L is the flat length and $R_w$ the radius of the wheel.

The vertical profile deviation of the wheel flat (Z) is defined as follows [41]:

$$Z = -\frac{D}{2}\left(1 - \cos\frac{2\pi x}{L}\right).H(x - (2\pi R_w - L)), \quad 0 \leq x \leq 2\pi R_w$$

where H represents the Heaviside periodic function, and x is the coordinate aligned with the track longitudinal direction.

Wheel-rail contact force values are significantly affected by imperfections in a real-condition environment, where the rails are not completely smooth. Although these irregularities are very small, they should be considered in the numerical analyses. Four real unevenness track profiles are taken into consideration in this study. The selected unevenness profiles of the rail are measured on the Northern Line of the Portuguese Railway network based on the track inspection vehicle EM120 and according to the details provided by Mosleh et al. [14]. The total length of the simulation was 1000 m.

To evaluate the proposed methodology, the accelerations on eight positions of the rail were evaluated in both baseline (undamaged) and damaged scenarios. The baseline condition corresponds to a train passing with healthy wheels, while the damaged scenarios correspond to the passage of trains with defective wheels. Table 1 summarizes the assumptions for damaged and baseline scenarios, as well as the number of numerical analyses performed for each scenario.

**Table 1.** Damaged and undamaged scenarios.

| | Baseline Scenarios | Damaged Scenarios |
|---|---|---|
| Train | Freight—Laagrss wagon | |
| Number of loading schemes | 6 | 1 (full capacity) |
| Unevenness profiles | 4 | 1 |
| Speeds (km/h) | 40–120 | 80 |
| Noise ratio | 5% | |
| Flat lengths (mm) | – | 50–100 |
| Number of numerical analyses | 100 | 30 |

Figure 3 presents the baseline scenario for which 113 simulations were performed considering a freight train comprising five wagons. Six different types of loading schemes were considered: (i) full-loaded train; (ii) half-loaded train, (iii) empty train, as well as trains with unbalanced loads in the transversal and longitudinal directions, namely (iv) UNB1, (v) UNB2 and (vi) UNB3. According to UIC loading guidelines [44] different unbalanced loading schemes were defined for the wagon model, where the cargo gravity center was offset in longitudinal and transversal directions.

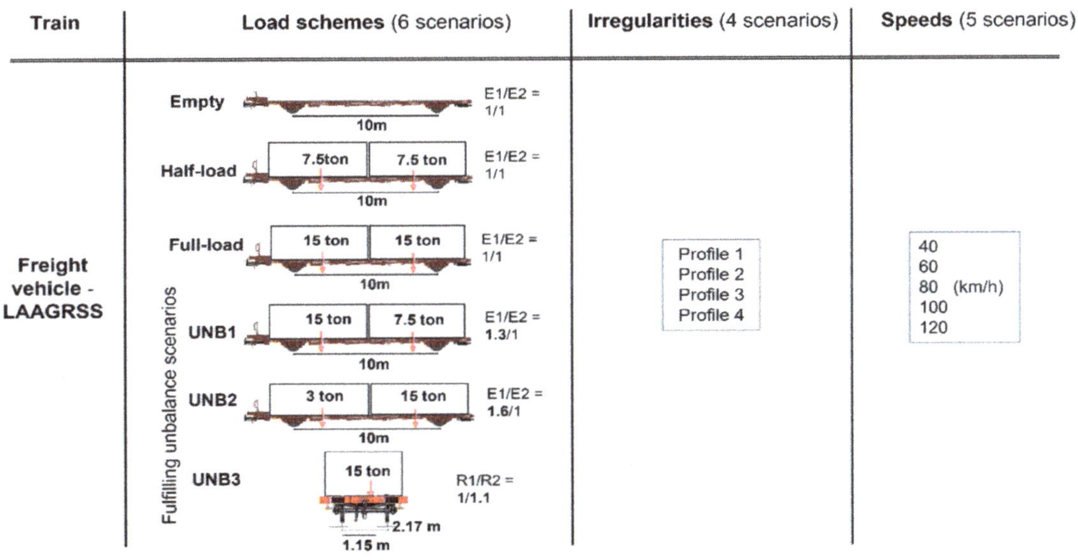


**Figure 3.** Baseline scenarios.

Figure 4 illustrates the damaged scenarios for 30 simulations which were implemented considering several combinations of flat properties for defective wheels. As mentioned before three defective cases were considered in this study, namely Damage 1, Damage 2 and Damage 3, which are located on the 1st, 3rd and 5th wagons, respectively. In total, 10 analyses were performed for each damaged wheel (Damages 1, 2 and 3) and the speed was considered equal to 80 km/h. Moreover, a sampling frequency of 10 kHz was used to evaluate acceleration signals for both baseline and damage scenarios.

The numerical signal was then polluted with artificial noise (5% of the amplitude) based on the maximum response of the signal for a more realistic reproduction of the measured rail response.

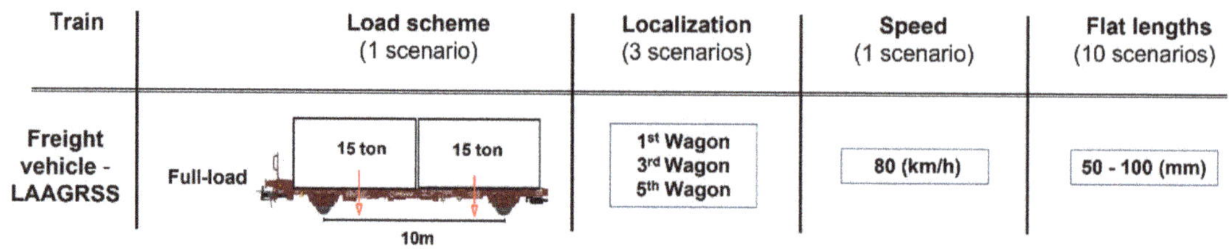


**Figure 4.** Damaged scenarios.

In Figure 5 are shown examples of acceleration time series for baseline scenarios obtained in sensor 3, located on the rail. These figures show the influence of different loading schemes, train speeds, and unevenness profiles on the track response. All-time series were filtered using a low-pass Chebyshev type II digital filter with a cut-off frequency of 500 Hz.

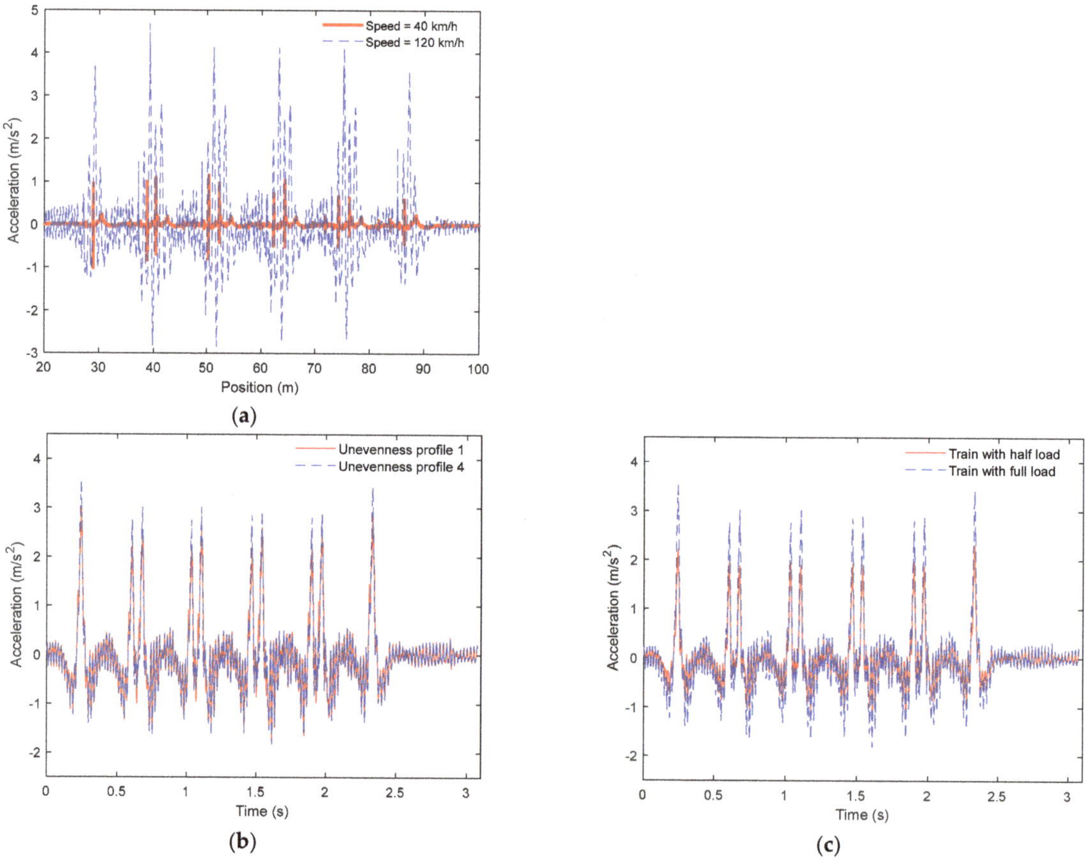


**Figure 5.** Acceleration time series for sensor 3 for a freight train considering a healthy wheel (baseline scenario): (**a**) influence of vehicle speed, (**b**) influence of the unevenness profile, (**c**) influence of the loading schemes.

Figure 5a demonstrates the relevant influence of the train speed on the evaluated acceleration, and the need to consider various train speeds for identifying wheel flats. Additionally, as shown in Figure 5b, both unevenness rail profiles induced similar acceleration responses. Finally, Figure 5c shows that both loading schemes affect the track responses particularly on the peak acceleration values.

### 3. Unsupervised Learning Methodology for Wheel Flat Detection

The purposed methodology for the automatic detection of wheel flats presented in Figure 6 includes four steps, particularly:

1. Features extraction: application of four advanced data-driven models, including the continuous wavelet transform (CWT), auto-regressive model (AR), principal component analysis (PCA), and ARX to extract the damage-sensitive features from the time series;
2. Feature normalization: normalization of the extracted features by the principal component analysis (PCA) method to increase the sensitivity to damage and remove environmental and operational variations (EOVs);
3. Data fusion: implementation of a Mahalanobis distance (MD) to merge the features derived from each sensor and detect wheel defects more effectively. In the first stage, the features from each sensor are merged and, in the second stage, the multi-sensor information is fused to enhance the sensibility to the damage [26,32];
4. Outlier analyses: upon completion of the previous step, a damage indicator (DI) is generated for each train passage; to distinguish each DI into a defective or a healthy wheel a statistical-based approach is used, in particular, an inverse cumulative distribution function that allows estimating a statistical confidence boundary (CB).

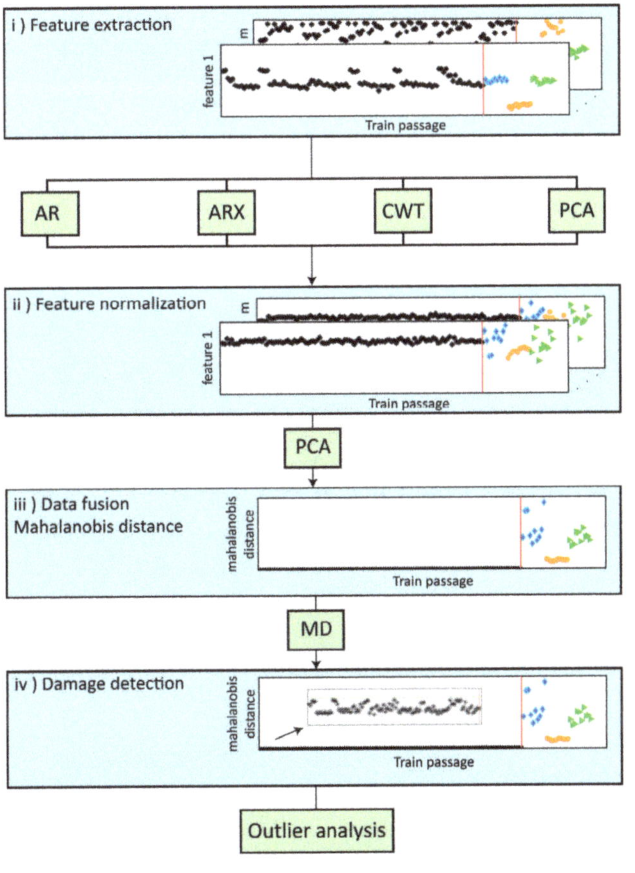


**Figure 6.** Methodology of the automatic detection of wheel flats.

The theoretical framework of each technique implemented within the methodology is available in the authors' previous publications [1,5,8,19,45].

## 4. Application of the Methodology of Wheel Flat Detection to a Freight Train

This section presents the application of the unsupervised learning methodology of wheel flat detection to the case of a freight train and considers different feature extraction methods, namely the AR, ARX, CWT and PCA. The purpose of this comparison was to assess the sensitivity to damage of each extraction method.

### 4.1. Feature Extraction

Damage-sensitive feature extraction from dynamic signals is the first step of the automatic damage detection methodology. The main goal of this step is to reduce the dimensions of the three-dimensional dynamic features matrices 143−by−q−by−n, in which, 143 is the total number of scenarios, including 113 baseline scenarios and 30 damage scenarios, q is the number of sensors (four sensors) and n is the dimension of dynamic time-histories (70,000). For this purpose, the extraction of features sensitive to the effects of wheel flats was performed by considering auto-regressive model (AR), principal component analysis (PCA), continuous wavelet transform (CWT), and auto-regressive model with exogenous input (ARX).

#### 4.1.1. AR Model

Several AR models were analyzed to determine the appropriate model order based on the Akaike information criteria (AIC), particularly the orders between 1 and 50. The AIC function for 30 damaged scenarios is shown in Figure 7. It can be concluded that, as the model's order increases, the AIC values tend to stabilize, which demonstrates that after a model order of 40, higher orders do not yield relevant information.

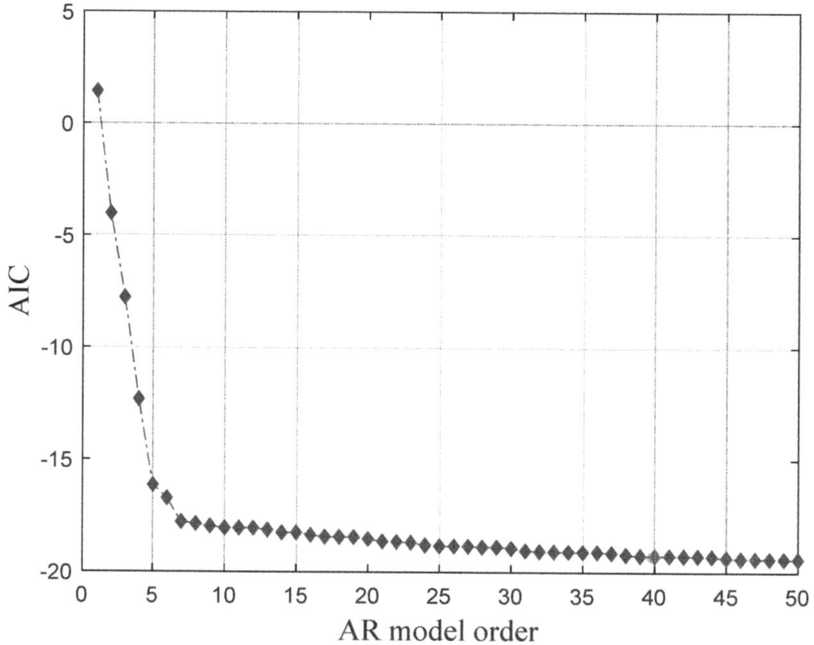


**Figure 7.** AR model order definition.

Extracted features from dynamic responses by implementing the AR method are obtained in 143−by−4−by−40 matrices which means that the number of features is significantly reduced from 70,000 to 40. Figure 8 illustrates five of the features obtained using the AR method for sensor 3. As shown in this figure, a particular sensitivity pattern is recognized for damaged scenarios, in such a way that, amplitude is sensitive to the side of the defect (right or left wheels). As an example, Figure 8b shows the amplitude of feature 19 for the 1st and 5th wagons with blue and green colors. Note that the defect is located on the right-side wheels for the 1st and 5th wagons, while for the 3rd wagon, the defect is placed on the left-side wheel which is presented in an orange color. It is noticeable that the amplitude is sensitive to the side of damage (left or right wheel). Additionally, in Figure 8d, due to the comparison of the amplitude variations between damage and baseline scenarios, it is possible to state a significant difference between healthy and damaged wheels. For other features, this difference is not so significant or visible, as is the case in Figure 8e.

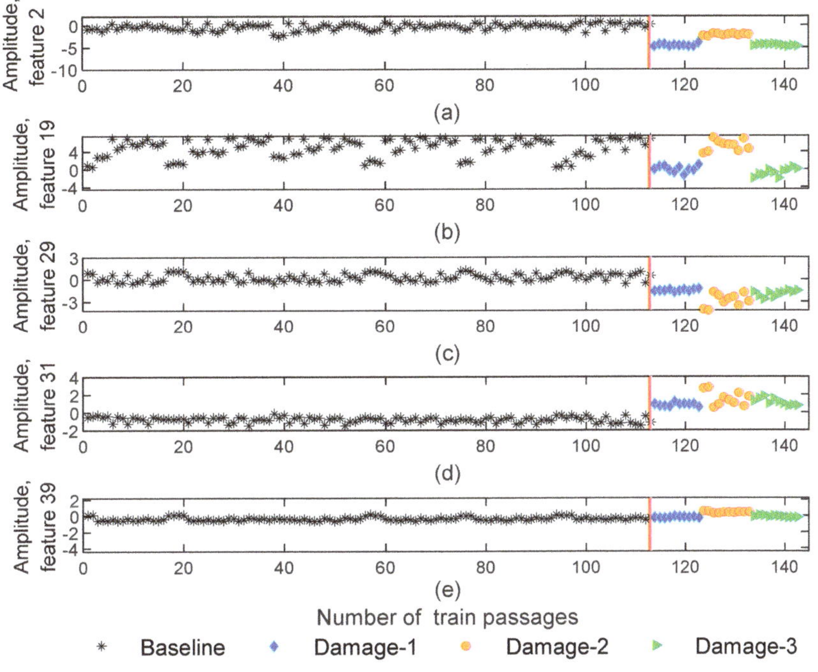


**Figure 8.** AR—feature extraction for all 143 baseline and damaged scenarios for accelerometer 3: (**a**) amplitude for feature 2, (**b**) amplitude for feature 19, (**c**) amplitude for feature 29, (**d**) amplitude for feature 31, (**e**) amplitude for feature 39.

#### 4.1.2. ARX Model

The auto-regressive model with exogenous input (ARX) is the second technique that was used to extract dynamic damage-sensitive features. This method of time-series analysis can perform a significant fusion while accurately generalizing the information contained in the data by adjusting the ARX (143−by−4−by−80) models. By using the ARX model the number of features is enlarged to 80 in comparison to the AR model. Figure 9 presents five of the features obtained by using the ARX method for sensor 3. As in the AR model, the damage scenarios features are also sensitive to the side of the wheel damage. As an example, in Figure 9b, the blue and green colors corresponding to a defective wheel on the right side of the 1st and 5th wagons, have similar amplitude values and are distinct from the ones associated with the defective wheel on the left side of the 3rd wagon, represented by the orange color. Moreover, as shown in Figure 9d, the difference between the amplitudes between healthy and defective wheels is evident.

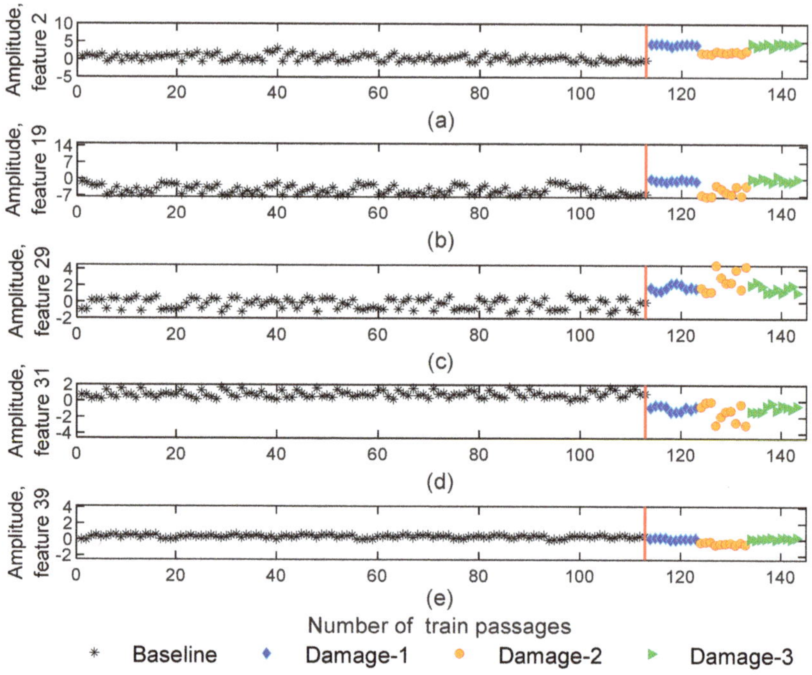


**Figure 9.** ARX—feature extraction for all 143 baseline and damage scenarios for accelerometer 3: (**a**) amplitude for feature 2, (**b**) amplitude for feature 19, (**c**) amplitude for feature 29, (**d**) amplitude for feature 31, (**e**) amplitude for feature 39.

#### 4.1.3. CWT

Another methodology that was implemented for feature extraction to reduce the size of the feature matrices was the continuous wavelet transform. By using CWT, the number of features is decreased from 70,000 to 468 and the obtained features matrices are of size 143−by−4−by−468. Figure 10 represents the extracted features for the CWT method, which shows sensitivity to the damage but not as much as the AR and ARX models. As an example, Figure 10a,b provides evidence that the features are sensitive to damage since their amplitude variation for damage scenarios is higher than for the healthy scenarios. However, for the features shown in Figure 10c,e, the amplitude variation is similar for healthy and defective wheels and the features are not sensitive to the damage. Furthermore, all the extracted features using CWT extraction are not sensitive to the side of the wheel defect.

#### 4.1.4. PCA

Data science frequently uses principal component analysis (PCA) to extract features based on the data projection into a new dimensionless subspace. PCA identifies the covariance matrix eigenvectors with the highest values [1,5,8,19]. In other words, the PCA method minimizes the number of features that effectively can capture the most significant of the original features. Thus, the number of extracted features is reduced to four and the matrices of damaged features are generated with 143−by−4−by−4 dimensions. The extracted features using PCA are represented in Figure 11. As shown, for features one and two, the dispersion of amplitude for healthy and defective wheels is almost imperceptible (Figure 11a,b). In turn, as Figure 11c shows for feature three, the amplitude variation of damaged scenarios is higher than baseline scenarios, and the amplitude difference between the damaged and healthy wheels is quite visible. This proves that only specific features have the potential to identify the damage.

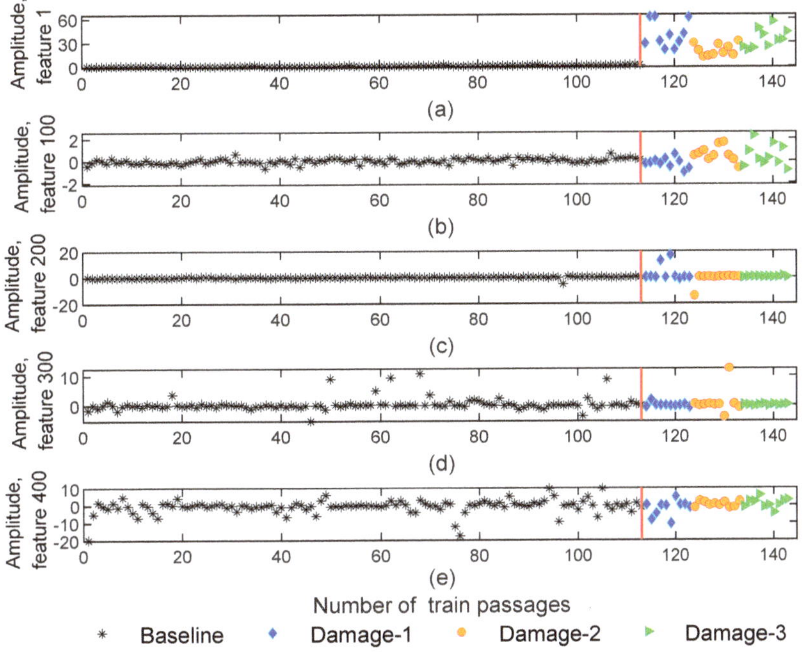


**Figure 10.** CWT—feature extraction for all 143 baseline and damage scenarios for accelerometer 3: (**a**) amplitude for feature 1, (**b**) amplitude for feature 100, (**c**) amplitude for feature 200, (**d**) amplitude for feature 300, (**e**) amplitude for feature 400.

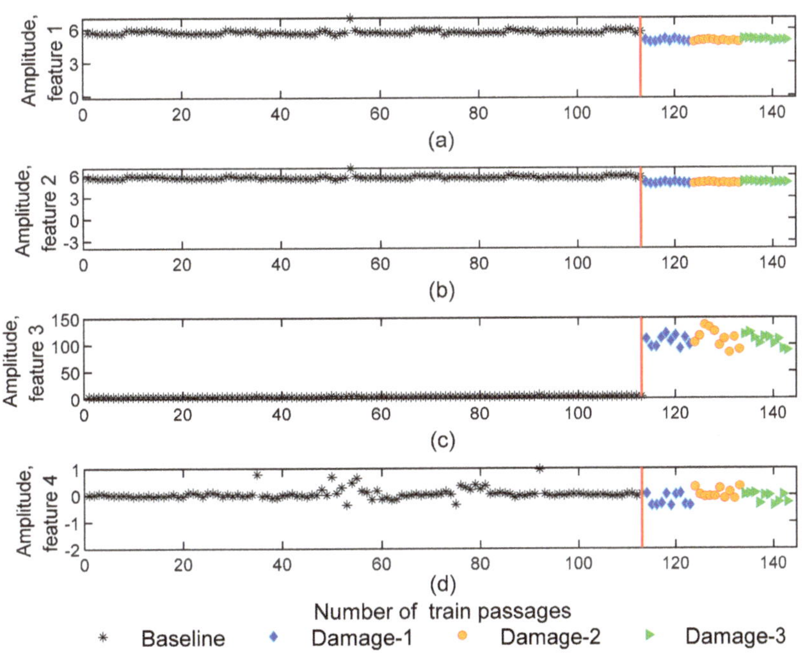


**Figure 11.** PCA—feature extraction for all 143 baseline and damage scenarios for accelerometer 3: (**a**) amplitude for feature 1, (**b**) amplitude for feature 2, (**c**) amplitude for feature 3, (**d**) amplitude for feature 4.

### 4.2. Feature Normalization

Data normalization allows to distinguish changes in the features acquired from sensor readings influenced by environmental and operational variations. One of the significant issues in damage detection is the difficulty of isolating environmental and operational disturbances from the observed dynamic properties to obtain features that are primarily sensitive to damage. Without the requirement to measure these actions, implementing a latent variable approach, such as PCA, to the retrieved features may effectively limit the effects of EOVs. In the feature normalization procedure, during the modeling phase, a cumulative percentage of the variance of components with a variance greater than 80% is removed [8].

#### 4.2.1. AR Model

By implementing the PCA method to AR parameters to normalize the features, for each train passage, a 4−by−40 matrix with PCA-based features was generated. Figure 12 represents 5 features out of 40 for all the 143 baseline and damage scenarios using the AR model. As shown in Figure 12a,d, after removing EOVs, features remain sensitive to damage and significant variations in amplitude can occur between the baseline and damage scenarios. Additionally, it is noteworthy that, after normalization, the extracted features by using the AR model remain sensitive to the side where the wheel defect occurs. As an example, in Figure 12a,e, the variation in amplitude for the 1st and 5th wagons (blue and green colors) are differentiable from the 3rd wagon (orange color).

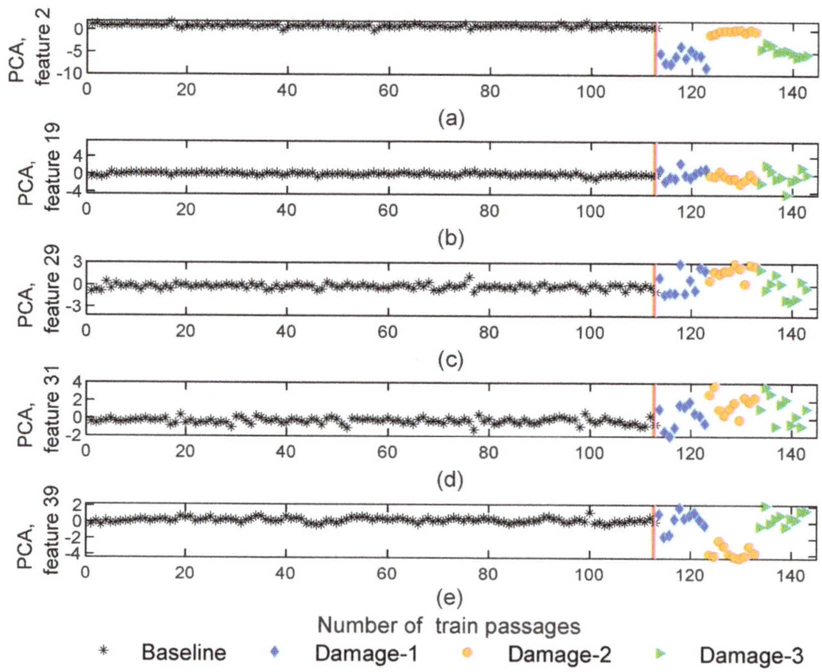


**Figure 12.** AR—feature normalization for all 143 baseline and damage scenarios for accelerometer 3: **(a)** PCA for feature 2, **(b)** PCA for feature 19, **(c)** PCA for feature 29, **(d)** PCA for feature 31, **(e)** PCA for feature 39.

#### 4.2.2. ARX Model

Feature normalization was also applied to the features extracted from the ARX method, and as a result, a matrix with dimension 4−by−80 was obtained individually for each train passage. Figure 13 shows that after implementing normalization the ARX features show specific sensitivity to damage. As seen in the examples shown in Figure 13c,e, the wheel defects have a noticeable effect on the variation in the features' amplitude. Moreover, as with the AR feature, the sensitivity to the side of the damage (left or right defective wheel) is still recognizable in some features after the elimination of the environmental effects, as stated in Figure 13a,e.

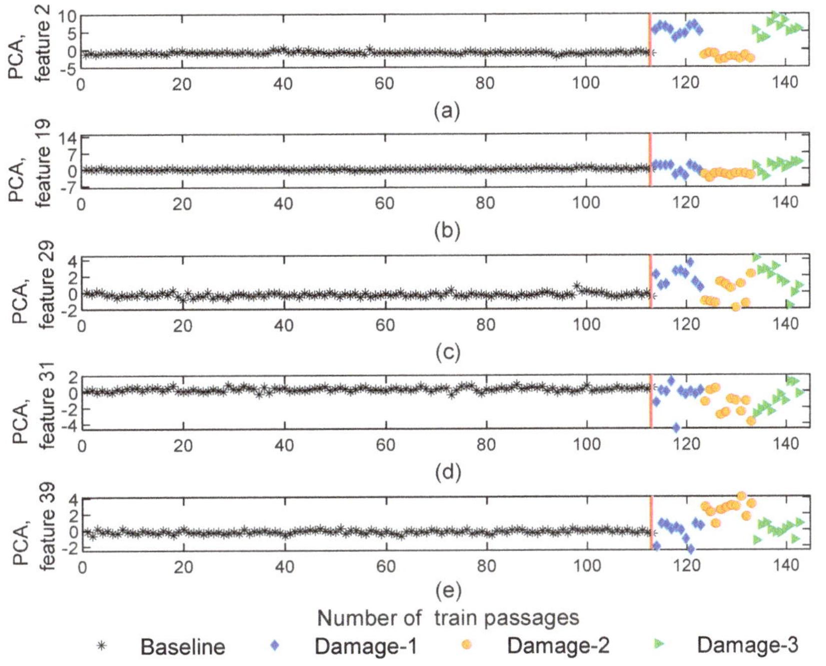


**Figure 13.** ARX—feature normalization for all 143 baseline and damage scenarios for accelerometer 3: (**a**) PCA for feature 2, (**b**) PCA for feature 19, (**c**) PCA for feature 29, (**d**) PCA for feature 31, (**e**) PCA for feature 39.

#### 4.2.3. CWT

Figure 14 shows five of the normalized features which are obtained by using CWT. In contrast to the AR and ARX feature normalization, the CWT normalization has an adverse effect on the sensitivity of the features to damage, and therefore, after normalization, the features are not sensitive enough to wheel defects. As an example, Figure 14a,c,e shows that the PCA-based normalized features lose sensitivity to the defects. Therefore, the different damages cause negligible variations in the amplitude of the feature, and no clear distinction is achieved in relation to the baseline. Additionally, the sensitivity in relation to the side of the damage is not recognizable for the CWT normalized features.

#### 4.2.4. PCA

Figure 15 shows that the PCA normalized features are influenced by environmental and operational effects, as shown in the compression of the amplitude's variation in comparison to the situation before normalization (Figure 11). Moreover, as shown in Figure 15b,d, the variation in amplitude for the damaged scenarios is quite distinguishable from the baseline scenarios, and features after normalization are sensitive to the defects. On the other hand, the PCA normalized features are not sensitive to the side of the wheel defect.

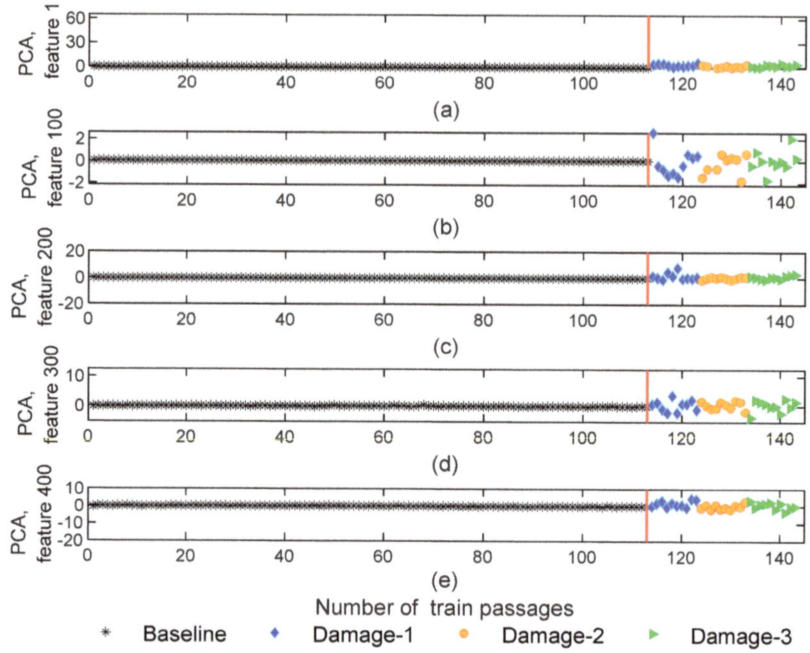


**Figure 14.** CWT—feature normalization for all 143 baseline and damage scenarios for accelerometer 3: (**a**) PCA for feature 1, (**b**) PCA for feature 100, (**c**) PCA for feature 200, (**d**) PCA for feature 300, (**e**) PCA for feature 400.

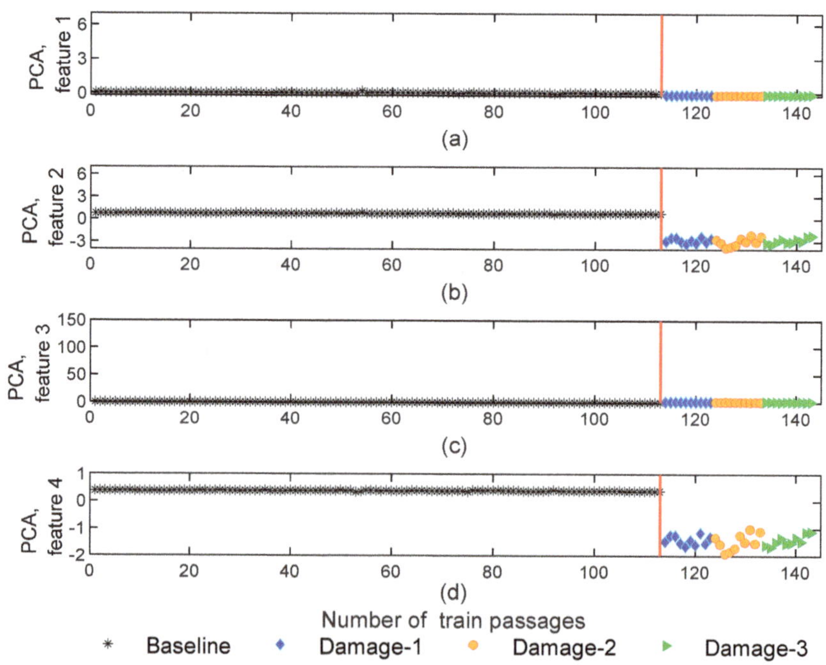


**Figure 15.** PCA—feature normalization for all 143 baseline and damage scenarios for accelerometer 3: (**a**) PCA for feature 1, (**b**) PCA for feature 2, (**c**) PCA for feature 3, (**d**) PCA for feature 4.

### *4.3. Data Fusion*

The results of Section 4.2 show that after the elimination of the environmental and operational effects, the difference between the baseline and damaged scenarios is not sufficient to distinguish healthy from damaged wheels. Therefore, the data fusion process was performed to increase the sensitivity of the features to the defect, and, as a result, a damage index (DI) was achieved for each simulation. Mahalanobis distance (MD) was used to reduce multivariate data into one single DI. To determine the similarities between the damaged and baseline features, the Mahalanobis distance (MD) calculates the distance between defective and healthy wheels, in which shorter distances represent higher similarities. In this step, the MD was obtained for each measurement point and train passage, and therefore, can transform all features into one single damage-sensitive feature. Thus, as a result, a distances vector with dimension 143−by−1 was calculated for every four sensors associated with each feature extraction method.

#### 4.3.1. AR Model

Figure 16 shows the values for the Mahalanobis distance for accelerometers 1–4 (see Figure 2). It is noticeable that the MD is sensitive to the defects and the variation in MD for defective wheels is higher than for healthy ones. Additionally, the MD is clearly sensitive to the side of the damage, as stated by the train passages of the defective wheel in the 3rd wagon (orange color) which have less amplitude compared with defective wheels on the right side of the 1st and 5th wagons (blue and green colors, respectively). Thus, as illustrated in this figure, it is possible to distinguish the damaged scenarios based on the side of the wheel defect.

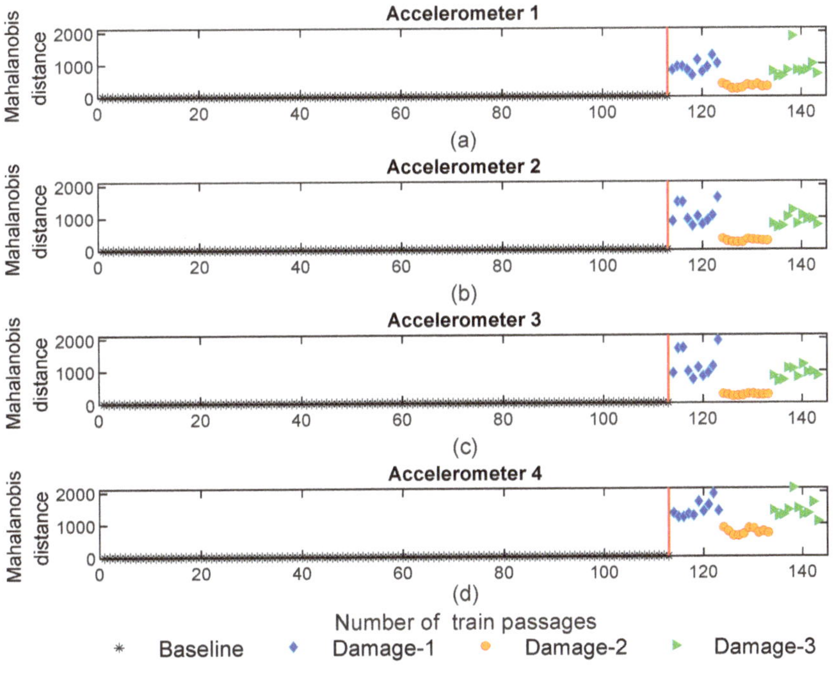


**Figure 16.** AR—data fusion for all 143 baseline and damage scenarios: (**a**) MD for accelerometer 1, (**b**) MD for accelerometer 2, (**c**) MD for accelerometer 3, (**d**) MD for accelerometer 4.

#### 4.3.2. ARX Model

The MD values for the ARX-normalized features are presented in Figure 17. It can be observed that the fusion of the features significantly increases the sensitivity to damage, and after the fusion the influence of damages is recognized, as stated by the amplitude of the variation for the damage scenarios which reaches a magnitude of 10,000. From this point, it can also be concluded that defective wheels can be distinguished from healthy ones. Furthermore, as shown in Figure 17, the MD is sensitive to the side of the damage and the defective wheels on the right side (blue and green colors) can be distinguished from the ones on the left side (orange color). Additionally, based on the amplitude values, it is possible to conclude that the features extracted by the ARX model are more sensitive than the ones derived from the AR model.

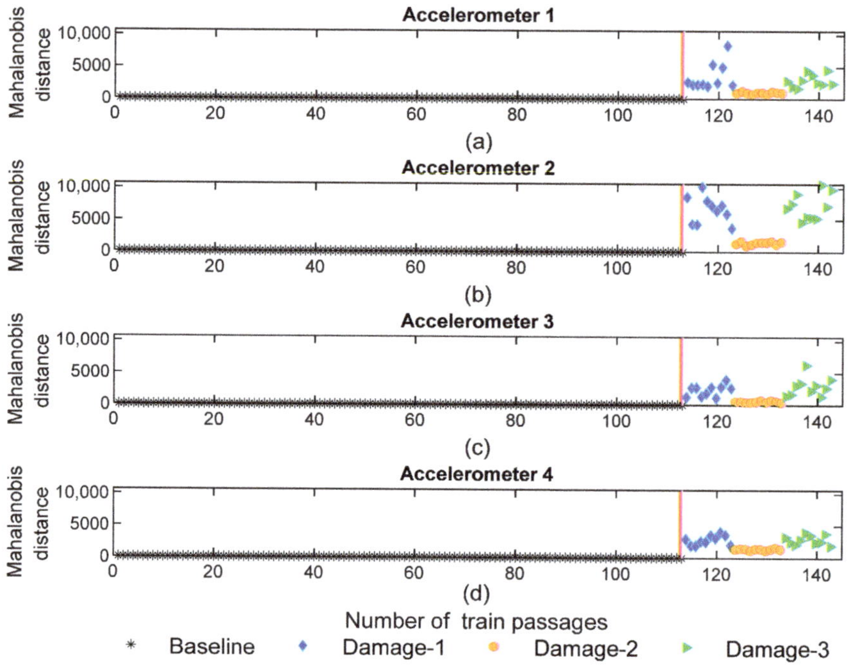


**Figure 17.** ARX—data fusion for all 143 baseline and damage scenarios: (**a**) MD for accelerometer 1, (**b**) MD for accelerometer 2, (**c**) MD for accelerometer 3, (**d**) MD for accelerometer 4.

#### 4.3.3. CWT

Figure 18 shows the MD for sensors 1–4 using CWT feature fusion. As presented in this figure, the amplitude of variation for the damaged scenarios is higher than for baseline scenarios; however, the maximum amplitude of the MD is 600 which is less than the value obtained for AR and ARX. On the other hand, it is noteworthy that, the amplitude of the MD for the defective wheel on the right side has the same range as the damaged wheel on the left side, which means that the MD based on the CWT features is not sensitive to the side of the wheel defect.

#### 4.3.4. PCA

Figure 19 shows the Mahalanobis distance based on the PCA extraction method. As shown in Figure 19c, the sensitivity for the MD based on the PCA is less than the AR and ARX models. Additionally, it should be mentioned that like CWT and in opposition to AR and ARX, the MD is not sensitive to the side of the wheel defect since the variation in the amplitude for the MD does not change between the three distinct damage scenarios.

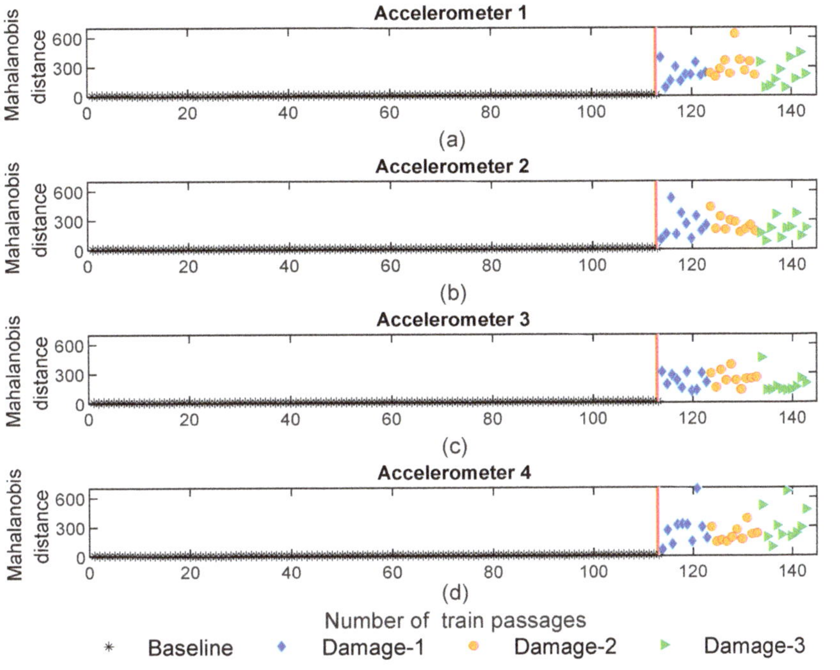


**Figure 18.** CWT—data fusion for all 143 baseline and damage scenarios: (**a**) MD for accelerometer 1, (**b**) MD for accelerometer 2, (**c**) MD for accelerometer 3, (**d**) MD for accelerometer 4.

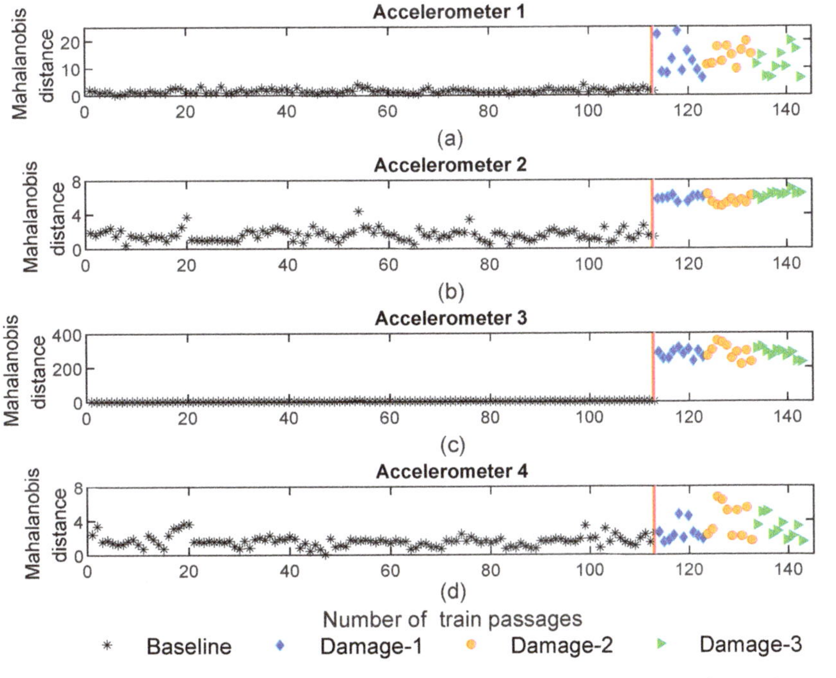


**Figure 19.** PCA—data fusion for all 143 baseline and damage scenarios: (**a**) MD for accelerometer 1, (**b**) MD for accelerometer 2, (**c**) MD for accelerometer 3, (**d**) MD for accelerometer 4.

### 4.4. Outlier Analysis

Outlier analysis allows the assessment of how effectively the suggested methodologies distinguish healthy wheels from defective ones for all feature extraction methods without human intervention. In general, the literature presupposes that a chi-squared distribution in n−dimensional space can approximate the Mahalanobis-squared distance. Therefore, a Gaussian distribution can roughly represent the Mahalanobis distance, and an outlier analysis based on a statistical threshold can be performed. The threshold's significance level is established as equal to 1% [46]. According to this theory, a confidence boundary (CB) for identifying a damage index consisting of an outlier is calculated using the Gaussian inverse cumulative distribution function (ICDF), considering the mean value, $\bar{\mu}$, and standard deviation, $\sigma$, of the baseline feature vector. Finally, feature damage indicators equal or greater than the CB are considered outliers (the null hypothesis is rejected).

#### 4.4.1. AR Model

Figure 20 depicts the results of the automatic damage detection system that considers the responses from accelerometers 1–4 using the AR model. This figure indicates that damage detection can be effectively performed using only sensors installed on the rail (between or above sleepers). As an example, according to Figure 20a–c, the damaged wheels are efficiently detected without the occurrence of false-positive cases, and so the healthy wheels can be robustly separated from damage scenarios. Moreover, the distance between the damaged wheels and the CB is sufficiently high; however, for the baseline scenarios this distance is sometimes very close to the CB. On the other hand, in the case of sensor 4 (Figure 20d), located on the sleeper, the damage detection implies some false-positive cases, which means that damage detection is not accurate enough based on the data exclusively derived from accelerometers on the sleeper. Furthermore, by using the AR-derived features, it is possible to observe a distinction between the behavior of indicators from wheel flats on the right and left sides. It is relevant to mention that only one sensor is adequate to detect a defective wheel using the AR-derived features.

#### 4.4.2. ARX Model

The results of the automatic damage detection for the ARX-derived features are presented in Figure 21. It can be observed that the extracted features can effectively detect all the damage scenarios without the occurrence of any false positives or negatives. Additionally, from the accelerometers located on the sleeper it is possible to detect the damages (Figure 21d). This conclusion is particularly relevant since it is a clear advantage in relation to the performance of the AR model, and because installing sensors on the sleeper is easier than installing on the rail. Additionally, the ARX method is also promising in terms of its ability to distinguish between damaged wheels on the left or right sides. Another advantage of using the ARX method is that this technique can detect defective wheels without any false positives or negatives, regardless of the sensor's position. Finally, in the case of ARX, it should be mentioned that installing one sensor is sufficient to distinguish a healthy wheel from a defective one.

#### 4.4.3. CWT

Figure 22 illustrates the damage detection assessment based on CWT-derived features. It is possible to infer that, automatic damage detection can provide an accurate distinction between the baseline and damaged scenarios without any false positives or negatives. Moreover, by locating the accelerometers on the sleeper only, damage detection using CWT is possible, in addition to the simplicity of installation. However, it can be concluded that damage detection by implementing the CWT-derived features is not sensitive to the side of the defect. Nevertheless, as in previous features, only one sensor is necessary to distinguish a defective wheel from a healthy one.

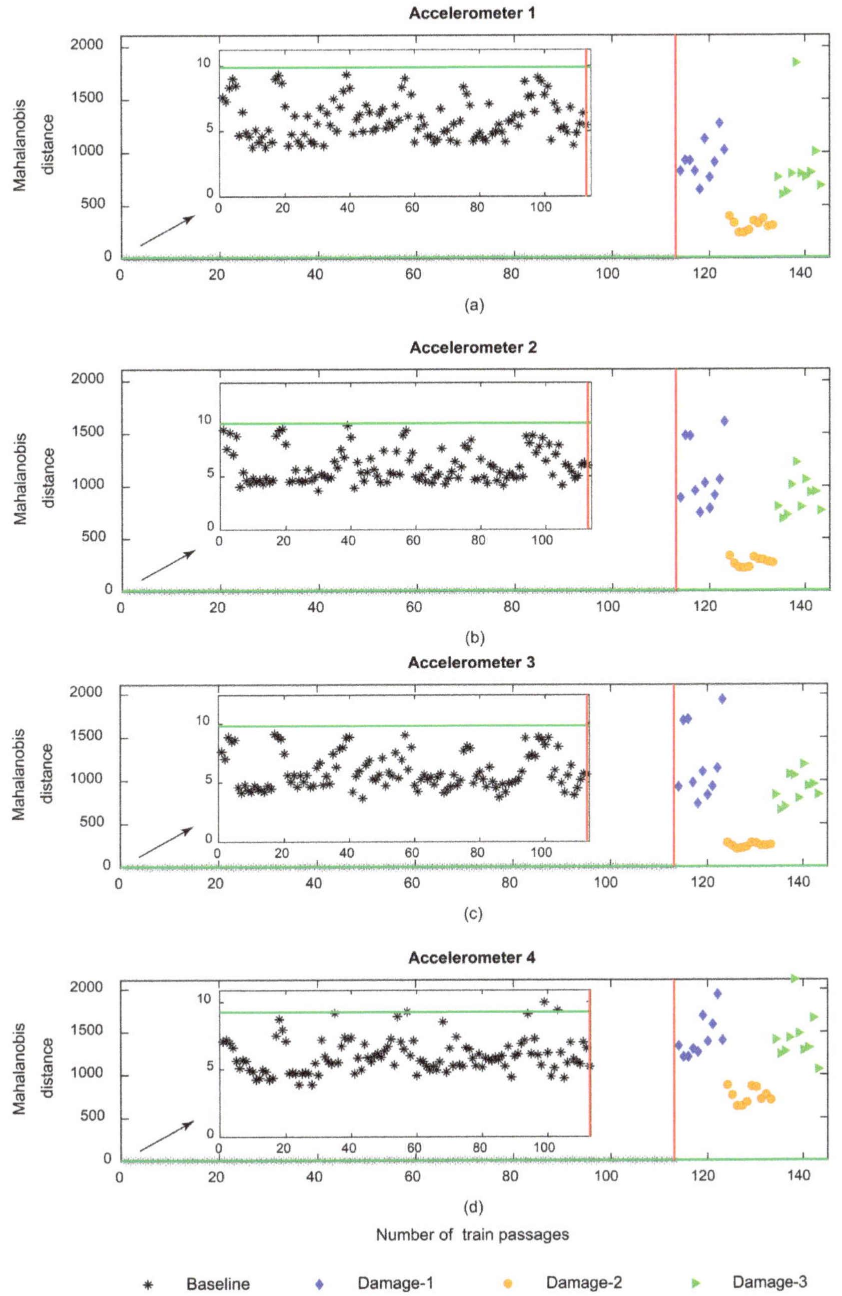


**Figure 20.** AR—automatic wheel flat damage detection considering the responses from accelerometers 1–4: (**a**) MD for accelerometer 1, (**b**) MD for accelerometer 2, (**c**) MD for accelerometer 3, (**d**) MD for accelerometer 4.

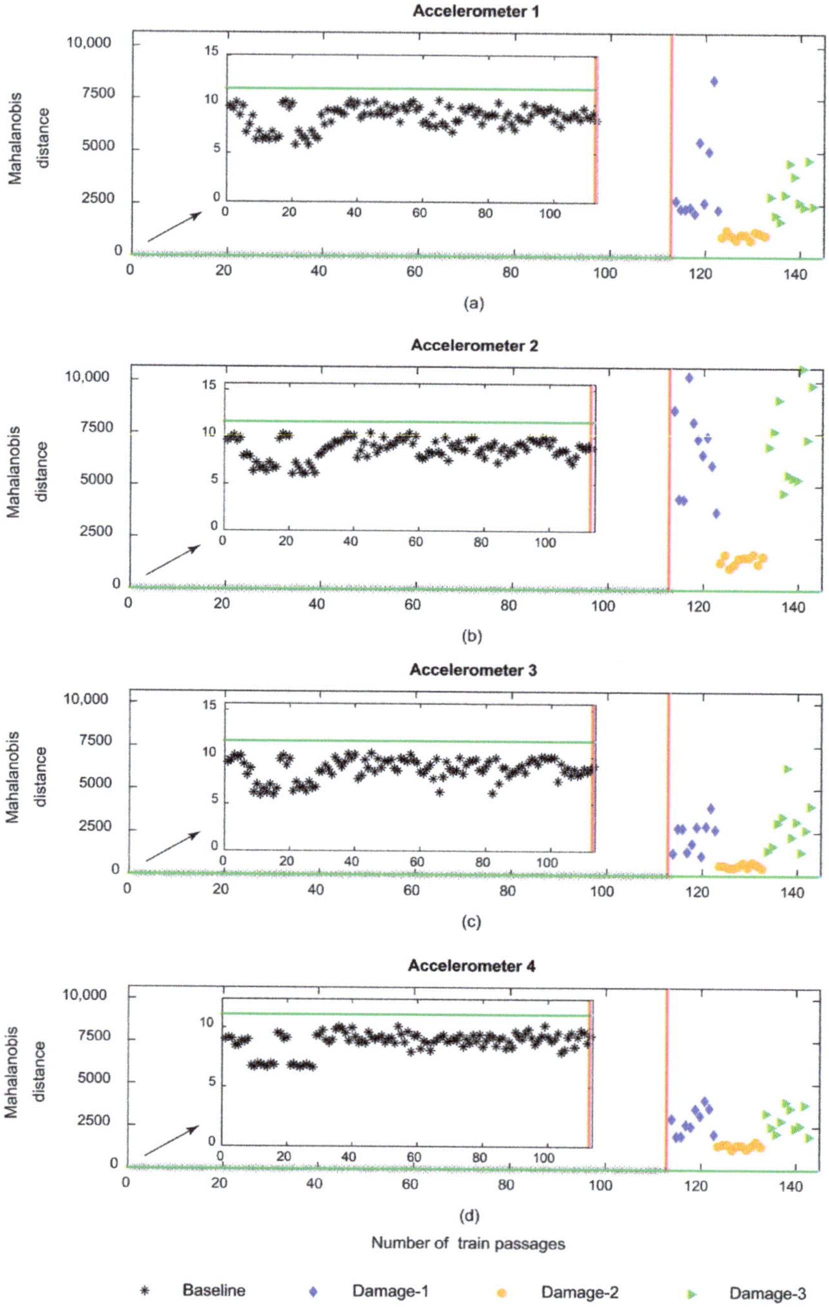


**Figure 21.** ARX—automatic wheel flat damage detection considering the responses from accelerometers 1–4: (**a**) MD for accelerometer 1, (**b**) MD for accelerometer 2, (**c**) MD for accelerometer 3, (**d**) MD for accelerometer 4.

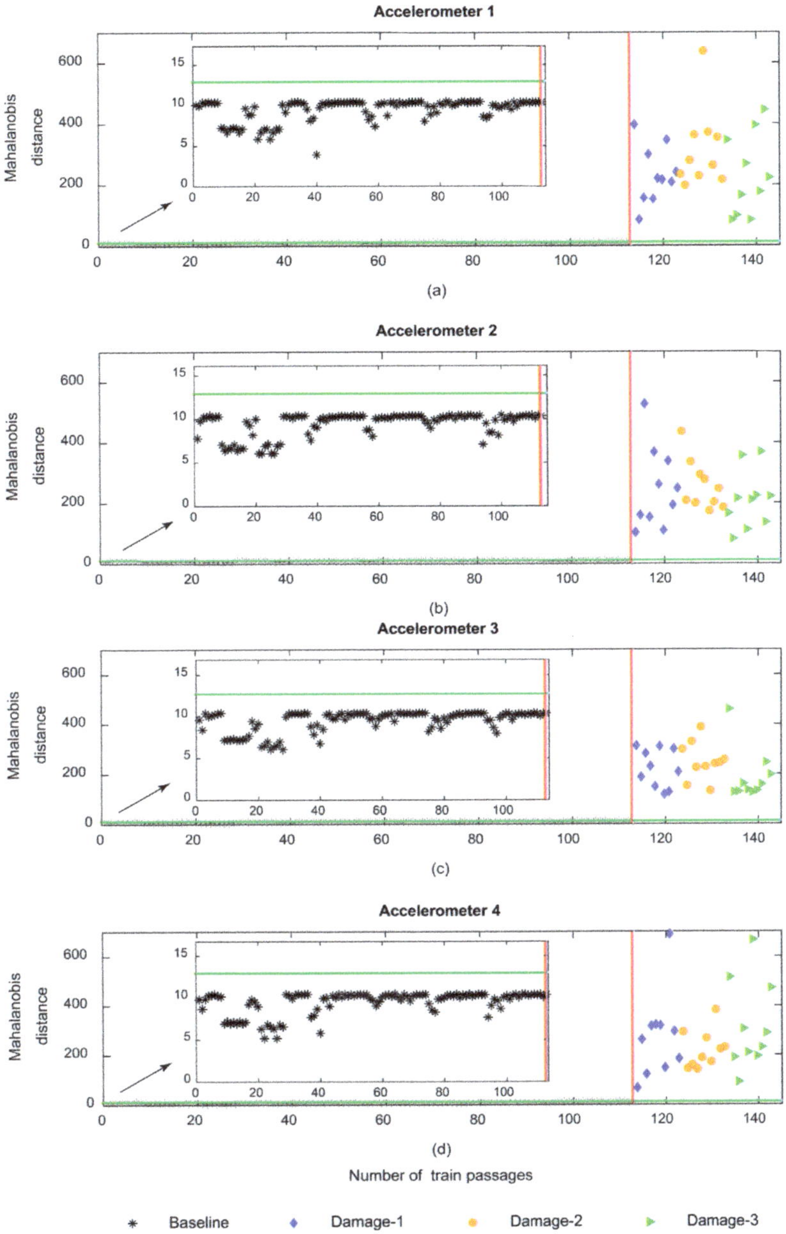


**Figure 22.** CWT—automatic wheel flat damage detection considering the responses from accelerometers 1–4: (**a**) MD for accelerometer 1, (**b**) MD for accelerometer 2, (**c**) MD for accelerometer 3, (**d**) MD for accelerometer 4.

#### 4.4.4. PCA

Figure 23 represents the automatic damage detection based on PCA-derived features for sensors 1–4. As shown in this figure, damage detection comes with at least two false positives. The output of damage detection based on the PCA-derived features lacks robustness and is not able to properly detect damaged wheels. In comparison to the AR-, ARX- and CWT-derived features, the PCA has less accuracy in damage detection.

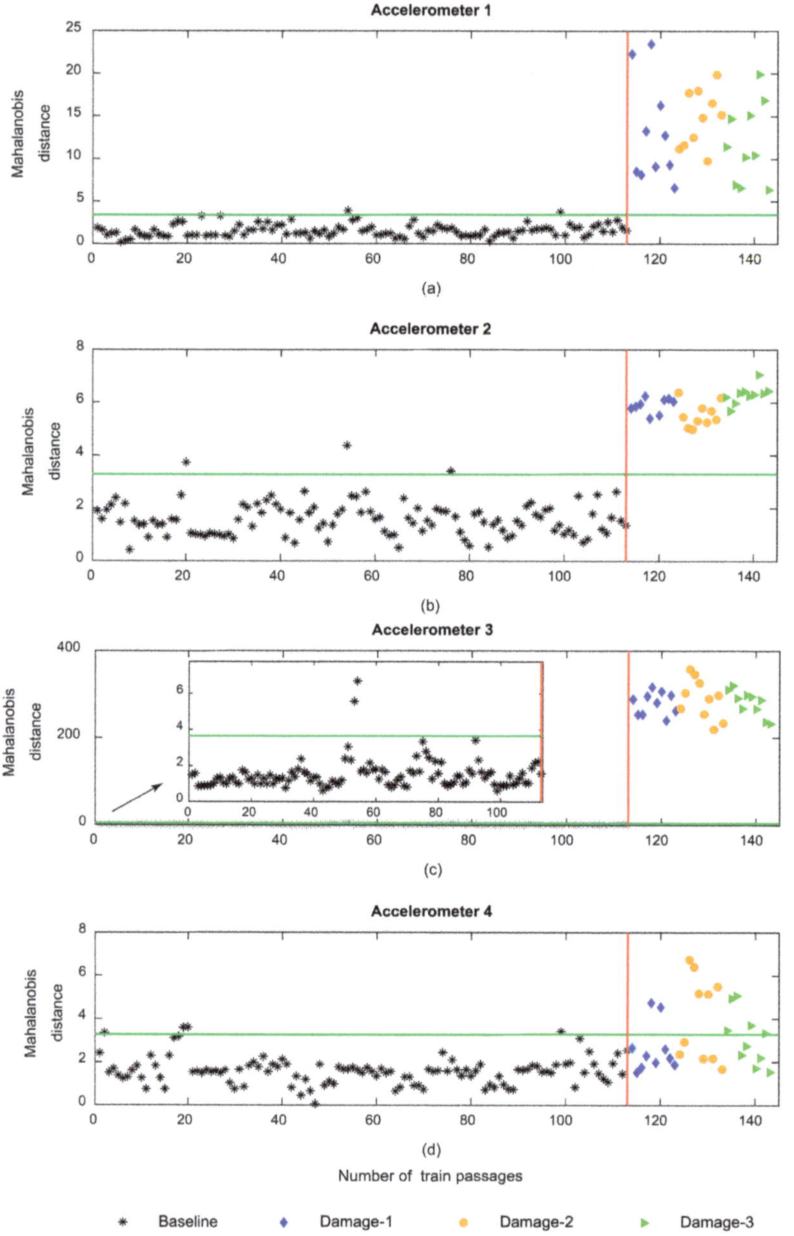


**Figure 23.** PCA—automatic wheel flat damage detection considering the responses from accelerometers 1–4: (**a**) MD for accelerometer 1, (**b**) MD for accelerometer 2, (**c**) MD for accelerometer 3, (**d**) MD for accelerometer 4.

From Figure 23, it can be concluded that the automatic damage detection based on PCA-derived features lacks robustness and the output comes with false positives and negatives. Therefore, to enhance the sensitivity to defects, the second stage of the data fusion consisting of multi-sensor fusion is implemented, by using data from sensors on both sides of the track. As shown in Figure 24, it is visible that after the second stage of data fusion the PCA-derived features come without any false negatives. On the other hand, the number of false positives is reduced to only two.

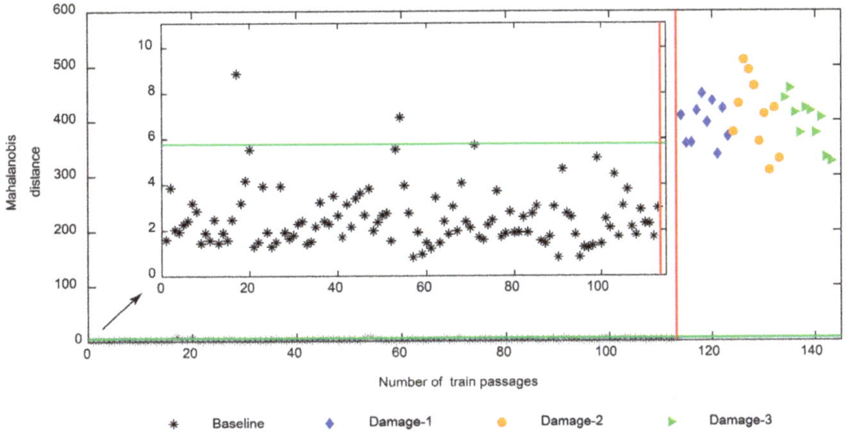


**Figure 24.** PCA—automatic wheel flat damage detection considering the multi-sensor data fusion.

## 5. Conclusions

This study aimed to compare the accuracy of an unsupervised data-driven methodology, based on four distinct features (AR, ARX, CWT, and PCA), for the automatic detection of wheel flats and based on time–history accelerations on the track elements (rails and sleepers).

The proposed methodology includes (i) feature extraction from acquired responses using dedicated feature extraction methods; (ii) feature normalization based on principal component analyses (PCA); (iii) data fusion to merge features derived from each sensor and (iv) damage detection by performing an outlier analysis using a specific confidence boundary.

From the research presented herein, it is possible to draw the following conclusions:

- the AR and ARX methods are the most accurate feature extraction methods for wheel flat damage detection as they can robustly detect defects; these two methods are sensitive to the side of the damage being the most promising to automatically distinguish an existing defective wheel on the right side from the left side in future works;
- the CWT method is only capable of detecting damaged wheels and is not sensitive to the side of the defect;
- the accuracy of the PCA method to detect the defective wheel is low and damage detection using this method lacks reliability;
- the ARX method is the only method that can robustly detect the wheel flat with accelerometers placed in the sleepers.
- One of the novelties of this research in relation to previous works [5,8] is the comparison of the accuracy of four different feature extraction techniques using an unsupervised learning methodology to automatically detect a defective wheel, which is a clear step forward in terms of the effectiveness of the proposed method, and allows full implementation for real-world applications.

Such results clearly show the great potential of this innovative application of data mining in the railway industry, particularly for infrastructure managers. Future work includes a field trial to validate the proposed methodology based on on-site measurements. Furthermore, for the final development of the proposed methodology, it is imperative to develop a novel feature, or eventually upgrade the actual methodology, to additionally classify the severities of the wheel flats.

**Author Contributions:** Conceptualization, A.M. (Araliya Mosleh), C.V., D.R., A.M. (Andreia Meixedo), P.M.; methodology, A.M. (Andreia Meixedo), P.M.; software, A.M. (Andreia Meixedo), P.M.; validation, M.M.; formal analysis, M.M.; investigation, M.M.; resources, C.V., D.R.; writing—original draft preparation, M.M.; writing—review and editing, A.M. (Araliya Mosleh), C.V., D.R.; supervision, A.M. (Araliya Mosleh), C.V., D.R.; project administration, C.V., D.R.; funding acquisition, C.V., D.R. All authors have read and agreed to the published version of the manuscript.

**Funding:** This work was financially supported by Base Funding-UIDB/04708/2020 and Programmatic Funding-UIDP/04708/2020 of the CONSTRUCT—Instituto de Estruturas e Construções, funded by national funds through the FCT/MCTES (PIDDAC). The paper reflects research developed in the ambit of the project Way4SafeRail, NORTE-01-0247-FEDER-069595, founded by Agência Nacional de Inovação S.A., program P2020 | COMPETE—Projetos em Copromoção. The second author acknowledges Grant no. 2021.04272.CEECIND from the Stimulus of Scientific Employment, Individual Support (CEECIND)—4th Edition provided by FCT—Fundação para a Ciência e Tecnologia.

**Institutional Review Board Statement:** Not applicable.

**Informed Consent Statement:** Not applicable.

**Data Availability Statement:** All data: models: and code generated or used during the study appear in the paper.

**Acknowledgments:** We gratefully appreciate the financial support from the Base Funding-UIDB/04708/2020 and Programmatic Funding-UIDP/04708/2020 of the CONSTRUCT—Instituto de Estruturas e Construções, funded by national funds through the FCT/MCTES (PIDDAC). The paper reflects research developed in the ambit of the project Way4SafeRail, NORTE-01-0247-FEDER-069595, founded by Agência Nacional de Inovação S.A., program P2020 | COMPETE—Projetos em Copromoção. The second author acknowledges Grant no. 2021.04272.CEECIND from the Stimulus of Scientific Employment, Individual Support (CEECIND)—4th Edition provided by FCT—Fundação para a Ciência e Tecnologia.

**Conflicts of Interest:** The authors declare no conflict of interest.

## References

1. Meixedo, A.; Santos, J.; Ribeiro, D.; Calçada, R.; Todd, M. Damage detection in railway bridges using traffic-induced dy-namic responses. *Eng. Struct.* **2021**, *238*, 112189. [CrossRef]
2. Mohammadi, M.; Mosleh, A.; Razzaghi, M.; Alves Costa, P.; Calçada, R. Stochastic analysis of railway embankment with uncer-tain soil parameters using polynomial chaos expansion. *Struct. Infrastruct. Eng.* **2022**, 1–20. [CrossRef]
3. Pintão, B.; Mosleh, A.; Vale, C.; Montenegro, P.; Costa, P. Development and Validation of a Weigh-in-Motion Methodology for Railway Tracks. *Sensors* **2022**, *22*, 1976. [CrossRef] [PubMed]
4. Barke, D.W.; Chiu, W.K. A Review of the Effects of Out-Of-Round Wheels on Track and Vehicle Components. *Proc. Inst. Mech. Eng. Part F J. Rail Rapid Transit* **2005**, *219*, 151–175. [CrossRef]
5. Mosleh, A.; Meixedo, A.; Ribeiro, D.; Montenegro, P.; Calçada, R. Automatic clustering-based approach for train wheels condition monitoring. *Int. J. Rail Transp.* **2022**, 1–26. [CrossRef]
6. Vale, C. Wheel Flats in the Dynamic Behavior of Ballasted and Slab Railway Tracks. *Appl. Sci.* **2021**, *11*, 7127.
7. General Contract of Use for Wagons—GCU, Edition dated 1 January 2021. Available online: https://gcubureau.org/wp-content/uploads/Contract/2021/20210101_GCU_EN_full_version.pdf (accessed on 29 December 2022).
8. Mosleh, A.; Meixedo, A.; Ribeiro, D.; Montenegro, P.; Calçada, R. Early wheel flat detection: An automatic data-driven wavelet-based approach for railways. *Veh. Syst. Dyn.* **2022**, 1–30. [CrossRef]
9. Nielsen, J.C.O.; Johansson, A. Out-of-round railway wheels-a literature survey. *Proc. Inst. Mech. Eng. Part F J. Rail Rapid Transit* **2000**, *214*, 79–91. [CrossRef]
10. Bosso, N.; Gugliotta, A.; Zampieri, N. Wheel flat detection algorithm for onboard diagnostic. *Measurement* **2018**, *123*, 193–202. [CrossRef]
11. Cavuto, A.; Martarelli, M.; Pandarese, G.; Revel, G.; Tomasini, E. Train wheel diagnostics by laser ultrasonics. *Measurement* **2016**, *80*, 99–107. [CrossRef]
12. Alexandrou, G.; Kouroussis, G.; Verlinden, O. A comprehensive prediction model for vehicle/track/soil dynamic response due to wheel flats. *Proc. Inst. Mech. Eng. Part F J. Rail Rapid Transit* **2016**, *230*, 1088–1104.
13. Amini, A.; Entezami, M.; Huang, Z.; Rowshandel, H.; Papaelias, M. Wayside detection of faults in railway axle bearings using time spectral kurtosis analysis on high-frequency acoustic emission signals. *Adv. Mech. Eng.* **2016**, *8*, 1687814016676000. [CrossRef]

14. Mosleh, A.; Costa, P.A.; Calçada, R. A new strategy to estimate static loads for the dynamic weighing in motion of railway vehicles. *Proc. Inst. Mech. Eng. Part F J. Rail Rapid Transit* **2019**, *234*, 183–200. [CrossRef]
15. Jiang, H.; Lin, J. Fault diagnosis of wheel flat using empirical mode decomposition-Hilbert envelope spectrum. *Math. Probl. Eng.* **2018**, *2018*, 8909031 . [CrossRef]
16. Mosleh, A.; Montenegro, P.A.; Costa, P.A.; Calçada, R. Railway Vehicle Wheel Flat Detection with Multiple Records Using Spectral Kurtosis Analysis. *Appl. Sci.* **2021**, *11*, 4002. [CrossRef]
17. Krummenacher, G.; Ong, C.S.; Koller, S.; Kobayashi, S.; Buhmann, J.M. Wheel Defect Detection with Machine Learning. *IEEE Trans. Intell. Transp. Syst.* **2018**, *19*, 1176–1187. [CrossRef]
18. Ni, Y.-Q.; Zhang, Q.-H. A Bayesian machine learning approach for online detection of railway wheel defects using track-side monitoring. *Struct. Health Monit.* **2020**, *20*, 1536–1550. [CrossRef]
19. Meixedo, A.; Santos, J.; Ribeiro, D.; Calçada, R.; Todd, M.D. Online unsupervised detection of structural changes using train–induced dynamic responses. *Mech. Syst. Signal Process.* **2022**, *165*, 108268. [CrossRef]
20. Alves, V.; Meixedo, A.; Ribeiro, D.; Calçada, R.; Cury, A. Evaluation of the performance of different damage indicators in railway bridges. *Procedia Eng.* **2015**, *114*, 746–753.
21. Javed, K.; Gouriveau, R.; Zerhouni, N.; Nectoux, P. Enabling Health Monitoring Approach Based on Vibration Data for Accurate Prognostics. *IEEE Trans. Ind. Electron.* **2015**, *62*, 647–656. [CrossRef]
22. Shin, S.; Yun, C.B.; Futura, H.; Popovics, J.S. Nondestructive evaluation of crack depth in concrete using PCA-compressed wave transmission function and neural networks. *Exp. Mech.* **2008**, *48*, 225–231.
23. Yan, A.M.; Kerschen, G.; De Boe, P.; Golinval, J.C. Structural damage diagnosis under varying environmental conditions—Part I: A linear analysis. *Mech. Syst. Signal Process.* **2005**, *19*, 847–864.
24. Oh, C.K.; Sohn, H.; Bae, I.-H. Statistical novelty detection within the Yeongjong suspension bridge under environmental and operational variations. *Smart Mater. Struct.* **2009**, *18*, 125022.
25. Figueiredo, E.; Cross, E. Linear approaches to modeling nonlinearities in long-term monitoring of bridges. *J. Civ. Struct. Health Monit.* **2013**, *3*, 187–194.
26. Figueiredo, E.; Park, G.; Farrar, C.R.; Worden, K.; Figueiras, J. Machine learning algorithms for damage detection under operational and environmental variability. *Struct. Health Monit.* **2010**, *10*, 559–572. [CrossRef]
27. Qian, L.; Zhang, L.; Bao, X.; Li, F.; Yang, J. Supervised sparse neighbourhood preserving embedding. *IET Image Process.* **2017**, *11*, 190–199. [CrossRef]
28. Liu, Y.; Shi, Y.; Mu, F.; Cheng, J.; Li, C.; Chen, X. Multimodal MRI volumetric data fusion with convolutional neural networks. *IEEE Trans. Instrum. Meas.* **2022**, *71*, 1–15.
29. Pan, Y.; Sun, Y.; Li, Z.; Gardoni, P. Machine learning approaches to estimate suspension parameters for performance degradation assessment using accurate dynamic simulations. *Reliab. Eng. Syst. Saf.* **2022**, *230*, 108950.
30. Kontolati, K.; Loukrezis, D.; dos Santos, K.R.; Giovanis, D.G.; Shields, M.D. Manifold learning-based polynomial chaos expansions for high-dimensional surrogate models. *Int. J. Uncertain. Quantif.* **2022**, *12*, 39–64. [CrossRef]
31. Wang, J.; Xie, J.; Zhao, R.; Zhang, L.; Duan, L. Multisensory fusion based virtual tool wear sensing for ubiquitous manufacturing. *Robot. Comput. Integr. Manuf.* **2017**, *45*, 47–58. [CrossRef]
32. Bull, L.A.; Worden, K.; Fuentes, R.; Manson, G.; Cross, E.J.; Dervilis, N. Outlier ensembles: A robust method for damage detection and unsupervised feature extraction from high-dimensional data. *J. Sound Vib.* **2019**, *453*, 126–150.
33. Li, Y.; Zuo, M.J.; Lin, J.; Liu, J. Fault detection method for railway wheel flat using an adaptive multiscale morphological filter. *Mech. Syst. Signal Process.* **2017**, *84*, 642–658. [CrossRef]
34. Nick, W.; Asamene, K.; Bullock, G.; Esterline, A.; Sundaresan, M. A study of machine learning techniques for detecting and classifying structural damage. *Int. J. Mach. Learn. Comput.* **2015**, *5*, 313. [CrossRef]
35. Addin, O.; Sapuan, S.M.; Mahdi, E.; Othman, M. A Naïve-Bayes classifier for damage detection in engineering materials. *Mater. Des.* **2007**, *28*, 2379–2386. [CrossRef]
36. Vitola, J.; Pozo, F.; Tibaduiza, D.A.; Anaya, M. Distributed Piezoelectric Sensor System for Damage Identification in Structures Subjected to Temperature Changes. *Sensors* **2017**, *17*, 1252. [CrossRef]
37. Montenegro, P.A.; Neves, S.G.M.; Calçada, R.; Tanabe, M.; Sogabe, M. Wheel–rail contact formulation for analyzing the lateral train–structure dynamic interaction. *Comput. Struct.* **2015**, *152*, 200–214.
38. Hertz, H. Ueber die Berührung fester elastischer Körper. *J. Für Die Reine Und Angew. Math.* **1882**, *92*, 156–171. [CrossRef]
39. Kalker, J.J. *Book of Tables for the Herzian Creep-force Law*; Faculty of Technical Mathematics and Informatics, Delft University of Technology: Delft, The Netherlands, 1996.
40. *MATLAB®*, version R2022a; The MathWorks Inc.: Natick, MA, USA, 2022.
41. *ANSYS®*; Release 19.2; Academic Research: Canonsburg, PA, USA, 2018.
42. Mosleh, A.; Montenegro, P.; Alves Costa, P.; Calçada, R. An approach for wheel flat detection of railway train wheels using envelope spectrum analysis. *Struct. Infrastruct. Eng.* **2021**, *17*, 1710–1729. [CrossRef]
43. Mosleh, A.; Montenegro, P.A.; Costa, P.A.; Calçada, R. Approaches for weigh-in-motion and wheel defect detection of railway vehicles. In *Rail Infrastructure Resilience*; Elsevier: Amsterdam, The Netherlands, 2022; pp. 183–207.
44. UIC. *Code of Practice for the Loading and Securing of Goods on Railway Wagons*; UIC: Paris, France, 2022.

45. Meixedo, A.; Ribeiro, D.; Santos, J.; Calçada, R.; Todd, M.D. Real-Time Unsupervised Detection of Early Damage in Railway Bridges Using Traffic-Induced Responses. In *Structural Health Monitoring Based on Data Science Techniques*; Springer: Berlin/Heidelberg, Germany, 2022; pp. 117–142.
46. Tomé, E.S.; Pimentel, M.; Figueiras, J. Damage detection under environmental and operational effects using cointegration analysis–application to experimental data from a cable-stayed bridge. *Mech. Syst. Signal Process.* **2020**, *135*, 106386.

sensors

MDPI

Article

# Semantic Segmentation of Terrestrial Laser Scans of Railway Catenary Arches: A Use Case Perspective

**Bram Ton [1,*], Faizan Ahmed [1,2] and Jeroen Linssen [1]**

1 Ambient Intelligence, Saxion University of Applied Sciences, 7513 AB Enschede, The Netherlands
2 Formal Methods and Tools, University of Twente, 7522 NB Enschede, The Netherlands
* Correspondence: b.t.ton@saxion.nl

**Abstract:** Having access to accurate and recent *digital twins* of infrastructure assets benefits the renovation, maintenance, condition monitoring, and construction planning of infrastructural projects. There are many cases where such a digital twin does not yet exist, such as for legacy structures. In order to create such a digital twin, a mobile laser scanner can be used to capture the geometric representation of the structure. With the aid of *semantic segmentation*, the scene can be decomposed into different object classes. This decomposition can then be used to retrieve CAD models from a CAD library to create an accurate digital twin. This study explores three deep-learning-based models for semantic segmentation of point clouds in a practical real-world setting: PointNet++, SuperPoint Graph, and Point Transformer. This study focuses on the use case of catenary arches of the Dutch railway system in collaboration with Strukton Rail, a major contractor for rail projects. A challenging, varied, high-resolution, and annotated dataset for evaluating point cloud segmentation models in railway settings is presented. The dataset contains 14 individually labelled classes and is the first of its kind to be made publicly available. A modified PointNet++ model achieved the best mean class Intersection over Union (IoU) of 71% for the semantic segmentation task on this new, diverse, and challenging dataset.

**Keywords:** semantic segmentation; point cloud; railway infrastructure; deep learning; terrestrial laser scanner; catenary arch

**Citation:** Ton, B.; Ahmed, F.; Linssen, J. Semantic Segmentation of Terrestrial Laser Scans of Railway Catenary Arches: A Use Case Perspective. *Sensors* **2023**, *23*, 222. https://doi.org/10.3390/s23010222

Academic Editors: Araliya Mosleh, Diogo Ribeiro, Abdollah Malekjafarian, Maria D. Martínez-Rodrigo

Received: 18 November 2022
Revised: 20 December 2022
Accepted: 22 December 2022
Published: 26 December 2022



## 1. Introduction

Renovation, maintenance, condition monitoring, and construction of infrastructural projects demand assessments of the current situation [1]. These processes are necessary for the evaluation of the existing situation, possibly leading to advice for re-designing aspects such as structural integrity, optimisation of traffic flow, and safety. In addition, the introduction of BIM (*Building Information Modelling*) and 3D design in general have created an increased need for accurate, up-to-date, 3D information of existing infrastructure.

Three-dimensional information can easily become outdated as the actual constructed infrastructure can deviate from the original design plans or can currently exist in an altered state [2]. Furthermore, blueprints do not always exist with a sufficient level of detail, are not available in a digital format, or only exist in 2D. These factors underline the need for accurate, up-to-date, 3D information.

At present, assessments and the subsequent translation to 3D are performed mostly manually, which is a time-consuming and error-prone task. This has given rise to technology aimed at automating the digitisation of infrastructure, such as photogrammetry and mobile laser scanning [3]. Laser scanning is a method that provides immediate 3D geometric information without any elaborate processing, which is in contrast with photogrammetry; further, accuracy-wise, laser scanning has better performance over photogrammetry [4]. Another benefit of laser scanning is its independence of illumination, which means specialised measuring trains can operate at night when the utilisation of the rail network is lower. A downside of laser scanning is the high cost of measurement devices compared to vision-based systems and the unstructured nature of the data. We believe the benefits of laser scanning outweigh its downsides; therefore, this technique was chosen for this application.

This paper evaluates several state-of-the-art approaches to semantic segmentation for the digitisation of infrastructure through a use case of railway catenary arches in the Netherlands. Catenary arches are the supporting structures above the railway track that carry the power lines for the trains, see Figure 1. The catenary system in the Netherlands consists of a variety of new and legacy arches, with custom and standardised components being mixed. Digitising the physical arches into their 3D, digital counterparts is an ongoing task. As part of this undertaking, mobile laser scans have been made of a small piece of railway track in the Netherlands (see Section 3).

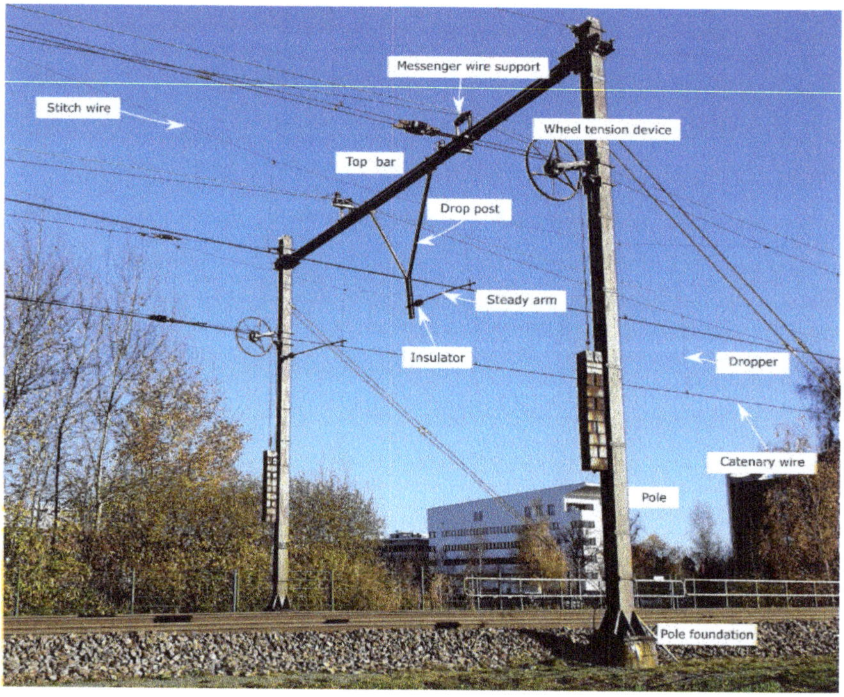


**Figure 1.** An example of a catenary arch (not in dataset) that shows the labels of the majority of the classes (own work).

Supervised semantical segmentation of the scanned scene provides a starting point for digitisation. In turn, these segments can be matched to individual components from a CAD library to create a full digital twin of the scene. This paper considers three state-of-the-art approaches, which semantically segment a dataset of catenary arches in the Netherlands, and compares their efficacy. Specifically, we address variations on PointNet++ [5], an implementation of SuperPoint Graph [6], and Point Transformer [7]. PointNet++ is considered as it has been a milestone in applying deep learning to point clouds. A catenary arch can be seen as a set of geometrically placed objects, which fits well with the objective of SuperPoint Graph to encode geometrical relations. The third method chosen is Point Transformer, as this was the first method to break the 70% mean Intersection over Union (mIoU) threshold on the S3DIS dataset [8]. Finally, implementation availability of code was taken into account when selecting these methods.

The remainder of this article is organised as follows. First, existing work on semantic segmentation in point clouds is described (Section 2). This is followed by a description of the dataset (Section 3). Next, the methodology for comparing the semantic segmentation techniques is described (Section 4). Thereafter, the results and discussions (Section 5) are provided. The discussion contains valuable pointers for future work and highlights the importance of explainable artificial intelligence. Finally, the conclusion and outlook for future work (Section 6) complete the article.

## 2. Related Work

Point clouds have vast applications in different areas of science and engineering such as the construction industry [9], digital photogrammetry [10], surveying [11], and robotics [12]. Therefore, many survey papers [13–18] have been written to compare different point-cloud-based machine learning models both technically and empirically.

Liu et al. [16] compared various deep-learning-based algorithms for different point cloud tasks, such as classification, segmentation, and object detection. The algorithms were divided into two categories, namely, raw point-cloud-based methods and tree-based deep learning. The raw point-cloud-based methods use the points directly as an input for training a deep learning model. The tree-based algorithm first forms a $k$-dimensional tree [19] (or $k$d-tree in short) representation of the raw point clouds. Local and global cues imposed by this tree structure can be exploited to progressively learn representation vectors [20]. An extensive empirical comparison of the performance of these models for a large number of benchmark datasets was also reported [16]. However, these datasets do not include railway catenary systems.

In a recent paper, Guo et al. [15] provided a comprehensive survey of deep learning methods for different point cloud tasks. The methods were categorised according to the three tasks associated with point clouds, namely, shape classification, object detection and tracking, and segmentation. The methods for each of these classes were further classified into different categories such as projection-based, point-based, object detection, object tracking, scene flow estimation, semantic segmentation, instance segmentation, and part segmentation ([15], Figure 1). Guo et al. not only briefly described the datasets but also commented on evaluation metrics. Furthermore, a chronological overview of the methods for all three categories was also given. The algorithms are empirically compared via different standardised metrics using benchmark datasets.

Although the surveys described above are comprehensive in describing the algorithmic advancement concerning various tasks related to point cloud data, they lack the aspect of one dataset, namely, datasets related to railway infrastructure. In the remainder of this section, the literature on point clouds related to railway infrastructure is surveyed. The reader should be warned that the performance metric used by various authors is not consistent—F1-score, accuracy, and mIoU are all used.

Arastounia used a high-density point cloud covering 550 m of Austrian railroad for segmenting individual catenary components [21]. A heuristic method was employed for this task based on the local neighbourhood structure, the shape of objects, and the topological relationship between objects. Each of the objects being segmented had a different model. The first step towards segmentation was detecting the track bed, which acted as a reference base for detecting the other components such as tracks, poles, and wires. In total, six different objects were segmented, with an average accuracy of 96.4% being obtained.

Chen et al. [22] used a more data-driven approach towards segmentation of catenary arches. Their approach starts by extracting line primitives at three different scales from the point cloud data. These line primitives are then used for training a hierarchical Conditional Random Field (CRF) model. A total of ten different objects were segmented with an overall accuracy of 99.67%.

Soilán et al. compared two deep-learning-based approaches, PointNet [23] and KPConv [24], for the task of segmenting railway tunnels [25]. Four classes were defined: tunnel lining, tracks, wires, and ground. The PointNet model achieved an average F1-score of 86.7% and the KPConv approach achieved an average F1-score of 87.2%. It is surprising that no data augmentation methods were used because the number of samples is small. An additional surprise is the low F1-score on segmenting the tracks, which have a very consistent geometric shape. This low score is attributed to labelling errors by the original authors.

The works of Chen et al. and Lin et al. share the same dataset [26,27]. This dataset was collected using a mobile, 2D laser scanning device mounted on a cart moving along the railway track to collect data. The data consist of sequences of 2D slices, which are perpendicular to the direction of the railway track. The sensor location is constrained to the track, making it very easy to define a constrained search area. We hypothesise that the variation within the captured railway catenary system is small, resulting in highly accurate results for both works. Due to the sequential nature of the data, Chen et al. opted to use a *Recursive Neural Network* (RNN) [26]. First, each of the slices is partitioned into non-overlapping regions of points using an iterative point partitioning algorithm. After this, PointNet [23] is used to derive local features from these regions. Thereafter, an RNN based on *Long Short-Term Memory* (LSTM) architecture is used to segment the points. Seventeen classes of catenary components are defined for the segmentation task. Obtained accuracies in terms of mIoU were extremely high, even smaller components such as droppers and suspension insulators achieved scores of 90.8% and 97.8%, respectively. The approach of Lin et al. [27] to segment the point clouds into individual catenary components is by first classifying each of the slices into one of the following categories: wires, droppers, or poles [27]. After this, adjacent slices with the same category are grouped together. For each of the groups, a different deep-learning-based segmentation model is trained. In total, eight classes were segmented, and a mean accuracy of 97.01% was achieved.

We hypothesise that the majority of the work on semantic segmentation of catenary systems rely on data from scenes with little variation. To determine the robustness of the segmentation models, the work presented here depends on a dataset with a large variety of catenary arches. In addition, a large number of components (14) is segmented.

## 3. Catenary Arch Dataset

This section details the process of creating the catenary arch dataset. First, the acquisition of the raw data is described. Thereafter, arch localisation, cropping, and labelling are addressed. Finally, a summary of the dataset, both visual and textual, is provided.

### 3.1. Acquisition

To the best of our knowledge, there are no publicly available point cloud datasets of railway catenary arches. Therefore, our work is based on a dataset provided by Strukton Rail, containing an 800 m stretch of railway track near Delft, the Netherlands containing 15 catenary arches, which has been digitised into a point cloud. The point cloud data were collected with a Trimble TX8 laser scanner using the level 2 operation mode. This model has a scan duration of three minutes and a point spacing of 11.3 mm at 30 m. Points are referenced within the Rijksdriehoeksstelsel [28] coordinate system, a national standard coordinate system of the Netherlands.

### 3.2. Arch Localisation

The scanned stretch of railway track was made available by the data provider in four chunks of data. A semi-automated method was used to detect the location of the catenary arches within these chunks of data. This method follows a similar approach as described by Zhu et al. [29] and Corongiu et al. [30]. The method is based on the assumption that poles are represented by a dense volume in the z-direction.

Our method first downsamples each chunk of data using a voxel filter with a cell size of 10 cm. This cell size enables the detection of poles and reduces the computational load. After that, the scene is flattened to a two-dimensional grid by summing in the z-direction. The grid size used is 20 cm. This two-dimensional representation is written to disk as a greyscale image. Within these images, pole locations are clearly visible because of their high-intensity values. The procedural steps of arch localisation within a larger scene are depicted in Figure 2. Other elements such as trees or signalling posts also produce high-intensity regions in the image. Therefore, pixel coordinates of the outer catenary poles are manually selected and are used to define a rectangular crop region with a padding of 2 m around the catenary arch. The major axis of the rectangular crop coincides with the line being defined by the poles of the catenary arch. Each of the arches is cropped from the larger chunk of data and stored individually in an LAS file [31]. After this, each of the samples is manually labelled into 14 different classes. Labelling was performed by five students and one senior researcher. The classes labelled are as follows: top bar, pole, drop post, top tie, bracket, pole foundation, steady arm, contact wire, stitch wire, wheel tension device, dropper, messenger wire support, insulator, and unlabelled.

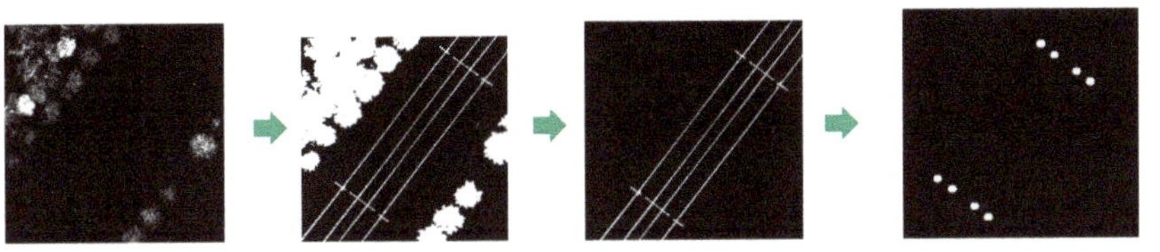


**Figure 2.** Processing steps for locating catenary arches within a large scene.

### 3.3. Data Summary

A summary of the data is provided in Table 1. It shows that the number of points in a catenary arch ranges between 1.6 and 11 M points. In total, the dataset contains roughly 55.4 M points. Not all classes are always present in each sample; for instance, tension wheels are only needed every few arches and, thus, occur less frequently in the dataset.

**Table 1.** Statistical description of the dataset.

| Arch | Name | Points | Classes (out of 14) |
|---|---|---|---|
| 0 | 01_01 | 1,586,927 | 13 |
| 1 | 01_02 | 2,147,546 | 13 |
| 2 | 01_03 | 2,664,907 | 13 |
| 3 | 02_01 | 11,112,574 | 13 |
| 4 | 02_02 | 2,415,930 | 11 |
| 5 | 02_03 | 4,362,055 | 11 |
| 6 | 02_04 | 5,257,501 | 11 |
| 7 | 03_01 | 2,787,253 | 12 |
| 8 | 03_02 | 6,782,568 | 10 |
| 9 | 03_03 | 1,973,730 | 6 |
| 10 | 03_04 | 6,582,344 | 11 |
| 11 | 04_01 | 2,271,179 | 11 |
| 12 | 04_02 | 1,673,804 | 11 |
| 13 | 04_03 | 1,598,090 | 11 |
| 14 | 04_04 | 2,183,600 | 12 |

A graphical overview of the entire dataset is provided in Figure 3. Some arches still have the track bed present; this is due to the fact that different individuals labelled the data. Some samples had the ground removed using an approximate progressive morphological filter [32]. The overview also clearly shows the large variation of catenary arches. For example, some arches span two adjacent tracks whilst others span four adjacent tracks.

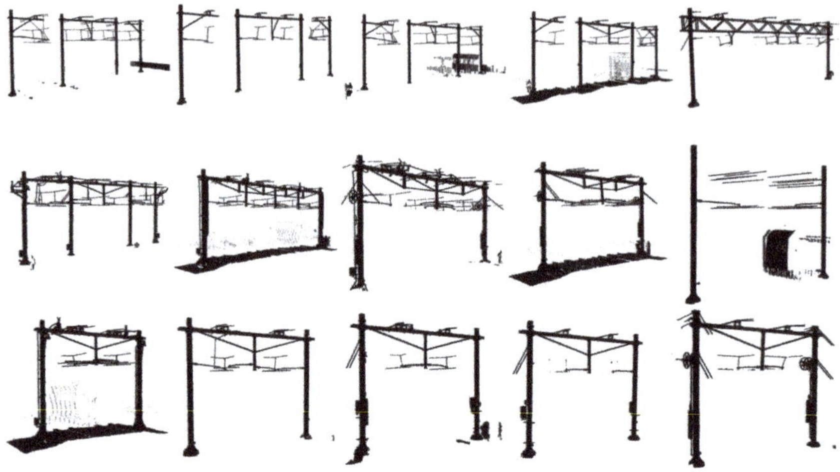

**Figure 3.** Overview of the dataset. Note the large variation of catenary arch types present in the dataset.

The dataset has a large imbalance in the distribution of the classes, which is inherent to the type of object, see Figure 4. The three largest classes (unlabelled, pole, and top bar) jointly constitute 72.3% of the points in the dataset. On the other hand, the three smallest classes (dropper, stitch wire, and wheel tension device) constitute only 1% of the dataset.

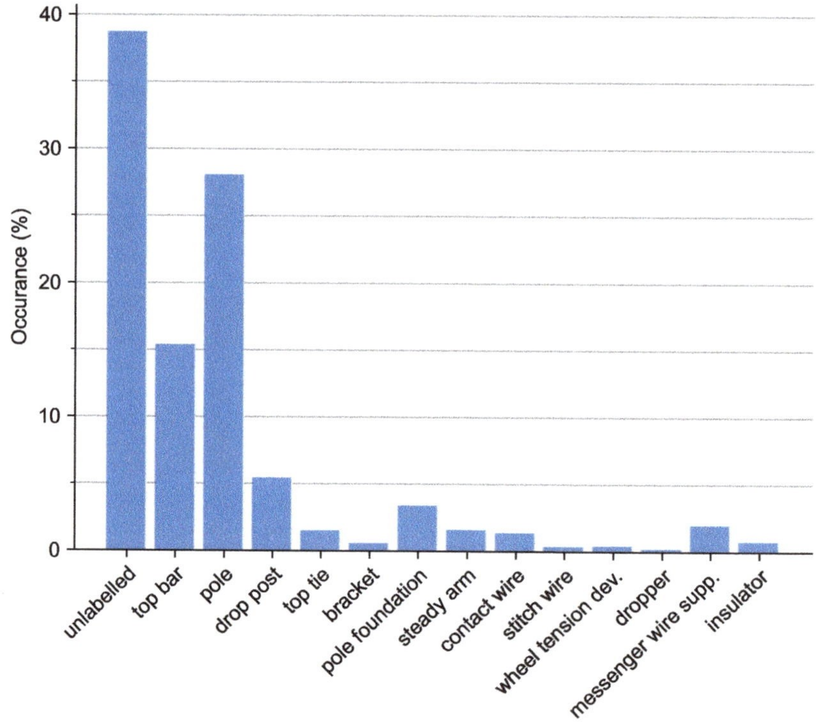


**Figure 4.** Normalised class distribution after the voxel centroid nearest neighbour filter is applied.

## 4. Methodology

Three different deep learning models are evaluated with regards to the semantic segmentation task of point clouds. The models evaluated are PointNet++, SuperPoint Graph, and PointTransformer. The first subsection describes the general pre-processing of the data, augmentation procedures, and the metric used for evaluation. The subsections following this describe the details of each of the three individual models.

### *4.1. General Pre-Processing*

Point clouds collected using a mobile laser scanner have a non-uniform density where the density decreases as the distance from the laser scanner becomes larger. In pursuit of a uniform density within each sample, each of the samples is downsampled using a voxel centroid nearest neighbour filter with a cell size of 1 cm. A centroid nearest neighbour approach is used to preserve the local point density distribution within a cell.

Each of the samples is normalised by centring the scene to the midpoint of the data span. In addition, a scaling factor is used to limit the range of coordinate values between $-1$ and 1. Taking into consideration the maximum dimension of a catenary arch in the dataset as 24 m and adding 3 m of safety margin, the resulting maximum dimension would be 27 m. Therefore, the appropriate scaling factor to limit coordinates between $-1$ and 1 is set to 13.5 m (half of the maximum dimension).

To increase the robustness of the models and to artificially increase the variations of the data seen by the models, various data augmentation techniques are used. The following three augmentations are sequentially applied to the input point cloud.

1. Uniform random rotation between $-180°$ and $180°$ of the points around the $z$-axis;
2. Uniform random translation of the point coordinates between $-1$ m and 1 m in all directions;
3. Adding random noise to the points. The random noise is selected from a truncated normal distribution with a mean of zero, a standard deviation of 2 cm, and truncated at $\pm 5$ cm.

The parameters for the additive noise are chosen based on intuition and the facts that the laser beam has a width of 10 mm at 30 m and the smallest object (insulator) for segmentation has a maximum dimension of $\approx$30 cm.

A summary of the processing steps described previously is provided in Figure 5.

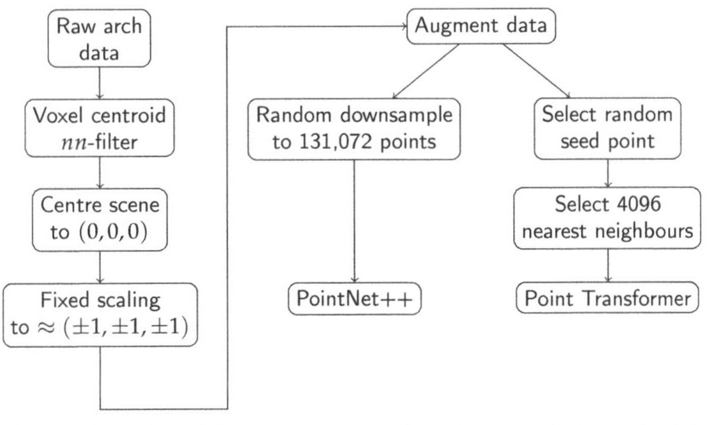


**Figure 5.** Overview of the processing steps, from the raw arch data to the data fed to the machine learning model.

As the number of samples of the dataset is limited, a leave-one-out cross-validation procedure is used. In total, 14 different components are segmented within the scene. The *mean Intersection over Union* (mIoU) per class is used as a performance metric. This metric weighs all the classes equally and is independent of the class size. It should be noted that not all components are always present in each sample. As an additional overall metric, the mean of the mIoU across all samples is reported.

As the dataset has a large class imbalance, each of the models is also trained with a loss function that incorporates the class weights—that is, classes that are rare will have a higher weight compared with the more common classes.

The models are trained on a desktop computer with 64 GB of RAM, an AMD Ryzen Threadripper 2950X CPU, and two NVIDIA TITAN V GPUs with 12 GB of memory each.

### 4.2. PointNet++

The data pre-processing for the PointNet++ models follows the procedure outlined in the previous section. To ensure a fixed number of points per sample, the pre-processed point cloud is randomly downsampled to 131,072 ($2^{17}$) points. This number of points still shows sufficient density for the smaller objects in the scene such as insulators, and also allows for a small batch size of four. This means the benefits of batch normalisation [33] can still be reaped. A limiting factor to the number of points sampled is the available memory on the GPU. The normalisation method described in the previous section deviates from the original PointNet++ work, which scales each individual sample to a unit circle. A fixed scaling factor is used since the model does not need to learn to be scale-independent. As a baseline, a vanilla PointNet++ [5] model is trained to perform the semantic segmentation task. After the baseline is established, the PointNet++ model is modified to enhance the segmentation of smaller objects within the scene.

This modification consists of adding an additional *set abstraction* level to the model. The parameters are set such that they are in line with the sequence of the other parameters. The number of points is set to 2048, the radius is set to 0.05, and the size of the multi-layer perceptron is set to $[16, 16, 32]$.

The steps per epoch are set to 4, which ensures that the model has encountered all training samples at least once during each epoch. The training parameters are selected based on preliminary explorative research. The model is trained for 400 epochs with a learning rate of 0.01 followed by 200 epochs with a learning rate of 0.001. The choice was made to use a fixed number of epochs because with the leave-one-out approach there is no validation set, which can be used to determine an early stopping trigger.

### 4.3. SuperPoint Graph

The SuperPoint Graph (SPG) [6] method first geometrically partitions the input cloud into individual segments using an unsupervised global energy model. These segments are referred to as superpoints in the article and represent the nodes of the graph. Nodes are connected by edges that have feature attributes such as volume and surface ratios of the two nodes being connected. The features and the superpoints are then used as input to a neural network.

Calculations of some of these edge features make use of the Quickhull algorithm [34]. Due to calculation imprecision and given the complex structure of the input point clouds, the algorithm is not always able to calculate the convex hull of the point cloud. In our case, it was not possible to create the graph in five out of the fifteen arches. An envisioned solution to overcome the imprecision issue is by enabling the 'joggle input' option of the Quickhull tool. This option randomly perturbs the input point cloud before running the Quickhull algorithm.

Considering that only a small part of the dataset can be used, the derived results would not be representative. Together with the fact that the generation of the graphs is computationally expensive, the decision was made not to include the results of the SuperPoint Graph in this article.

### 4.4. Point Transformer

Self-attention networks [35] are a major milestone in the area of deep learning. These networks made a significant impact in the areas of computational linguistics [36] and computer vision [37]. The goal of the Point Transformer model is to apply the concept of self-attention to point clouds.

In contrast to the PointNet++ model, this model is trained on subsets of the catenary arch. A subset is created by selecting a random point within the arch together with its nearest neighbours. A *k*-d tree is used to efficiently query the nearest neighbours of a point—in this implementation, 4096. During each training step, an arch is selected at random from which to draw the subset of points. To compensate for the fact that not all arches contain the same number of points, the probability of choosing an arch is proportional to the number of points in the arch.

The training parameters are selected based on preliminary explorative research. The model is trained for 100 epochs with 100 steps per epoch. The initial learning rate is 0.1, which is decreased to 0.01 after 60 epochs and decreased once more to 0.001 after 80 epochs. A batch size of 32 is used.

## 5. Results and Discussion

The mIoU per class for both the PointNet++ model and the Point Transformer model can be seen in Table 2. The numbers in bold indicate the highest scores. The classes are ordered subjectively based on size from large to small.

**Table 2.** Per-class mIoU compared for the vanilla PointNet++ model, the modified PointNet++ model, and the Point Transformer model (standard deviations between parentheses). The term *nw* refers to non-weighted losses and *iw* refers to inversely weighted losses.

| Class | PointNet++ | | | | Point Transformer | |
|---|---|---|---|---|---|---|
| | Vanilla | | Modified | | Vanilla | |
| | nw | iw | nw | iw | nw | iw |
| unlabelled | 0.63 | 0.63 | 0.69 | 0.67 | **0.73** | 0.43 |
| top bar | 0.73 | 0.73 | **0.80** | 0.78 | 0.78 | 0.70 |
| pole | 0.81 | 0.81 | 0.83 | 0.82 | **0.89** | 0.76 |
| drop post | 0.77 | 0.77 | **0.81** | 0.79 | 0.80 | 0.64 |
| top tie | 0.42 | 0.59 | **0.83** | 0.79 | 0.32 | 0.20 |
| bracket | 0.59 | 0.74 | **0.88** | 0.82 | 0.33 | 0.26 |
| pole foundation | 0.60 | 0.60 | 0.67 | 0.66 | **0.74** | 0.48 |
| steady arm | 0.54 | 0.54 | 0.58 | 0.58 | **0.70** | 0.63 |
| contact wire | 0.65 | 0.65 | 0.69 | 0.68 | **0.71** | 0.69 |
| stitch wire | 0.60 | 0.67 | **0.71** | 0.68 | 0.58 | 0.60 |
| wheel tension device | 0.52 | 0.44 | 0.70 | **0.76** | 0.07 | 0.09 |
| dropper | 0.31 | 0.31 | 0.51 | 0.46 | **0.54** | 0.39 |
| messenger wire supp. | 0.45 | 0.52 | 0.69 | 0.64 | **0.73** | 0.50 |
| insulator | 0.33 | 0.38 | 0.48 | 0.46 | **0.76** | 0.58 |
| class mean | 0.57 | 0.60 | **0.71** | 0.69 | 0.62 | 0.50 |
| | (0.15) | (0.14) | (0.12) | (0.12) | (0.22) | (0.20) |
| sample mean | 0.58 | 0.60 | **0.68** | 0.66 | 0.65 | 0.50 |
| | (0.10) | (0.10) | (0.12) | (0.11) | (0.15) | (0.11) |

It is challenging to objectively compare the performance of both models as they are trained using two different methods of feeding in the input data. The PointNet++ model trains on the downsampled version of the entire arch, whereas the Point Transformer trains on subsets of individual arches.

Overall, the modified PointNet++ model has the best performance in terms of mean class and mean sample accuracy. On the other hand, when counting the best performing metrics, the Point Transformer model is clearly superior. It performs best for eight out of the fourteen classes, with its performance entirely pulled down by the following three classes: top tie, bracket, and wheel tension device.

It is surprising that even with a small number of samples during training, good results can be obtained. This can be attributed to the fact that most of the catenary arch components have a well-defined geometrical structure, which does not vary between instances. This might also explain the fact that applying class weights to the loss function does not give a significant performance boost for the case of the PointNet++ model. In contrast, the Point Transformer model shows a large gap in performance when comparing the weighted and non-weighted results. It is unknown why this is the case.

One of the difficulties of using this dataset is the large differences in the sizes of objects, which require segmentation. For instance, an insulator measures approximately 30 cm and a top-bar might measure approximately 24 m, which is a factor of 80 difference with respect to the size of the insulator. This difficulty translates into the fact that large objects are segmented more accurately compared with smaller objects. The modified PointNet++ model shows an improvement in terms of class mIoU for the smaller components such as the droppers, messenger wire supports, and insulators.

The dataset contains an `unlabelled` class, which contains all points that do not fall into one of the other categories. Even though the models are able to correctly classify this class to a certain extent, it is difficult to understand how it does so. Does it learn to recognise the large variety of features associated to the unlabelled class? Or does it learn the process of elimination—that is, if a point does not belong to one of the thirteen other classes, then must it be an unlabelled point? These questions highlight the necessity of *explainable artificial intelligence* [38] to discover the hidden, underlying functionality of such models.

Exploring the explainability of a deep learning model is a challenging task on its own [39–42]. As a preliminary step, we focused on the shape and location of an object since these two aspects are most important for the semantic segmentation task. We applied transformations such as translation and rotation to the example dataset and measured the segmentation performance of the PointNet++ based deep learning model. The preliminary results indicate that shape has an almost negligible effect on the segmentation performance. Changing the object's location significantly affects the performance (these results can be found in the student paper [43]). We are currently devising more experiments to explore the explainability and robustness of the model.

Another interesting question that arises when trying to segment a large number of classes is whether the performance degrades when the number of classes is large. Additionally, it is hypothesised that having dedicated models for each individual class is beneficial.

When comparing the per-class mIoU of both models, the wheel tension device stands out, which the PointNet++ model is still able to segment reasonably well. On the other hand, the Point Transformer model has poor performance for this class. This could be caused by the unique shape of the wheel tension device, which is circular, flat, and symmetrical. This is another example where explainability of the model can aid in the understanding.

## 6. Conclusions

This work evaluated three deep-learning-based point cloud segmentation methods (PointNet++, SuperPoint Graph, and Point Transformer) in a real-world scenario. A custom dataset containing high-resolution point cloud scans of catenary arches was collected for this application. The arches were manually labelled into 14 different classes. To the best of our knowledge, this is the first high-resolution point cloud dataset of catenary arches that is available to the public.

Overall, the modified PointNet++ model performed best, achieving an average class mIoU of 71%. However, when counting the number of best-performing metrics, the non-weighted Point Transformer is superior. Its mean class performance was dragged down by just a few classes. The SuperPoint Graph model was not deemed appropriate for this use case as it was very prone to calculation imprecision and had high computational demands.

To counter the substantial class imbalance of the dataset, the models were also trained using class weights. Surprisingly, this had a negligible effect on the result for the PointNet++ models, yet for the Point Transformer model it was destructive.

*Outlook*

Semantic segmentation of railway scenes provides a crucial stepping stone towards automated condition monitoring. For instance, the work of Burton and Heuckelbach is focused on vegetation monitoring [44]. Their work uses point cloud data to assess the risk of trees falling on the railway track. With the help of semantic segmentation, the track, masts, wires, relay cabinets, and other assets can be identified and the risk of a falling tree can be evaluated per object type.

Other opportunities arise when wires are also part of the semantic segmentation process. This enables monitoring parameters such as sag and stagger [45,46] of wires.

If catenary masts are part of the segmentation process, their tilt can automatically be determined [47]. Maintenance can be planned if certain thresholds for tilt are exceeded. Measuring the same piece of track at multiple epochs leads to an even more advanced maintenance paradigm: predictive maintenance. For instance, in the case of mast tilt, it would be possible to determine a tilt velocity. This can be used to make projections in the future and can aid the creation of optimal maintenance plans.

A bottleneck for semantic segmentation in real-world scenarios is the availability of labelled data. Creating such a dataset is tedious, time-consuming, and prone to human errors as the classes are manually labelled. These datasets also tend to be inflexible. For instance, adding a new class would require another iteration of manual labelling. To address this issue, the possibility of a more model-driven approach, where models from an existing CAD library are used, can be explored. For instance, Vock et al. proposed a robust method for template matching within point clouds [48]. In our case, the templates can be generated from existing CAD libraries [49]. Such an approach could be feasible as components of the catenary arch have strong geometric shapes. An alternative approach is to use an active learning paradigm [50] for reducing the labelling cost. It is possible to leverage the trained model in this feat for the human-in-loop approach. This technique has been applied successfully for image classification tasks [51] and also object detection in point clouds [52]. Given the additional computational challenge, exploring its applicability for point cloud segmentation opens new possibilities for research.

The current dataset was collected using a mobile laser scanner mounted on a tripod; such a solution would not be viable when moving towards a more production-ready solution. Therefore, the segmentation models should also be evaluated on data captured by a train-mounted mobile laser scanner. This poses new challenges such as lower resolution, shadow effects, and merging data together from multiple trajectories.

To further improve the segmentation quality, the possibility of adding additional features such as colour and intensity values should be explored. The current work only considers cropped out arches, further research should focus on segmenting entire railway scenes.

**Author Contributions:** Conceptualisation, B.T., F.A. and J.L.; methodology, B.T., F.A. and J.L.; software, B.T.; validation, B.T.; formal analysis, B.T.; investigation, B.T.; data curation, B.T.; writing—original draft preparation, B.T.; writing—review and editing, B.T., F.A. and J.L.; visualisation, B.T. and F.A.; supervision, J.L.; project administration, J.L.; funding acquisition, J.L. All authors have read and agreed to the published version of the manuscript.

**Funding:** This work was mainly supported by collaboration with Strukton Rail (https://strukton.com/en/rail (accessed on 25 December 2022)) in a project called 'Digitalisatie Bovenleidingen en Draagconstructies', funded by TechForFuture (https://techforfuture.nl (accessed on 25 December 2022)). The APC was funded by the University of Twente.

**Institutional Review Board Statement:** Not applicable.

**Informed Consent Statement:** Not applicable.

**Data Availability Statement:** The dataset related to this article can be found at https://dx.doi.org/10.4121/17048816, an online data repository hosted at 4TU.ResearchData [53].

**Acknowledgments:** Student work by Mani Salahmand, Egbert Dijkstra, Floris Verburg, Zino Vieth, Benjamin Bakir, Job Jonkers, Joey Teunissen, Thomas Tunc and Rehan Ahmed assisted the authors in attaining the current results.

**Conflicts of Interest:** The data was collected by Strukton Rail. The authors declare no further conflict of interest. The funders had no role in the design of the study; the analyses, or interpretation of data; in the writing of the manuscript; or in the decision to publish the results.

## References

1. Uddin, W.; Hudson, W.R.; Haas, R. *Public Infrastructure Asset Management*; McGraw-Hill Education: New York, NY, USA, 2013.
2. Tang, P.; Huber, D.; Akinci, B.; Lipman, R.; Lytle, A. Automatic reconstruction of as-built building information models from laser-scanned point clouds: A review of related techniques. *Autom. Constr.* **2010**, *19*, 829–843. . [CrossRef]
3. Baltsavias, E.P. A comparison between photogrammetry and laser scanning. *ISPRS J. Photogramm. Remote Sens.* **1999**, *54*, 83–94. [CrossRef]
4. Kalvoda, P.; Nosek, J.; Kuruc, M.; Volarik, T. Accuracy Evaluation and Comparison of Mobile Laser Scanning and Mobile Photogrammetry Data Accuracy Evaluation and Comparison of Mobile Laser Scanning and Mobile Photogrammetry Data. In *IOP Conference Series: Earth and Environmental Science*; IOP Publishing: Bristol, UK, 2020. [CrossRef]
5. Qi, C.R.; Yi, L.; Su, H.; Guibas, L.J. PointNet++: Deep Hierarchical Feature Learning on Point Sets in a Metric Space. In Proceedings of the 31st International Conference on Neural Information Processing Systems, NIPS'17, Long Beach, CA, USA, 4–9 December 2017; Curran Associates Inc.: Red Hook, NY, USA, 2017; pp. 5105–5114.
6. Landrieu, L.; Simonovsky, M. Large-Scale Point Cloud Semantic Segmentation with Superpoint Graphs. In Proceedings of the 2018 IEEE/CVF Conference on Computer Vision and Pattern Recognition, Salt Lake City, UT, USA, 18–23 June 2018; pp. 4558–4567. [CrossRef]
7. Zhao, H.; Jiang, L.; Jia, J.; Torr, P.H.; Koltun, V. Point Transformer. In Proceedings of the IEEE/CVF International Conference on Computer Vision (ICCV), Montreal, QC, Canada, 10–17 October 2021; pp. 16259–16268.
8. Armeni, I.; Sener, O.; Zamir, A.R.; Jiang, H.; Brilakis, I.; Fischer, M.; Savarese, S. 3D Semantic Parsing of Large-Scale Indoor Spaces Supplementary Material. In Proceedings of the 2016 IEEE Conference on Computer Vision and Pattern Recognition (CVPR), Las Vegas, NV, USA, 27–30 June 2016; pp. 1534–1543.
9. Wang, Q.; Kim, M.k. Applications of 3D point cloud data in the construction industry: A fifteen-year review from 2004 to 2018. *Adv. Eng. Inform.* **2019**, *39*, 306–319. [CrossRef]
10. Callahan, M.A.; LeBlanc, B.; Vreeland, R.; Bretting, G. *Close-Range Photogrammetry with Laser Scan Point Clouds*; Technical Report; SAE Technical Paper: Warrendale, PA, USA, 2012.
11. Valero, E.; Bosché, F.; Forster, A. Automatic segmentation of 3D point clouds of rubble masonry walls, and its application to building surveying, repair and maintenance. *Autom. Constr.* **2018**, *96*, 29–39. [CrossRef]
12. Mahler, J.; Matl, M.; Satish, V.; Danielczuk, M.; DeRose, B.; McKinley, S.; Goldberg, K. Learning ambidextrous robot grasping policies. *Sci. Robot.* **2019**, *4*, eaau4984. [CrossRef] [PubMed]
13. Bello, S.A.; Yu, S.; Wang, C.; Adam, J.M.; Li, J. Review: Deep Learning on 3D Point Clouds. *Remote Sens.* **2020**, *12*, 1729. [CrossRef]
14. Burume, D.M.; Du, S. Deep Learning Methods Applied to 3D Point Clouds Based Instance Segmentation: A Review. *Preprints* **2021**, 2021110228. [CrossRef]
15. Guo, Y.; Wang, H.; Hu, Q.; Liu, H.; Liu, L.; Bennamoun, M. Deep learning for 3D point clouds: A survey. *IEEE Trans. Pattern Anal. Mach. Intell.* **2020**, *43*, 4338–4364. [CrossRef]
16. Liu, W.; Sun, J.; Li, W.; Hu, T.; Wang, P. Deep Learning on Point Clouds and Its Application: A Survey. *Sensors* **2019**, *19*, 4188. [CrossRef]
17. Liu, S.; Zhang, M.; Kadam, P.; Kuo, C.C.J. Deep Learning-Based Point Cloud Analysis. In *3D Point Cloud Analysis*; Springer: Berlin/Heidelberg, Germany, 2021; pp. 53–86.
18. Zhang, J.; Zhao, X.; Chen, Z.; Lu, Z. A review of deep learning-based semantic segmentation for point cloud. *IEEE Access* **2019**, *7*, 179118–179133. [CrossRef]
19. Bentley, J.L. Multidimensional binary search trees used for associative searching. *Commun. ACM* **1975**, *18*, 509–517. [CrossRef]
20. Zeng, W.; Gevers, T. 3DContextNet: K-d tree guided hierarchical learning of point clouds using local and global contextual cues. In Proceedings of the European Conference on Computer Vision (ECCV) Workshops, Munich, Germany, 8–14 September 2018.
21. Arastounia, M. Automated Recognition of Railroad Infrastructure in Rural Areas from LiDAR Data. *Remote Sens.* **2015**, *7*, 14916–14938. [CrossRef]
22. Chen, L.; Jung, J.; Sohn, G. Multi-Scale HierarchicalCRF for Railway Electrification Asset Classification From Mobile Laser Scanning Data. *IEEE J. Sel. Top. Appl. Earth Obs. Remote Sens.* **2019**, *12*, 3131–3148. [CrossRef]
23. Charles, R.Q.; Su, H.; Kaichun, M.; Guibas, L.J. PointNet: Deep Learning on Point Sets for 3D Classification and Segmentation. In Proceedings of the 2017 IEEE Conference on Computer Vision and Pattern Recognition (CVPR), Honolulu, HI, USA, 21–26 July 2017; pp. 77–85. [CrossRef]

24. Thomas, H.; Qi, C.R.; Deschaud, J.E.; Marcotegui, B.; Goulette, F.; Guibas, L. KPConv: Flexible and Deformable Convolution for Point Clouds. In Proceedings of the 2019 IEEE/CVF International Conference on Computer Vision (ICCV), Seoul, Republic of Korea, 27 October–2 November 2019; pp. 6410–6419. [CrossRef]
25. Soilán, M.; Sánchez-Rodríguez, A.; del Río-Barral, P.; Perez-Collazo, C.; Arias, P.; Riveiro, B. Review of Laser Scanning Technologies and Their Applications for Road and Railway Infrastructure Monitoring. *Infrastructures* **2019**, *4*, 58. [CrossRef]
26. Chen, L.; Xu, C.; Lin, S.; Li, S.; Tu, X. A Deep Learning-Based Method for Overhead Contact System Component Recognition Using Mobile 2D LiDAR. *Sensors* **2020**, *20*, 2224. [CrossRef]
27. Lin, S.; Xu, C.; Chen, L.; Li, S.; Tu, X. LiDAR Point Cloud Recognition of Overhead Catenary System with Deep Learning. *Sensors* **2020**, *20*, 2212. [CrossRef]
28. Bruijne, A.d.; Buren, J.V.; Marel, H.V.D. *Geodetic Reference Frames in the Netherlands*; NCG, Nederlandse Commissie voor Geodesie, Netherlands Geodetic Commission: Delft, The Netherlands, 2005; pp. 1–117.
29. Zhu, L.; Hyyppa, J. The Use of Airborne and Mobile Laser Scanning for Modeling Railway Environments in 3D. *Remote Sens.* **2014**, *6*, 3075–3100. [CrossRef]
30. Corongiu, M.; Masiero, A.; Tucci, G. Classification of Railway Assets in Mobile Mapping Point Clouds. *Int. Arch. Photogramm. Remote Sens. Spat. Inf. Sci.* **2020**, *XLIII-B1-2*, 219–225. . [CrossRef]
31. American Society for Photogrammetry and Remote Sensing. *LAS Specification Version 1.4-R13*; Technical Report; ASPRS: Bethesda, MD, USA, 2013.
32. Zhan, K.; Chen, S.; Whitman, D.; Shyu, M.; Yan, J.; Zhang, C. A progressive morphological filter for removing nonground measurements from airborne LiDAR data. *IEEE Trans. Geosci. Remote Sens.* **2003**, *41*, 872–882. [CrossRef]
33. Ioffe, S.; Szegedy, C. Batch Normalization: Accelerating Deep Network Training by Reducing Internal Covariate Shift. In Proceedings of the 32nd International Conference on International Conference on Machine Learning—JMLR.org, ICML'15, Lille, France, 7–9 July 2015; Volume 37, pp. 448–456. [CrossRef]
34. Barber, C.B.; Dobkin, D.P.; Huhdanpaa, H. The quickhull algorithm for convex hulls. *ACM Trans. Math. Softw.* **1996**, *22*, 469–483. [CrossRef]
35. Vaswani, A.; Shazeer, N.; Parmar, N.; Uszkoreit, J.; Jones, L.; Gomez, A.N.; Kaiser, L.; Polosukhin, I. Attention is All You Need. In Proceedings of the 31st International Conference on Neural Information Processing Systems, NIPS'17, Long Beach, CA, USA, 4–9 December 2017; Curran Associates Inc.: Red Hook, NY, USA, 2017; pp. 6000–6010.
36. Devlin, J.; Chang, M.; Lee, K.; Toutanova, K. BERT: Pre-training of Deep Bidirectional Transformers for Language Understanding. In Proceedings of the 2019 Conference of the North American Chapter of the Association for Computational Linguistics: Human Language Technologies, NAACL-HLT, Minneapolis, MN, USA, 2–7 June 2019; Burstein, J., Doran, C., Solorio, T., Eds.; Association for Computational Linguistics: Minneapolis, MN, USA, 2019; pp. 4171–4186. [CrossRef]
37. Wang, X.; Girshick, R.; Gupta, A.; He, K. Non-local Neural Networks. In Proceedings of the 2018 IEEE/CVF Conference on Computer Vision and Pattern Recognition (CVPR), Salt Lake City, UT, USA, 18–22 June 2018; IEEE Computer Society: Los Alamitos, CA, USA, 2018; pp. 7794–7803. [CrossRef]
38. Samek, W.; Müller, K.R. Towards Explainable Artificial Intelligence. In *Explainable AI: Interpreting, Explaining and Visualizing Deep Learning*; Samek, W., Montavon, G., Vedaldi, A., Hansen, L.K., Müller, K.R., Eds.; Springer International Publishing: Cham, Switzerland, 2019; pp. 5–22. [CrossRef]
39. Linardatos, P.; Papastefanopoulos, V.; Kotsiantis, S. Explainable AI: A review of machine learning interpretability methods. *Entropy* **2021**, *23*, 18. [CrossRef] [PubMed]
40. Pan, H.; Wang, Z.; Zhan, W.; Tomizuka, M. Towards Better Performance and More Explainable Uncertainty for 3D Object Detection of Autonomous Vehicles. In Proceedings of the 2020 IEEE 23rd International Conference on Intelligent Transportation Systems (ITSC), Rhodes, Greece, 20–23 September 2020; pp. 1–7. [CrossRef]
41. Zhang, M.; You, H.; Kadam, P.; Liu, S.; Kuo, C.C.J. PointHop: An Explainable Machine Learning Method for Point Cloud Classification. *IEEE Trans. Multimed.* **2020**, *22*, 1744–1755. [CrossRef]
42. Matrone, F.; Paolanti, M.; Felicetti, A.; Martini, M.; Pierdicca, R. BubblEX: An Explainable Deep Learning Framework for Point-Cloud Classification. *IEEE J. Sel. Top. Appl. Earth Obs. Remote Sens.* **2022**, *15*, 6571–6587. . [CrossRef]
43. Verburg, F.M. Exploring Explainability and Robustness of Point Cloud Segmentation Deep Learning Model by Visualization. Bachelor's Thesis, University of Twente, Enschede, The Netherlands, 2022.
44. Burton, T.; Heuckelbach, D. Fugro vegetation control: A remote solution for lineside vegetation management. *Perm. Way Inst.* **2020**, *138*, 34–37.
45. Gutiérrez-Fernández, A.; Fernández-Llamas, C.; Matellán-Olivera, V.; Suárez-González, A. Automatic extraction of power cables location in railways using surface lidar systems. *Sensors* **2020**, *20*, 6222. [CrossRef]
46. Zhang, L.; Wang, J.; Shen, Y.; Liang, J.; Chen, Y.; Chen, L.; Zhou, M. A Deep Learning Based Method for Railway Overhead Wire Reconstruction from Airborne LiDAR Data. *Remote Sens.* **2022**, *14*, 5272. [CrossRef]
47. Marwati, A.; Wang, C.K. Automatic retrieval of railway masts tilt angle from Mobile Laser Scanning data. In Proceedings of the 42nd Asian Conference on Remote Sensing, ACRS 2021, Can Tho City, Vietnam, 22–24 November 2021; Asian Association on Remote Sensing (AARS): Can Tho, Vietnam, 2021.
48. Vock, R.; Dieckmann, A.; Ochmann, S.; Klein, R. Fast template matching and pose estimation in 3D point clouds. *Comput. Graph.* **2019**, *79*, 36–45. [CrossRef]

49. Vieth, Z.J. Point Cloud Classification and Segmentation of Catenary Systems. Bachelor's Thesis, University of Twente, Enschede, The Netherlands, 2022.
50. Sayin, B.; Krivosheev, E.; Yang, J.; Passerini, A.; Casati, F. A review and experimental analysis of active learning over crowd sourced data. *Artif. Intell. Rev.* **2021**, *54*, 5283–5305. [CrossRef]
51. Budd, S.; Robinson, E.C.; Kainz, B. A survey on active learning and human-in-the-loop deep learning for medical image analysis. *Med. Image Anal.* **2021**, *71*, 102062. [CrossRef]
52. Meng, Q.; Wang, W.; Zhou, T.; Shen, J.; Jia, Y.; Van Gool, L. Towards a weakly supervised framework for 3D point cloud object detection and annotation. *IEEE Trans. Pattern Anal. Mach. Intell.* **2021**, *44*, 4454–4468. [CrossRef]
53. Strukton, R.; Ton, B. High resolution labelled point cloud dataset of catenary arches in the Netherlands. *4TU.ResearchData* **2021**. [CrossRef]

sensors

MDPI

*Article*

# An Ensemble Learning Aided Computer Vision Method with Advanced Color Enhancement for Corroded Bolt Detection in Tunnels

**Lei Tan [1,2,3,*], Tao Tang [1,2] and Dajun Yuan [4]**

1 State Key Laboratory of Rail Traffic Control and Safety, Beijing Jiaotong University, Beijing 100044, China
2 Beijing Municipal Engineering Research Institute, Beijing 100037, China
3 School of Electronic and Information Engineering, Beijing Jiaotong University, Beijing 100044, China
4 School of Civil Engineering, Beijing Jiaotong University, Beijing 100044, China
* Correspondence: tanlei@bjtu.edu.cn

**Abstract:** Bolts, as the basic units of tunnel linings, are crucial to safe tunnel service. Caused by the moist and complex environment in the tunnel, corrosion becomes a significant defect of bolts. Computer vision technology is adopted because manual patrol inspection is inefficient and often misses the corroded bolts. However, most current studies are conducted in a laboratory with good lighting conditions, while their effects in actual practice have yet to be considered, and the accuracy also needs to be improved. In this paper, we put forward an Ensemble Learning approach combining our Improved MultiScale Retinex with Color Restoration (IMSRCR) and You Only Look Once (YOLO) based on truly acquired tunnel image data to detect corroded bolts in the lining. The IMSRCR sharpens and strengthens the features of the lining pictures, weakening the bad effect of a dim environment compared with the existing MSRCR. Furthermore, we combine models with different parameters that show different performance using the ensemble learning method, greatly improving the accuracy. Sufficient comparisons and ablation experiments based on a dataset collected from the tunnel in service are conducted to prove the superiority of our proposed algorithm.

**Keywords:** corroded bolt detection; computer vision; color enhancement; ensemble learning

**Citation:** Tan, L.; Tang, T.; Yuan, D. An Ensemble Learning Aided Computer Vision Method with Advanced Color Enhancement for Corroded Bolt Detection in Tunnels. *Sensors* **2022**, *22*, 9715. https://doi.org/10.3390/s22249715

Academic Editors: Abdollah Malekjafarian, Diogo Ribeiro, Araliya Mosleh and Maria D. Martínez-Rodrigo

Received: 23 November 2022
Accepted: 10 December 2022
Published: 11 December 2022



## 1. Introduction

Railway transportation has become the main mode of land transport with its remarkable carrying capacity and fast speed [1,2]. As an important branch, subway systems have developed rapidly in recent years [3], becoming the preferred traveling way for city dwellers. The lining, which is fixed and arranged by bolts, supports the tunnel structure and guarantees the operation of metros. However, the bolts are exposed to the open air, usually influenced by moisture and air pollutants, and the steel material thus tends to become corroded [4–6]. When it comes to maintenance and repair, human-based visual inspection still dominates the tunnel industry, which is also limited by training level. Patrol inspectors have to check all bolts during non-running times such as night and early morning. However, commonly, an inspection team composed of 10 to 15 trained maintainers could check two to three kilometers during a maintenance period of about three hours, which is costly and inefficient. Besides, quite a few bolts are misdiagnosed as normal or corroded due to the poor light in tunnels and fatigue caused by night work. Hence, some researchers have tried many approaches to design an automatic, high-accuracy, and fast detection speed method for practical engineering projects.

Computer Vision (CV), which overcomes the limitations of visual inspection by trained human resources and the ability to detect structural damage in images remotely [7,8], has become a prioritized technique for corroded bolt detection. However, the traditional CV algorithms require the manual design of filter modules, which has poor robustness and low accuracy. Deep learning-based CV bolt corrosion detection becomes available for engineering as deep learning develops [9–11]. For instance, Cha et al. [12] developed an autonomous structural visual inspection method via Region-based Convolutional Neural Networks (RCNNs) for real-time damage detection covering concrete cracks, steel and bolt corrosion, and steel delamination. Ta et al. [13] monitored and identified the corrosion levels of corroded bolts in a lab-scale steel structure with good illumination using a Mask-RCNN. Suh et al. [14] adopted a Faster RCNN-based model to detect and locate damage types, including bolt corrosion. These RCNNs search the target area with selective search and generate nearly 2000 eigenvectors for each figure. They are mostly applied in the precise pixel-level detection task. However, it is not easy to deploy RCNN models in practical applications compared to end-to-end models. Plus, it is not necessary to precisely distinguish the target pixels on the corrosion bolt in practice at the expense of speed and cost. Another branch of deep learning target detection algorithms, You Only Look Once (YOLO), reinterprets the principle of object detection tasks from classifications to regressions, speeding up the training and detecting processes [15–17]. We select YOLOv5 nano (YOLOv5n) as the basis of our proposed model caused of its speed, end-to-end characteristics, and high precision compared with the two-stage detectors.

Although the YOLOv5n shows its superior performance in computing speed and resource consumption, the complex corrosion targets still require improvements in accuracy. Using multiple models with different preferences, ensemble learning makes a better and more comprehensive decision to avoid the wrong prediction created by weak classifiers. For example, Xu et al. [18] applied ensemble deep learning technology to learn and extract features of forest fires. Mohammad et al. [19] presented an ensemble deep-learning approach to recognize structural corrosion in drone images. Seijo-Pardo et al. [20] concluded ensemble learning of homogeneous and heterogeneous approaches, showing the availability of integrating models with different parameters. Inspired by these works, we put forward ensemble learning with YOLOv5n (YOLOv5n-EL) to raise accuracy without slowing down the computing speed too much.

In addition to the corroded bolt detector, tunnels are usually damp and dim, weakening the tunnel scan image to low definition, poor contrast, and color distortion. These problems bring big troubles to the task of corroded bolt detection in such tunnels, which require figures to be pre-processed to make the features of the image more apparent for better corroded bolt detection. It has been proved that the Retinex theory (a color-invariance-based principle) is effective for low-light image enhancement like night and underwater [21–23]. Retinex mainly consists of three basic algorithms—Single Scale Retinex (SSR), MultiScale Retinex (MSR), and MultiScale Retinex with Color Restoration (MSRCR). Compared with SSR and MSR, MSRCR shows better image quality improvement and the ability to avoid the color distortion caused by the imbalance of each color channel proportion after convolution computation. However, the performance still degrades in the dim tunnel environment caused by its Gaussian Blur, which reduces the sharpness of edges while brightening the dark areas. Thus, we proposed the Improved-MSRCR (IMSRCR) algorithm to solve the problem of fuzzy bolt edges in low-illumination tunnel images using auto-matched dynamic filters and $L_0$ regularization. Through a combination scheme of the IMSRCR and the YOLOv5n-EL, our model appears to have excellent performance at bolt corrosion detection. Our main contributions can be summarized as follows.

1. We optimized the MSRCR color enhancement algorithm based on auto-matched dynamic filters and $L_0$ regularization to avoid blurring the image when brightening the dark areas.
2. We put forward ensemble learning with its fusion strategy combining models with different parameters to improve precision accuracy.
3. The experiments are conducted on actual data collected from a practical railway tunnel. We disclosed our labeled dataset, the first public corroded lining bolt dataset using a professional tunnel scanner.

The rest of this paper is organized as follows. Section 2 exhaustively describes the proposed approach covering the improved color-enhanced module and ensemble learning algorithm for bolt corrosion detection. Section 3 thoroughly exhibits the details of the experiments, including the dataset, experiment settings, comparison schemes, performance evaluations, and the analysis of the results. Section 4 gives a discussion about the method. Section 5 outlines our main results.

## 2. Methodology

Figure 1 depicts the flow chart of the corroded bolt detection scheme in a dim tunnel, including two main modules, i.e., the image color enhancement algorithm and the object detection module. Considering the difficulty of distinguishing corroded and normal bolts in a dim environment, an improved MSRCR (IMSRCR) is proposed to sharpen the contrast between the rust-infected area and the background, enhancing the appearance of image features. Then, for essential prediction speed and training efficiency in the object detection module, YOLOv5n is introduced to finish the object detection and location of corroded bolts on the color enhancement image, which is an end-to-end train and predict structure. For a further step up in accuracy, we propose YOLOv5n-EL based on YOLOv5n. Specifically, we train a series of models with different parameters and adopt ensemble learning to integrate all model outputs.

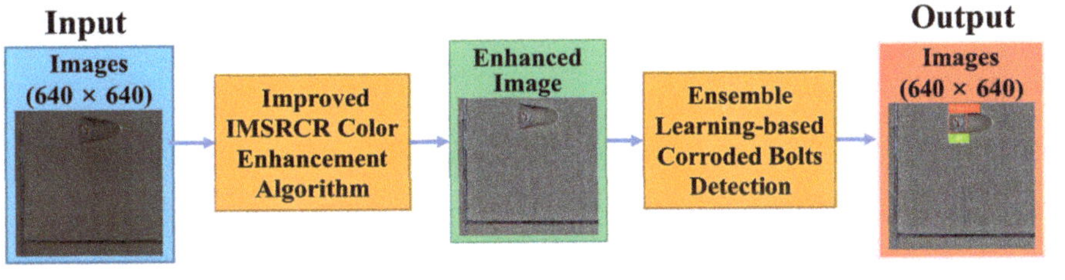


**Figure 1.** Flow chart of corroded bolt detection scheme in a dim tunnel.

### *2.1. The Improved IMSRCR Color Enhancement Algorithm*

As is well-known, the illumination is poor, so the tunnel images gathered are dim and unclear. Thus, we need to enhance the contrast between the bolts and the background. MSRCR is developed on MSR and SSR based on Retinex theory, which has been approved as an effective color enhancement method. However, MSRCR has a limited effect in dark areas and the edges of the dark areas. In our work, we propose IMSRCR to enhance the bolts features in dark areas. According to Retinex, the observed image $I(x,y)$ can be divided into the reflection component $R(x,y)$ carrying target information and the irradiation component $L(x,y)$ of ambient light is

$$I(x,y) = L(x,y) \times R(x,y). \tag{1}$$

Therefore, image enhancement aims to get rid of the irradiated component and extract a reflective part that carries information about the object. By simple mathematical transformation, we can get the expression of $R(x,y)$ with

$$\log R(x,y) = \log I(x,y) - \log L(x,y). \tag{2}$$

$L(x,y)$ can be estimated through low-pass Gaussian center function $F(x,y)$ and the observed image $I(x,y)$ as

$$\log L(x,y) = \log[F(x,y) \otimes I(x,y)], \tag{3}$$

where $F(x,y)$ is defined by

$$F(x,y) = \lambda e^{-\frac{x^2+y^2}{2c^2}}. \tag{4}$$

Meanwhile, $F(x,y)$ should satisfies

$$\iint F(x,y)dxdy = 1. \tag{5}$$

As a result, the expression of SSR can be obtained from (2)–(4) to

$$r_{ssr}(x,y) = \log R(x,y) = \log I(x,y) - \log[F(x,y) \otimes I(x,y)]. \tag{6}$$

The parameter $c$ in (4) is strongly related to the scale of image enhancement. However, the enhancements of SSR are not always satisfactory because the parameter $c$ is not suitable for all kinds of images. In response to the above question, MSR imports Gaussian center function at different scales as

$$r_{msr}(x,y) = \sum_{k}^{K} \omega_k \log I(x,y) - \log[F_k(x,y) \otimes I(x,y)], \tag{7}$$

where $\omega_k$ and $F_k(x,y)$ meets the Equations (8)–(10).

$$\sum_{k}^{K} \omega_k = 1, \tag{8}$$

$$F_k(x,y) = \lambda_k e^{-\frac{x^2+y^2}{2c_k{}^2}}, \tag{9}$$

$$\iint F_k(x,y)dxdy = 1. \tag{10}$$

Although MSR enhances image features at both low and high scales, color distortion will occur as the parameters are different for each color channel. Thus, the color recover factor C is added in MSRCR to keep the appearance true through

$$r_{msrcr}(x,y) = C_i \sum_{k}^{K} \omega_k \log I_i(x,y) - \log[F_k(x,y) \otimes I_i(x,y)], \tag{11}$$

where $i$ represents the $i_{\text{th}}$ color channel and $C_i$ can be expressed by

$$\begin{aligned} C_i &= f[I_i'(x,y)] \\ &= \beta \log[\alpha I_i'(x,y)] \\ &= \beta \log[\alpha \frac{I_i(x,y)}{\sum_{j=1}^{N} I_j(x,y)}] \\ &= \beta \log[\alpha I_i(x,y)] - \beta \log[\sum_{j=1}^{N} I_j(x,y)], \end{aligned} \tag{12}$$

in which $\alpha$ denotes controlled nonlinear treatment strength and $\beta$ is the gain constant.

Although MSRCR performs better in image enhancement comparing MSR and SSR, the edge of the enhanced image is still inconspicuous, which makes the performance of MSRCR degrade in a dim environment. Accordingly, we propose an IMSRCR algorithm to solve the problem of fuzzy bolt edges in low-illumination tunnel images. Our algorithm uses Automatic Guide Filtering (AGF) to estimate the illumination image first and then calculate the reflected image according to the Retinex theory mentioned above. Residual image is extracted by the norm. Finally, the color restoration is carried out on the fused image. The flow path of our algorithm is shown in Figure 2.

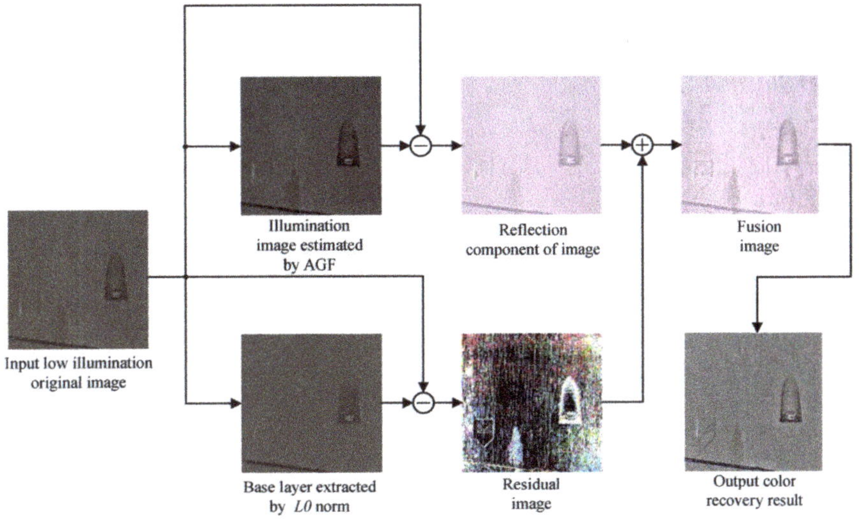


**Figure 2.** Flow path of IMSRCR.

2.1.1. Illumination Estimation

In order to reduce the edge blur problem of the Gaussian filter, the Illumination Estimation is powered by AGF, which is different from traditional MSRCR using a Gaussian filter. The illuminance images estimated by AGF and Gaussian filter are shown in Figure 3.

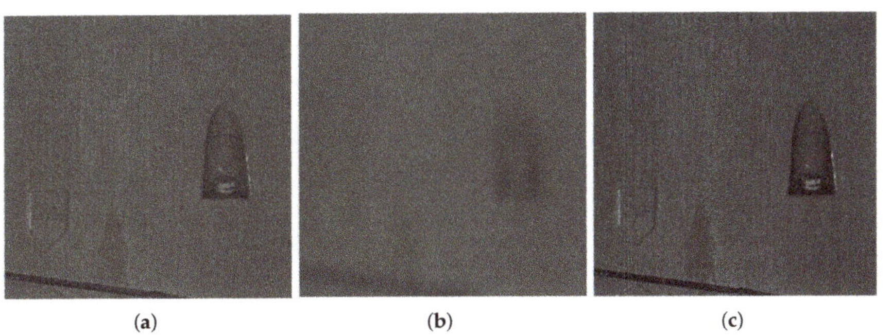


**Figure 3.** Original and illumination images. (**a**) Original image; (**b**) Illumination image estimated by Gaussian filter; (**c**) Illumination image estimated by AGF.

Guided filter is a local linear model with smooth edge preserving characteristics [24,25] which is defined as

$$g_t = a_k\, G_t + b_k \ , \ \forall t \epsilon \Omega_k, \tag{13}$$

where $g$ is the output image after guided filtering and $G$ is the guided image, $a_k$ and $b_k$ are the linear coefficients at the sub-windows $\Omega_k$, $\Omega_k$ represents the sub-window with scale $r$, and $t$ is the index of pixels in $\Omega_k$. We specify to input image $I$ as the guided image $Q$. $a_k$ and $b_k$ could be defined according to Guiding filtering-related theory as

$$\begin{aligned} a_k &= \frac{\sigma_k^2}{\sigma_k^2 + \varepsilon}, \\ b_k &= \mu_k (1 - a_k). \end{aligned} \tag{14}$$

The scale $r$ of the guided filter is set to three values referring to the process of the MSRCR algorithm. The range of three values of scale $r$ is $[1, r_{min}]$, $[r_{min}, r_{mid}]$ and $[r_{mid}, r_{max}]$ respectively [26]. $r_{min}$, $r_{mid}$ and $r_{max}$ could be determined as

$$\begin{aligned} r_{min} &= \left\lfloor \frac{\min(m,\, n)}{2^N} \right\rfloor, \\ r_{max} &= \left\lfloor \frac{\min(m,\, n)}{2} - 1 \right\rfloor, \\ r_{mid} &= \left\lfloor \frac{r_{min} + r_{max}}{2} \right\rfloor, \end{aligned} \tag{15}$$

where $m$ and $n$ are the width and height of the image, and $N$ is the number of selected scales. To balance the smoothing and edge-preserving effects of guided filtering, an Auto multi-scale selection algorithm is expressed by

$$\begin{aligned} r_1 &= \left\lfloor \frac{1 + r_{min}}{2^N} \right\rfloor, \\ r_2 &= \left\lfloor \frac{r_{min} + r_{mid}}{2} \right\rfloor, \\ r_3 &= \left\lfloor \frac{r_{mid} + r_{max}}{2} \right\rfloor. \end{aligned} \tag{16}$$

The illumination estimation result applies AGF to each channel of the input image. The reflection component in the logarithmic domain could be defined according to the Retinex theory

$$F_{AGF} = \sum_{j=1}^{3} \omega_j [logI_i(x, y) - logg_i(x, y)], \tag{17}$$

where $F_{AGF}$ is the reflected image channel corresponding to the AGF.

### 2.1.2. Residual Fusion

In order to overcome the problem of $F_{AGF}$ detail loss, we used $L_0$ norm in IMSRCR [27]. Residual results extracted by $L_0$ norm is shown in Figure 4.

(**a**) (**b**) (**c**)

**Figure 4.** Original and residual images. (**a**) Original image; (**b**) Base layer extracted by $L_0$ norm; (**c**) Residual image.

$L_0$ norm can be expressed as the number of non-zero elements in a vector. The $L_0$ norm of image gradient can be expressed as

$$C(f) := \#\{p || f_p - f_{p+1} | \neq 0\}, \tag{18}$$

where $p$ and $p+1$ are adjacent elements in the image. $|f_p - f_{p+1}|$ is the image gradient which is the forward difference of the image. # represents the number of pixels in the image that satisfied $|f_p - f_{p+1}| \neq 0$. $C(f)$ is the $L_0$ norm of the image gradient.

Taking one-dimensional signal as an example, the objective function can be defined as

$$\min_f \sum_p (f_p - g_p)^2 \quad \text{s.t.} \quad C(f) = k. \tag{19}$$

It must be converted into unconstrained problems for two-dimensional images. We set smoothing parameter $\lambda$ to 0.01 in combination with our use scene

$$\min_f \sum_p (f_p - g_p)^2 + \lambda \cdot C(f). \tag{20}$$

The number of gradients in the horizontal and vertical directions of the image needs to be constrained in the two-dimensional images. The objective function and its constraints are expressed as

$$\begin{gathered} \min_f \sum (f_p - g_p)^2 + \lambda \cdot C(\partial_x f, \partial_y f), \\ C(\partial_x f, \partial_y f) = \#\{p \| \partial_x f_p| + |\partial_y f_p| \neq 0\}. \end{gathered} \tag{21}$$

Since the $L_0$ norm is non-differentiable, the variable splitting method is used here to relax it into two quadratic programming problems. Finally, the iterative method is used to find the global optimum. We rewrite the objective function as

$$\min_f \sum_p (f_p - g_p)^2 + \lambda \cdot C(\partial_x f, \partial_y f) + \beta \cdot \sum^p \left( (\partial_x f_p - h_p)^2 + (\partial_y f_p - v_p)^2 \right). \tag{22}$$

The iterative solution result of the objective function is expressed as

$$h_p, v_p = \begin{cases} (0,0) & (\partial_x f_p)^2 + (\partial_y f_p)^2 \leq \frac{\lambda}{\beta} \\ (\partial_x f_p, \partial_y f_p) & \text{otherwise} \end{cases} \tag{23}$$

As presented in Figure 5, the image processed by IMSRCR is more apparent and has higher color contrast based on subjective visual judgment. And the edge of the bolts is more clear compared with the enhanced image processed by SSR, MSR, and MSRCR. Hence, IMSRCR is developed for the detection module to ensure that the pictures inputted to YOLOv5n have distinct visual features.

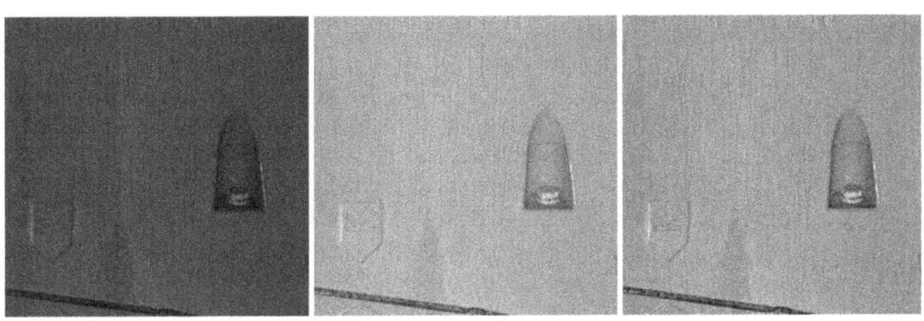

(**a**) Original image (**b**) SSR processed (**c**) MSR processed

**Figure 5.** *Cont.*

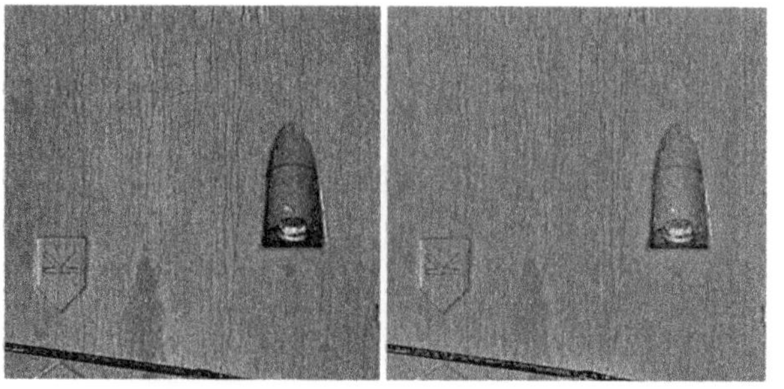

(d) MSRCR processed (e) IMSRCR processed

**Figure 5.** Effect comparison of different image enhancement algorithm.

### 2.2. *Ensemble Learning-Based Corroded Bolts Detection*

CV modules with different stages, mainly one and two, are used for object detection tasks. One-stage end-to-end algorithms give the prediction results (type and location) directly through the backbone, while two-stage methods form a series of sample boxes first, then classify and locate the object inside the boxes. So the non-end-to-end structure requires much more time than the one-stage method to train and detect separately, slowing the speed in real corroded bolt detection work. YOLOv5n, a fast and accurate one-stage CV model, is chosen as the baseline of our ensemble learning.

#### 2.2.1. Ensemble Learning Method

Usually, a target detection task is based on one given model to train and learn for a good performance in detection results. As far as we know, there are some excellent models to resolve the detection task, such as YOLO and FCNN. However, the performance of the models mentioned above can still be improved. Adjusting HyperParameters of training is a common technique to improve the model performance. However, it has a limited effect as the structure of the model restricts a better performance. Ensemble learning is a machine learning method that integrates the prediction of multiple deep learning models to improve robustness and detection performance. It processes the multiple model outputs as a decision question. If a mistake occurs on one of the multiple models and the others are right, the final output of ensemble learning will correct the error considering the whole model's outputs. Compared with the single model, ensemble learning combining multiple models will improve the accuracy heavily.

Ensemble learning can be divided into two categories according to training methods: Boosting and Bagging. Boosting constructs a series of object detectors through serial learning, which means the new detector is improved based on the adjustment to the mistake detection data weight in the last detector. In contrast, Bagging is a parallel learning method that utilizes the independence of different detectors to improve performance, while a single detector cannot extract whole features. In our work, Bagging is adopted as the ensemble learning method while we integrate different kinds of models which are independent of each other. The structure of ensemble learning is shown in Figure 6. It is worth noticing that our proposed integrated learning model is a parallel structure, corresponding to the use of multi-threaded parallel learning operations to avoid bringing excessive consumption of model training and inference time.

Bagging draws training data from the whole dataset at random and the drawn training data will be put back before the next round of extraction. This process will be continued for $k$ rounds, so we can get $k$ independent sub-datasets. Every sub-dataset is adopted to train a basic model. As a result, we can get $k$ independent basic models.

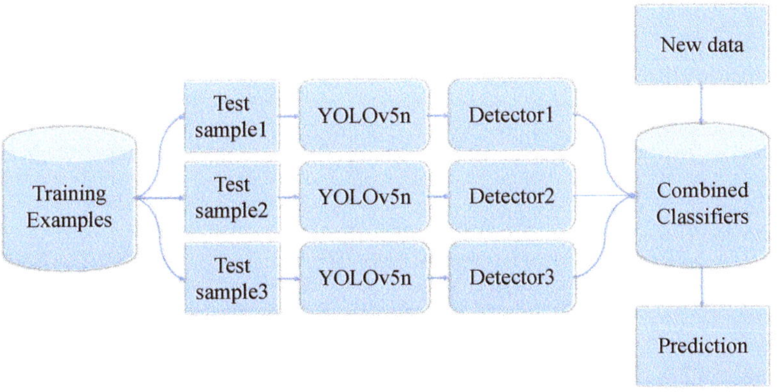


**Figure 6.** Structure of Ensemble Learning.

2.2.2. Fusion Strategy in Ensemble Learning

Fusion strategy is fundamental in ensemble learning. With an excellent fusion strategy, ensemble learning can combine the strengths of each model and get a better result comparing any single model without ensemble learning. We adopt a probabilistic ensemble method to combine the independent basic models in our work. Assume that we have an object with a label $y$ and two outputs of the basic models $x_1$ and $x_2$ ( it can easily be expanded to more outputs ). As Bagging mentioned above, the basic models are independent, so the measurements are also conditionally independent, which can be formulated as

$$p(x_1, x_2|y) = p(x_1|y)p(x_2|y). \tag{24}$$

This is also can be expressed as $p(x_1|y) = p(x_1|x_2, y)$ as the independence between $x_1$ and $x_2$ exists, which means that the $x_2$ will not be changed if we give the value of $y$. Our purpose is to get the value of $y$, which can be expressed as

$$p(x_1, x_2 \mid y) = \frac{p(x_1 \mid x_2, y)p(y)}{p(x_1, x_2)} \propto p(x_1 \mid x_2, y)p(y). \tag{25}$$

As the independence mentioned above, the probabilistic relation can be written as

$$p(y \mid x_1, x_2) \propto p(x_1 \mid y)p(x_2 \mid y)p(y) \propto \frac{p(x_1 \mid y)p(y)p(x_2 \mid y)p(y)}{p(y)} \propto \frac{p(y \mid x_1)p(y \mid x_2)}{p(y)}. \tag{26}$$

Utilizing the probabilistic relation, we can calculate the score of $y$. Given the existence of conditional independence, it can be considered the optimal fusion scheme. The calculation can be formulated as

$$p(y \mid \{x_i\}_{i=1}^{M}) \propto \frac{\prod_{i=1}^{M} p(y|x_i)}{p(y)^{M-1}}. \tag{27}$$

The class prior $p(y)$ can be easily obtained by taking the statistics for $y$ from the dataset. Then, according to (27), the results of all basic models can be fused.

## 3. Experiment

### *3.1. Data Acquisition System and Dataset*

Figure 7 shows the data acquisition system named MS100 produced by South Surveying & Mapping Technology Co., Ltd. (Guangzhou, China). It can automatically move and scan with a motor at a speed of 1 km/h in disease-scanning mode. The experiments are performed on the corroded bolt dataset collected from a certain Beijing metro tunnel in service. The dataset consists of 1441 pictures in the size 640 × 640. All the targets are labeled with a 100 × 100 ground truth box. We split the dataset into the training set and the test set in a ratio of 8:2, with the test set also serving as the validation set.

**Figure 7.** MS100 3D scanner.

### 3.2. Experiment Settings

Experiments in the study have been implemented on an Intel® Core™ i7-11700K CPU (3.6 GHz, 32 GB RAM) and an NVIDIA GeForce RTX 3060 (CUDA version 11.6) with Python 3.9.12 (PyTorch 1.11.0) in 64 Bit Ubuntu 18.04.1 Long Term Support operating system.

To train the module properly, we set the input resolution to 640 × 640 and use Stochastic Gradient Descent (SGD) with 0.9 momenta as the optimizer. The learning rate is initialized to 0.001 and the cosine decay with warm-up is selected as the learning rate schedule. All models have been trained completely in the experiments.

As for data augmentation, we set the image rotation rate to 0.5 and the image translation rate to 0.1. Both the image scale rate and image shear rate are set to 0.5. We mainly used Mosaic to further enhance the performance of the detector, and the Mosaic rate is set to 1.0.

### 3.3. Evaluation Metrics

Taking the popular assessment in the CV detection field as a reference, the performance is evaluated by the average precision, recall rate, precision rate, and *F*1 score. We determined the predicted box as positive based on a common metric where the Intersection over Union ($IoU$) between the predicted box and the ground truth box is greater than 0.5. The definition of the targets are

$$Recall = \frac{X_{TP}}{X_{TP} + X_{FN}}, \tag{28}$$

$$Precision = \frac{X_{TP}}{X_{TP} + X_{FP}}, \tag{29}$$

$$\text{F1 score} = \frac{2 \times Recall \times Precision}{(Recall + Precision)}, \tag{30}$$

where $Recall$ and $Precision$ represent the recall and precision rate, respectively. $X_{TP}$ denotes the number of objects correctly identified as true. $X_{FP}$ denotes the number of misidentifications of false targets. $X_{FN}$ represents the number of objects that fail to be correctly detected. $F1$ score can be regarded as a weighted average of recall rate and precision rate to evaluate the model comprehensively. The engineering problem pays more attention to the $F1$ score. From the perspective of recall rate and precision rate, the experiments utilize AP to test the detection accuracy of our method.

### 3.4. Performance Comparisons

Table 1 shows the comparative results on the test set between our method and some state-of-the-art detection approaches. Faster-RCNN is a two-stage CNN-based object detector, which is a widely used non-end-to-end detection method [28]. YOLOv5n is a fast and powerful end-to-end detector and YOLOv5s denotes a larger size of YOLOv5n. YOLOv5n6 adds a detection head to YOLOv5n, which can have a larger focus scale on targets. Experiments of different color enhancement algorithms, detection structures, and YOLOs are fully taken into consideration in performance comparison.

**Table 1.** The experimental results.

| Method | *Precision* | *Recall* | F1 Score | AP@0.5 | AP@0.5:0.95 |
|---|---|---|---|---|---|
| Faster-RCNN | 0.690 | 0.917 | 0.790 | 0.832 | 0.316 |
| YOLOv5s | 0.876 | 0.950 | 0.912 | 0.957 | 0.509 |
| YOLOv5n6 | 0.883 | 0.927 | 0.904 | 0.945 | 0.506 |
| YOLOv5n | 0.889 | 0.974 | 0.930 | 0.970 | 0.525 |
| YOLOv5n-EL (baseline) | 0.895 | 0.974 | 0.938 | 0.970 | 0.530 |
| MSR + YOLOv5s | 0.886 | 0.924 | 0.905 | 0.959 | 0.510 |
| MSR + YOLOv5n6 | 0.858 | 0.948 | 0.901 | 0.952 | 0.495 |
| MSR + YOLOv5n | 0.864 | 0.969 | 0.913 | 0.961 | 0.506 |
| MSR + YOLOv5n-EL | 0.880 | 0.970 | 0.922 | 0.965 | 0.515 |
| MSRCR + YOLOv5s | 0.877 | 0.933 | 0.904 | 0.961 | 0.509 |
| MSRCR + YOLOv5n6 | 0.889 | 0.962 | 0.924 | 0.970 | 0.514 |
| MSRCR + YOLOv5n | 0.912 | 0.970 | 0.940 | 0.966 | 0.533 |
| MSRCR + YOLOv5n-EL | 0.917 | 0.970 | 0.943 | 0.972 | 0.534 |
| IMSRCR + YOLOv5s | 0.881 | 0.933 | 0.906 | 0.965 | 0.512 |
| IMSRCR + YOLOv5n6 | 0.915 | 0.972 | 0.943 | 0.972 | 0.520 |
| IMSRCR + YOLOv5n | 0.914 | 0.970 | 0.941 | 0.971 | 0.535 |
| IMSRCR + YOLOv5n-EL | **0.921** | **0.975** | **0.947** | **0.975** | **0.537** |

As shown in Table 1, compared with Faster-RCNN, YOLOv5s, and YOLOv5n, the $F1$ score of YOLOv5n-EL has been enhanced by 0.148, 0.026, and 0.008, respectively. From the perspective of AP, YOLOv5n-EL achieves 0.970 AP@0.5 and 0.530 AP@0.5:0.95, which is the best of Faster-RCNN (0.832 AP@0.5 and 0.316 AP@0.5:0.95), YOLOv5s (0.957 AP@0.5 and 0.509 AP@0.5:0.95) and YOLOv5n (0.969 AP@0.5 and 0.525 AP@0.5:0.95). This illustrates the advantage of YOLOv5n-EL as a corrosion bolt detector. In this problem, the corroded bolt is the target of fixed scale, and the detection head on a larger scale may cause redundancy of features.Therefore, YOLOv5n6 failed to improve the detection performance. Meanwhile, YOLOv5n6 (0.945 AP@0.5 and 0.506 AP@0.5:0.95), which own a larger size of parameters, get a lower AP than YOLOv5n-EL. The detection time consumption of contrastive models is shown in Table 2. It is clear that the Faster-RCNN costs nearly 10 times longer than the YOLOs in experiments caused by the non-end-to-end structure. Because the features of corroded bolts in the dataset are relatively simple, the model with large parameters may be more prone to overfitting in training. In this detection task, YOLOv5n-EL not only can avoid overfitting but also achieves better performance without wasting too much time (only 7 ms more than YOLOv5n, far less than the consumption of color enhancement). Besides, the FLOPs cost of YOLOv5n-EL is still lower than YOLOv5s, while the results are significantly better. The above analysis shows the correctness of choosing YOLOv5n-EL as the detector.

**Table 2.** The Time and FLOPs Consumption of Detectors.

| | Faster-RCNN | YOLOv5s | YOLOv5n | YOLOv5n-EL |
|---|---|---|---|---|
| FLOPs (G) | \ | 15.8 | 4.1 | 12.3 |
| Pre-process Time (ms) | \ | 0.677 | 0.614 | 0.643 |
| Inference Time (ms) | \ | 9.571 | 5.041 | 12.716 |
| NMS Time (ms) | \ | 1.475 | 1.405 | 1.008 |
| Total Time (ms) | 83.990 | 11.723 | 7.060 | 14.367 |

With the color feature enhancement module, Table 1 also shows that MSR makes the detection performance of YOLOv5s and YOLOv5n-EL even worse instead of the enhancement. That is due to MSR causing some color distortion, which makes the data processed deviate from real data distribution. However, we notice that MSR lightly improves the detection performance of YOLOv5n6 and YOLOv5n, which illustrates that, with MSR, the overfitting caused by more parameters is somewhat relieved.

We also compare the results of MSRCR and IMSRCR to evaluate the performance further. It can be seen from Table 1 that, compared with MSRCR, IMRCR enhances the performance of detectors. YOLOv5n-EL achieves 0.975 mAP@0.5 and 0.537 mAP@0.5:0.95 with IMSRCR. IMSRCR effectively enhances the darker areas in the image and improves the intensity of the target edge, which offers more help to the detector. This illustrates the effectiveness of the IMSRCR method. We show the effects of different color enhancement algorithms in Figure 8. In contrast, although MSR and MSRCR can also enhance the color features of the corroded parts, color distortion may occur on other occasions, and the edge is not clear in a dim environment. The IMSRCR can not only strengthen the features significantly but also avoid obscurity in a dim environment, which leads to an improvement in comprehensive detection effectiveness.

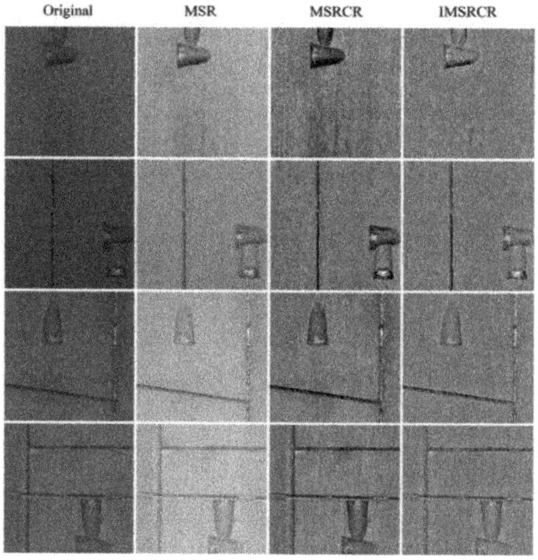


**Figure 8.** Examples of results from different color enhancement algorithm.

Furthermore, Table 3 shows the time consumption of different color enhancement methods. Since MSRCR uses Gaussian blur, the enhancement speed is significantly slowed down to undertake many numerical calculations. IMSRCR, however, avoids the shortcomings, and the speed increases by about a quarter.

**Table 3.** The Time Consumption of Color Enhancement.

| | MSR | MSRCR | IMSRCR |
|---|---|---|---|
| Time Cost (ms) | 47.576 | 93.308 | 69.870 |

Figures 9–12 show the visualization of some representative detection results in the test set. Compared to other YOLO detectors and the baseline YOLOv5n, YOLOv5n-EL offers better performance in detection like Figure 9b,c. Comparison between Figures 9–11 shows that different color enhancement algorithms can heighten the significance of features, changing the effect of the total model.

In summary, the experimental results indicate that YOLOv5n-EL is an efficient corroded bolt target detector. In addition, the ablation study demonstrates that the IMSRCR is helpful for the enhancement of the color features and improves the detection performance for corroded bolts both in speed and accuracy.

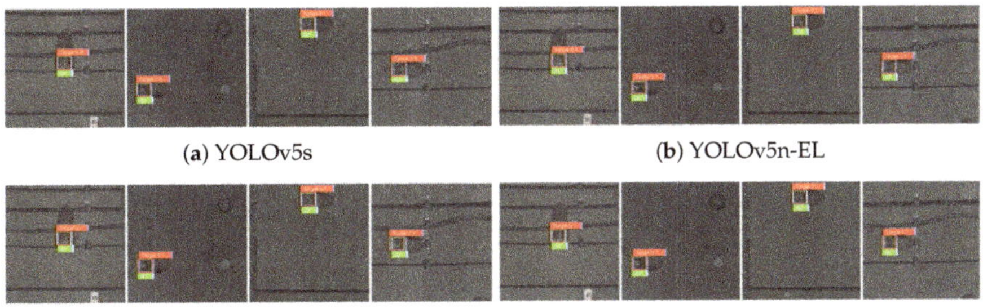

(**a**) YOLOv5s (**b**) YOLOv5n-EL
(**c**) YOLOv5n (**d**) YOLOv5n6

**Figure 9.** Examples of detection results without color enhancement.

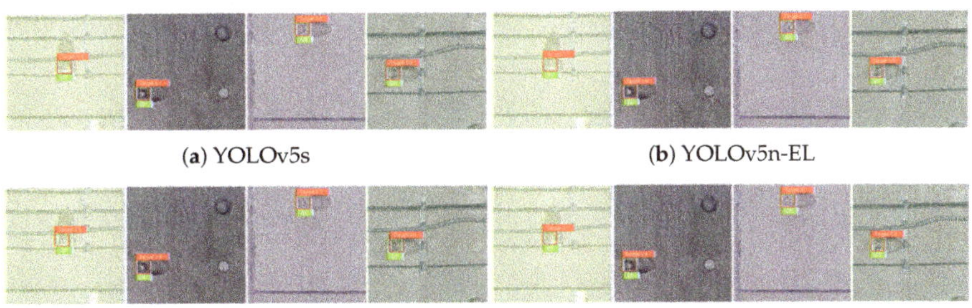

(**a**) YOLOv5s (**b**) YOLOv5n-EL
(**c**) YOLOv5n (**d**) YOLOv5n6

**Figure 10.** Examples of detection results with MSR color enhancement.

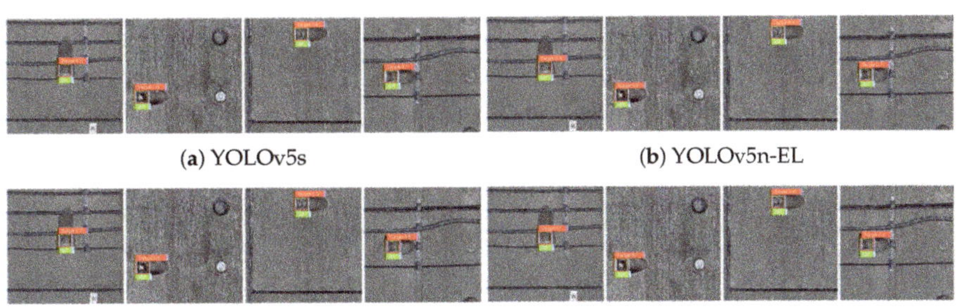

(**a**) YOLOv5s (**b**) YOLOv5n-EL
(**c**) YOLOv5n (**d**) YOLOv5n6

**Figure 11.** Examples of detection results with MSRCR color enhancement.

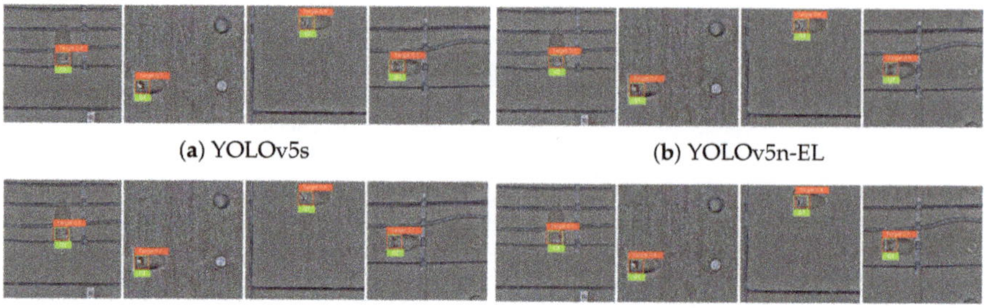

(**a**) YOLOv5s (**b**) YOLOv5n-EL

(**c**) YOLOv5n (**d**) YOLOv5n6

**Figure 12.** Examples of detection results with IMSRCR color enhancement.

## 4. Discussion

The method composed in this paper is a corroded bolt detection model, which combines the IMSRCR module and YOLOv5n-EL into one algorithm. The experimental results on the test set demonstrate that our method has good detection performance for corroded bolts. The parameter size of the YOLOv5n-EL basic model (YOLOv5n) is only about 14 MB, which is suitable for project deployment and real-time detection. Our method outperforms other comparative methods in both accuracy and speed. The color feature enhancement made by IMSRCR is helpful for the detector to detect corroded bolts with inconspicuous corrosion features.

## 5. Conclusions

In this paper, a method was put forward for tunnel corroded bolt detection. For this purpose, an efficient CV module with color enhancement and ensemble learning is proposed. Considering the low definition, poor contrast, and color distortion in the tunnel, IMSRCR enhances the color and edge appearance based on auto-matched dynamic filters and $L_0$ regularization. Moreover, YOLOv5n-EL also directly improves the accuracy of detection. To examine the effectiveness of our model, we collect corroded bolts with a professional tunnel scanner from a practical railway tunnel. It achieves a precision of 0.921 and a recall of 0.975 within 84.237 ms (14.367 + 69.870), which confirms that the IMSRCR + YOLOv5n-EL is the most suitable structure for the task.

**Author Contributions:** Data curation, L.T.; Investigation, L.T. and T.T.; Methodology, L.T. and D.Y.; Supervision, L.T.; Validation, L.T. and D.Y.; Writing-original draft, L.T. and T.T.; Writing-review and editing, L.T., T.T. and D.Y. All authors have read and agreed to the published verision of the manuscript.

**Funding:** This research was funded by Financial Project of Beijing Municipal Engineering Research Institute (YCZ202208B003).

**Institutional Review Board Statement:** Not applicable.

**Informed Consent Statement:** Not applicable.

**Data Availability Statement:** Not applicable.

**Conflicts of Interest:** The authors declare no conflict of interest.

## References

1. Hu, X.; Cao, Y.; Sun, Y.; Tang, T. Railway Automatic Switch Stationary Contacts Wear Detection Under Few-Shot Occasions. *IEEE Trans. Intell. Transp. Syst.* **2022**, *23*, 14893–14907. [CrossRef]
2. Hu, X.; Cao, Y.; Tang, T.; Sun, Y. Data-driven technology of fault diagnosis in railway point machines: Review and challenges. *Transp. Saf. Environ.* **2022**, *4*, tdac036. [CrossRef]
3. Wen, T.; Xie, G.; Cao, Y.; Cai, B. A DNN-Based Channel Model for Network Planning in Train Control Systems. *IEEE Trans. Intell. Transp. Syst.* **2022**, *23*, 2392–2399. [CrossRef]

4. Nikravesh, S.M.Y.; Goudarzi, M. A review paper on looseness detection methods in bolted structures. *Lat. Am. J. Solids Struct.* **2017**, *14*, 2153–2176. [CrossRef]
5. Reddy, M.S.B.; Ponnamma, D.; Sadasivuni, K.K.; Aich, S.; Kailasa, S.; Parangusan, H.; Ibrahim, M.; Eldeib, S.; Shehata, O.; Ismail, M.; et al. Sensors in advancing the capabilities of corrosion detection: A review. *Sens. Actuators A Phys.* **2021**, *332*, 113086. [CrossRef]
6. Xu, Y.; Li, D.; Xie, Q.; Wu, Q.; Wang, J. Automatic defect detection and segmentation of tunnel surface using modified Mask R-CNN. *Measurement* **2021**, *178*, 109316. [CrossRef]
7. Fan, Z.; Song, Z.; Xu, J.; Wang, Z.; Wu, K.; Liu, H.; He, J. Object Level Depth Reconstruction for Category Level 6D Object Pose Estimation From Monocular RGB Image. *arXiv* **2022**, arXiv:2204.01586.
8. Fan, Z.; Liu, H.; He, J.; Jiang, S.; Du, X. PointFPN: A Frustum-based Feature Pyramid Network for 3D Object Detection. In Proceedings of the 2020 IEEE 32nd International Conference on Tools with Artificial Intelligence (ICTAI), Baltimore, MD, USA, 9–11 November 2020; IEEE: Piscataway, NJ, USA, 2020; pp. 1129–1136. [CrossRef]
9. Zhang, X.; Sheng, Z.; Shen, H.L. FocusNet: Classifying better by focusing on confusing classes. *Pattern Recognit.* **2022**, *129*, 108709. [CrossRef]
10. Cao, S.Y.; Hu, J.; Sheng, Z.; Shen, H.L. Iterative Deep Homography Estimation. In Proceedings of the IEEE/CVF Conference on Computer Vision and Pattern Recognition (CVPR), New Orleans, LA, USA, 19–20 June 2022; pp. 1879–1888. [CrossRef]
11. Chang, S.; Zhang, R.; Ji, K.; Huang, S.; Feng, Z. A Hierarchical Classification Head based Convolutional Gated Deep Neural Network for Automatic Modulation Classification. *IEEE Trans. Wirel. Commun.* **2022**, *21*, 8713–8728. [CrossRef]
12. Cha, Y.J.; Choi, W.; Suh, G.; Mahmoudkhani, S.; Büyüköztürk, O. Autonomous structural visual inspection using region-based deep learning for detecting multiple damage types. *Comput.-Aided Civ. Infrastruct. Eng.* **2018**, *33*, 731–747. [CrossRef]
13. Ta, Q.B.; Huynh, T.C.; Pham, Q.Q.; Kim, J.T. Corroded Bolt Identification Using Mask Region-Based Deep Learning Trained on Synthesized Data. *Sensors* **2022**, *22*, 3340. [CrossRef] [PubMed]
14. Suh, G.; Cha, Y.J. Deep faster R-CNN-based automated detection and localization of multiple types of damage. In Proceedings of the Sensors and Smart Structures Technologies for Civil, Mechanical, and Aerospace Systems, Denver, CO, USA, 5–8 March 2018; SPIE: Bellingham, WA, USA, 2018; Volume 10598, pp. 197–204. [CrossRef]
15. Redmon, J.; Divvala, S.; Girshick, R.; Farhadi, A. You only look once: Unified, real-time object detection. In Proceedings of the IEEE Conference on Computer Vision and Pattern Recognition, Las Vegas, NV, USA, 27–30 June 2016; pp. 779–788. [CrossRef]
16. Song, Y.; Zhang, H.; Liu, L.; Zhong, H. Rail surface defect detection method based on YOLOv3 deep learning networks. In Proceedings of the 2018 Chinese Automation Congress (CAC), Xi'an, China, 30 November–2 December 2018; IEEE: Piscataway, NJ, USA, 2018; pp. 1563–1568. [CrossRef]
17. Guo, K.; He, C.; Yang, M.; Wang, S. A pavement distresses identification method optimized for YOLOv5s. *Sci. Rep.* **2022**, *12*, 3542. [CrossRef] [PubMed]
18. Xu, R.; Lin, H.; Lu, K.; Cao, L.; Liu, Y. A forest fire detection system based on ensemble learning. *Forests* **2021**, *12*, 217. [CrossRef]
19. Forkan, A.R.M.; Kang, Y.B.; Jayaraman, P.P.; Liao, K.; Kaul, R.; Morgan, G.; Ranjan, R.; Sinha, S. CorrDetector: A framework for structural corrosion detection from drone images using ensemble deep learning. *Expert Syst. Appl.* **2022**, *193*, 116461. [CrossRef]
20. Seijo-Pardo, B.; Porto-Díaz, I.; Bolón-Canedo, V.; Alonso-Betanzos, A. Ensemble feature selection: Homogeneous and heterogeneous approaches. *Knowl.-Based Syst.* **2017**, *118*, 124–139. [CrossRef]
21. Zhang, W.; Dong, L.; Xu, W. Retinex-inspired color correction and detail preserved fusion for underwater image enhancement. *Comput. Electron. Agric.* **2022**, *192*, 106585. [CrossRef]
22. Liu, R.; Ma, L.; Zhang, J.; Fan, X.; Luo, Z. Retinex-inspired unrolling with cooperative prior architecture search for low-light image enhancement. In Proceedings of the IEEE/CVF Conference on Computer Vision and Pattern Recognition, Nashville, TN, USA, 20–25 June 2021; pp. 10561–10570. [CrossRef]
23. Fan, M.; Wang, W.; Yang, W.; Liu, J. Integrating semantic segmentation and retinex model for low-light image enhancement. In Proceedings of the ACM International Conference on Multimedia, Seattle, WA, USA, 12–16 October 2020; pp. 2317–2325. [CrossRef]
24. He, K.; Sun, J.; Tang, X. Guided image filtering. *IEEE Trans. Pattern Anal. Mach. Intell.* **2012**, *35*, 1397–1409. [CrossRef] [PubMed]
25. Ochotorena, C.N.; Yamashita, Y. Anisotropic guided filtering. *IEEE Trans. Image Process.* **2019**, *29*, 1397–1412. [CrossRef]
26. Li, Z.; Song, X.; Chen, C.; Wang, C. Brightness level image enhancement algorithm based on retinex algorithm. *J. Data Acquisit. Process* **2019**, *2019*, 41–49.
27. Xu, L.; Lu, C.; Xu, Y.; Jia, J. Image smoothing via L 0 gradient minimization. In Proceedings of the 2011 SIGGRAPH Asia Conference, Hong Kong, China, 12–15 December 2011; pp. 1–12.
28. Ren, S.; He, K.; Girshick, R.; Sun, J. Faster r-cnn: Towards real-time object detection with region proposal networks. *Adv. Neural Inf. Process. Syst.* **2015**, *28*. [CrossRef] [PubMed]

*sensors*

MDPI

*Article*

# Virtual Axle Detector Based on Analysis of Bridge Acceleration Measurements by Fully Convolutional Network

**Steven Robert Lorenzen [1,*,†], Henrik Riedel [1,†], Maximilian Michael Rupp [1], Leon Schmeiser [1], Hagen Berthold [1], Andrei Firus [2] and Jens Schneider [1]**

1 Institute for Structural Mechanics and Design, Technical University of Darmstadt, 64287 Darmstadt, Germany
2 iSEA Tec GmbH, 88046 Friedrichshafen, Germany
* Correspondence: lorenzen@ismd.tu-darmstadt.de
† These authors contributed equally to this work.

**Abstract:** In the practical application of the Bridge Weigh-In-Motion (BWIM) methods, the position of the wheels or axles during the passage of a vehicle is a prerequisite in most cases. To avoid the use of conventional axle detectors and bridge type-specific methods, we propose a novel method for axle detection using accelerometers placed arbitrarily on a bridge. In order to develop a model that is as simple and comprehensible as possible, the axle detection task is implemented as a binary classification problem instead of a regression problem. The model is implemented as a Fully Convolutional Network to process signals in the form of Continuous Wavelet Transforms. This allows passages of any length to be processed in a single step with maximum efficiency while utilising multiple scales in a single evaluation. This allows our method to use acceleration signals from any location on the bridge structure and act as Virtual Axle Detectors (VADs) without being limited to specific structural types of bridges. To test the proposed method, we analysed 3787 train passages recorded on a steel trough railway bridge of a long-distance traffic line. Results of the measurement data show that our model detects 95% of the axles, which means that 128,599 out of 134,800 previously unseen axles were correctly detected. In total, 90% of the axles were detected with a maximum spatial error of 20 cm, at a maximum velocity of $v_{\max} = 56.3$ m/s. The analysis shows that our developed model can use accelerometers as VADs even under real operating conditions.

**Keywords:** moving load localisation; nothing-on-road; free-of-axle-detector; bridge weigh-in-motion; structural health monitoring; field validation; continuous wavelet transformation; machine learning; fully convolutional networks

**Citation:** Lorenzen, S.R.; Riedel, H.; Rupp, M.M.; Schmeiser, L.; Berthold, H.; Firus, A.; Schneider, J. Virtual Axle Detector Based on Analysis of Bridge Acceleration Measurements by Fully Convolutional Network. *Sensors* **2022**, 22, 8963. https://doi.org/10.3390/s22228963

Academic Editors: Abdollah Malekjafarian, Diogo Ribeiro, Araliya Mosleh and Maria D. Martínez-Rodrigo

Received: 25 October 2022
Accepted: 16 November 2022
Published: 19 November 2022



## 1. Introduction

All over the world, aging bridge infrastructure is facing the challenge of increasing traffic loads. For example, in the United States, there are more than 617,000 bridges, of which 42% are at least 50 years old and 7.5% are considered structurally deficient [1]. In Germany, more than 40% of the 25,710 railway bridges are older than 80 years, while the average lifespan is about 122 years [2,3]. The application of structural health monitoring (SHM) makes it possible to increase the operational availability and safety of these structures. Since knowledge of the actual operational loads is of high importance for the condition assessment of the structures, especially when it comes down to the assessment of fatigue failure and the evaluation of the remaining service life, the determination of the loads is a key aspect in the field of SHM. Since the direct measurement of loads is often technically difficult and usually requires significant financial resources [4–6], different methods for load identification based on measured structural responses have been developed [6–10]. In the case of bridges, these methods are referred to as Bridge Weigh-In-Motion (BWIM) [11–14].

For the majority of BWIM systems, information about the vehicle configuration (number of axles and axle spacing) and velocity is a prerequisite [14]. For this purpose, conventional axle detectors are used [5,15–17]. However, due to the impact loads of the wheels, axle detectors have a limited durability [18]. In addition, the installation of the axle detectors always implies road or railway track closures. The latter case especially requires a considerable amount of bureaucratic, logistic, and financial effort. To avoid these issues, modern BWIM systems use axle detection concepts that use only sensors installed under the bridge. These concepts are called nothing-on-road (NOR) or free-of-axle-detector (FAD) [19,20].

FAD technology uses two additional strain sensors at different positions on the bridge to determine vehicle configuration and speed [20]. Since FAD is only suitable for specific types of bridges [14], it was investigated whether the axle velocity and the axle spacing could be determined using global flexural strain or shear strain measurements [14,21]. For the proposed method in [21], which makes use of shear strains, the application of strain gauges at the level of the neutral axis is required. This is a challenge for complex structures, especially for railway bridges with ballasted tracks, because in such cases, the position of the neutral axis cannot be easily determined. However, the proposed method in [14] is only suitable for structures where the structural response is dominated by the quasi static response of the bridge, e.g., where the dynamic amplification is low. In addition, the second time derivative of the strain signals is used here; this makes the method sensitive to measurement noise, which entails the need for a suitable noise filter, depending on the application.

In [14], the method of virtual axles was proposed. It assumes a vehicle with many virtual axes, where all axles, except for the real ones, are weightless. The true axles and their weights are then determined by solving a constrained least square problem. As the authors stated, the method fails when there is significant noise in the signals. Since a significant amount of noise is present in field measurements and other practical applications, the method cannot be practically applied without a sophisticated regularisation method. Furthermore, the method uses experimentally determined lines of influence, so it is not applicable to cases with significant dynamic amplification in their structural response.

To the best of our knowledge, only Zhu et al. [22] have published a study on the accelerometer-based axle detection method so far. Here, a shallow Convolutional Neural Network (CNN) is used to detect potential axle sequences, which are then transformed with a continuous wavelet transform. Afterwards, the axles are detected in the transformed signals with the use of peak-finding methods. The method of Zhu et al. [22] requires accelerometers close to the supports. The acceleration signals of these sensors are dominated by the vehicle-induced impulses when entering and leaving the bridge, leading to the clear axle recognition in the time domain. In contrast to this method, our approach allows for the use of sensors at arbitrary positions, effectively reducing the number of sensors required by SHM applications.

An axle detection method based on acceleration measurements is desirable because the installation of acceleration sensors is much easier and less laborious compared to strain gauges. However, accelerometers are often already installed on the structure to determine the modal parameters and are not necessarily located close to the supports. Therefore, we propose a method that would enable accelerometers arbitrarily placed on a bridge to be used as VADs. In this way, the same acceleration sensors used for analysing the global structural behaviour (e.g., at midspan or quarter span of beam-like bridges) can also be employed in axle detection without having to install additional sensors in the proximity of the supports.

In the present work, Continuous Wavelet Transforms (CWTs) were used because they are generally an effective tool for analysing acoustic and visual signals [23]. In addition, previous work has shown that CWTs are an effective tool for axle identification [18,20–22,24]. The wavelet transformed signals are subsequently analysed using a Fully Convolutional Network (FCN) that is trained in a supervised manner to perform a binary classification task (Figure 1). As a result, the model outputs a pseudo-probability for each time step, whether a train axle is located above the sensor of the input signal at this time. A peak-finding algorithm is then used to classify the pseudo-probabilities into axle (peak) or no axle (no peak). This enables the processing of signals of arbitrary length without the need to divide them into time windows. Furthermore, analysis in this way is not limited to certain mother wavelets or specific scales, as in the previously mentioned work that used the CWT [18,20–22,24].

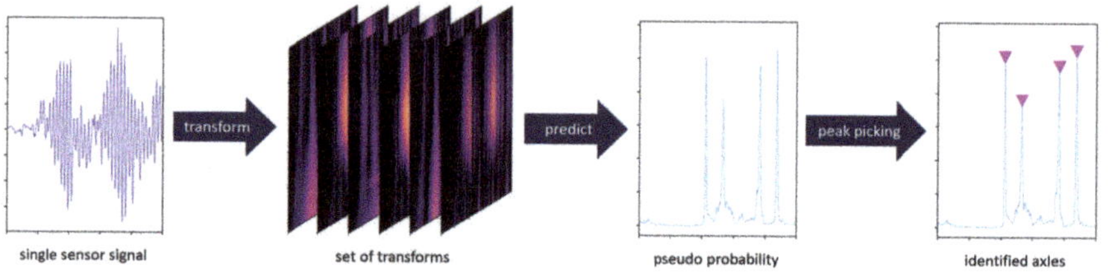


**Figure 1.** VAD process, from left to right: acceleration signals from a single sensor, set from different transformations of the signal, localization estimation as pseudo probabilities, and identified axles classified by a peak-finding algorithm. Signal section used is the same for each plot with horizontal axis in the samples.

To validate our method, we recorded a data set on a railway bridge with sensors distributed across the free span of the bridge on the main girders. The impulses of the wheel sets were superimposed with the vibrations of the bridge, which did not allow for their clear visual identification in the time domain. For many bridges and common sensor setups for monitoring purposes, similar to the ones used in this study, the method of Zhu et al. [22] would not be applicable. The VAD, on the other hand, can learn to distinguish the contribution of the structure-dependent natural vibration from the load-induced vibration and can thus be trained for sensors at any point on the bridge. This allows for the application of the method, independent of bridge type and accessibility to specific parts of the bridge.

This paper is structured as follows: In section two, the methods are presented. The first sub-section describes the data acquisition procedure in the field experiment and the subsequent data processing. The second sub-section contains the model definition. In the last sub-section, details on the training of the model are given. Section three presents and discusses the results. The paper ends with section four, in which the conclusions of the present study are drawn.

## 2. Methods

Since we opted to use a supervised learning approach for the VAD, a set of train passages with known axle distances and velocity was required to train the model. In the current research, this information was obtained from strain measurements at the rail level. For future practical applications, the information could be obtained from vehicles with a known axle configuration and through the use of a Differential Global Positioning System (DGPS). If such information is not available, a transfer learning approach based on simulated data could also be an option. In the application, the model can then identify the axles based on the transformed sensor signals. For this purpose, the model first gives a pseudo-probability for the presence of an axle at the longitudinal position of the sensor and for each time step of the analysed signal. In a subsequent step, a peak-finding algorithm is used to extract only the local maxima from the pseudo-probabilities. The extracted maxima represent the time points at which an axle is located above the sensor. As a result, the pseudo-probabilities are classified into axle (class 1) or no axle (class 0), without being limited to classical thresholds (Figure 1). A detailed description is given in the following subsections.

### 2.1. Data Acquisition

We recorded the measurement data used in the present study on a single-span steel trough railway bridge (Figure 2) located on a long-distance traffic line in Germany.

The single-track bridge with a ballasted superstructure was built in 1969 and has a total length of 18.4 m, with a free span of 16.4 m (Figure 3).

**Figure 2.** Photos of the investigated structure.

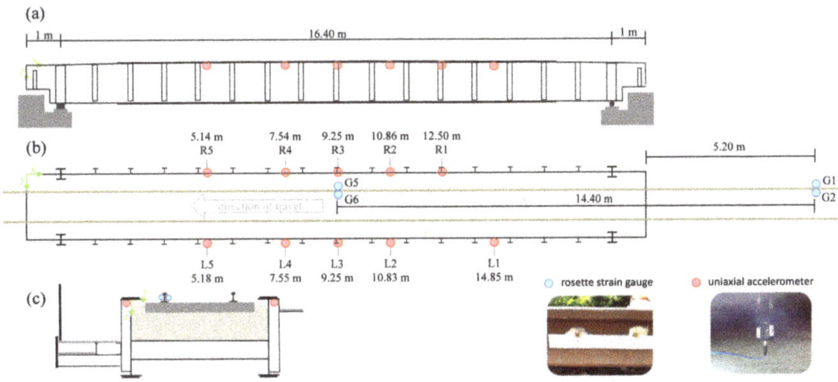


**Figure 3.** Bridge and sensor setup: (**a**) side view, (**b**) top view with sensor labels, accelerometer $x$-ordinates, and strain gauge distances, (**c**) cross section.

The measurement setup is shown in Figure 3. It can be seen that a total of ten seismic uniaxial accelerometers of the type PCB-39B04 (PCB Synotech GmbH, 41836 Hückelhoven, Germany ), with a sensitivity of 1000 mV/g ($\pm$10%), a broadband resolution of 0.000003 $g_{RMS}$, a measurement range of $\pm$5 $g_{pk}$, and a frequency range of 0.06 to 450 Hz ($\pm$5%) were installed. As previously mentioned, we chose an axle detection method based on acceleration measurements because the installation of accelerometers is much easier and less costly compared to strain gauges.

The measurements were triggered from the ring buffer via the rising slope of the wheel load measuring point G1 (Figure 3). The signal was recorded for 60 s, which started ten seconds before the trigger. All sensor signals were recorded with a sampling frequency of $f_s = 600$ Hz using the catmanAP software and the CX22 data recorder connected to an MX1601B universal amplifier and an MX1616B strain gauge amplifier (all products are from Hottinger Brüel & Kjaer GmbH, 64293 Darmstadt, Germany).

By means of two wheel load measuring points, the average velocity of each axle was determined, and from this, the actual position of the axles during the passage was deduced. Every measuring point involved the installation of at least one pair of rosette strain gauges (HBM 1-CXY41-6/350HE) on the rails. Each pair of strain gauges were placed at the level of the neutral axis of the UIC 60 rail profiles with a distance of 20 cm and allowed for the recording of bi-axial strains at an angle of 45° with respect to the neutral axis (Figure 4). Thus, shear strains were obtained. The difference of the shear strains allowed us to determine the acting wheel loads. For further details, please refer to [5]. To compensate for the influence of the lateral wheel loads, a pair of strain gauges was placed on each side of the rail, such that one wheel load measuring point retrieves two signals.

**Figure 4.** Top view of the bridge with the wheel load measuring points used at a distance of 14.4 m, with a detailed view of the rosette strain gauges and the weatherproof measuring point installed.

The peaks of the wheel load measurement signals were automatically identified (Figure 5a). All passages where the two wheel load measuring points had not detected the same number of peaks were discarded. This led to 3745 usable recorded passages out of a total of 3787, i.e., about 98.9%. Using the temporal differences of the peaks at the two measuring points and the known distance between the wheel load measurement points of 14.40 m, the mean velocity could be determined for each axle. The trains reached a maximum velocity of about 57 m/s (Figure 5b).

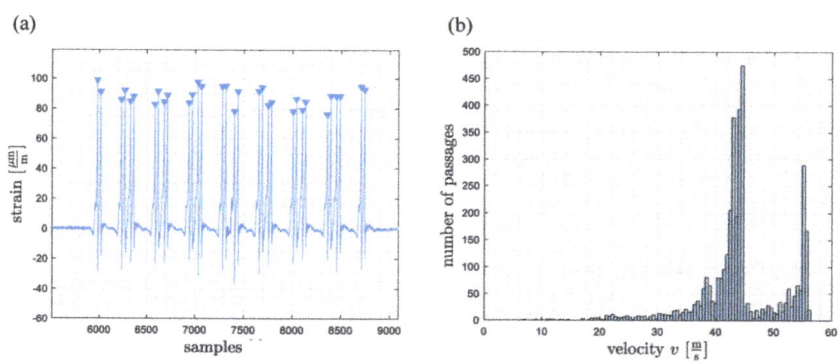


**Figure 5.** (**a**) Signal of the wheel load measuring point with detected peak values marked with blue triangles (**b**) Histogram of determined mean train velocities for all 3745 passages.

In the next step, by using the known distances—from the first wheel load measurement point to each of the ten accelerometers—and the mean velocity of each axle, the time at which the axle was at the same $x$-ordinate as the respective sensor could be calculated. Since the two strain gauges of one wheel load measuring point had a distance of 20 cm between them, the uncertainty with respect to the distance between the two wheel load measuring points $s_{\mathrm{WLM}} = 14.40$ m was assumed to be $\Delta s_{\mathrm{WLM}} = 0.2$ m. This propagated through the velocity determination. Together with an uncertainty in time of $\Delta t = \frac{1}{f_s} = \frac{1}{600}$ s, the absolute spatial error $\Delta x$ from the linear error propagation for each sensor (Figure 6) was calculated as follows:

$$\Delta x(v, \tilde{s}) = v\Delta t + \tilde{s}\left(|\frac{v}{s_{\mathrm{WLM}}}|\Delta t + |\frac{1}{s_{\mathrm{WLM}}}|\Delta s_{\mathrm{WLM}}\right) \quad (1)$$

This shows that the absolute position error is increased with increasing velocity and with the increasing distance of the sensor with respect to the first wheel load measurement point (G1/G2).

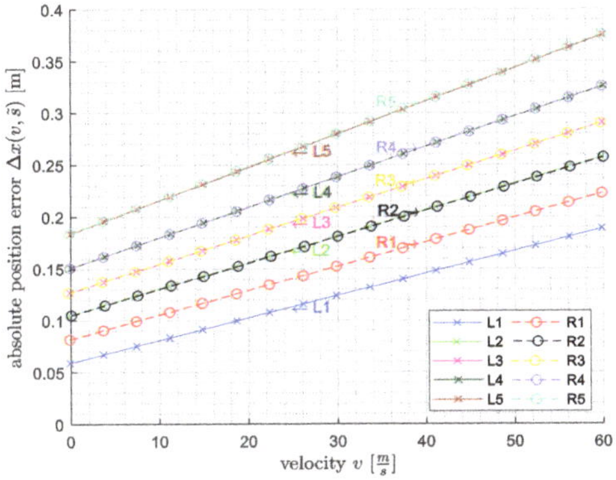


**Figure 6.** Absolute spatial error of the ground-truth per sensor as a function of velocity.

The acceleration signals were combined into one data matrix: $\mathbf{A}_{\mathrm{L}}^{36{,}000\times5}$ for the sensors L1–L5 and $\mathbf{A}_{\mathrm{R}}^{36{,}000\times5}$ for the sensors R1–R5 (Figure 3b), for each passage and without any further signal processing steps. Additionally, two data matrices, $\mathbf{L}_{\mathrm{L}}^{n_a\times5}$ and $\mathbf{L}_{\mathrm{R}}^{n_a\times5}$ ($n_a$: number of axles), containing the calculated indices at which an axle was at the respective sensor were created.

The complete data set as well as the processing code are available online [25].

### 2.2. Data Transformation

Transforming a signal into the frequency–time domain enables the localisation of frequency content in time [26]. In our case, low-frequency effects such as the bridge's natural vibration were separated from high-frequency effects such as measurement noise in the frequency domain, while the time domain was preserved. Therefore, the model could learn frequency-specific information, which should lead to faster training and more reliable results.

The most common choices for a frequency–time domain transformation are Short Time Fourier Transformation (STFT) and CWT. The multi-resolution approach of the CWT is particularly useful for complex signals since it adapts the window size to the frequency [27]. The STFT has a fixed resolution, which means that there is always a trade-off between a good time resolution and a good frequency resolution, depending on the window size [26]. As a result, we chose the CWT because it is more suitable for the analysis of acoustic and visual signals than the windowed Fourier transform [23]. The CWT has also been shown in previous work to be an effective tool for axle detection [18,20–22,24].

With respect to the signals, a section ranging from 150 samples before the first axle to 500 samples after the last axle was further processed and transformed with the PyWavelets package [28] using the determined settings (Table 1). Since the parameter space was too big to be tested on a large scale, the CWT were visualised and analysed for correlations between the axle positions (cyan dotted line) and the power of the transformed signal (Figure 7). As a result, we were able to find that within the range of the bridge's natural frequency of about 6.9 Hz for the first bending mode (Figure 7 left column), the influence of the bridge on the vibration was mainly visible, while a correlation between the train axles (dashed cyan lines) and the signal did not seem to be present. In the higher frequency range, a correlation became clearer, indicating that the influence of the axles were mainly located in the 64 Hz range (Figure 7 right column).

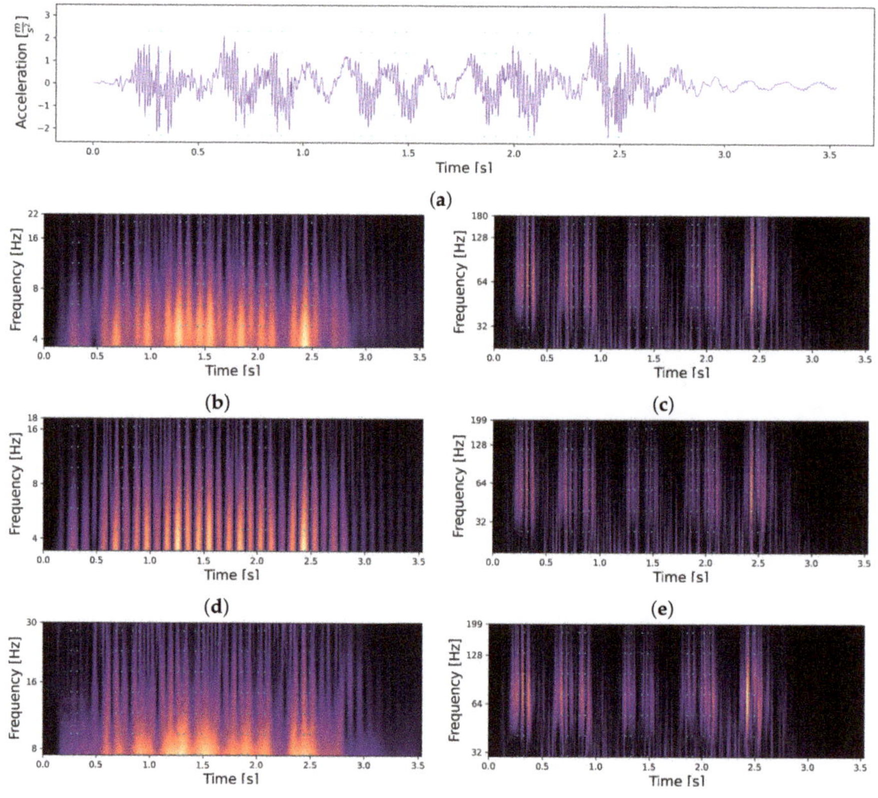


**Figure 7.** Set of continuous wavelet transformations (CWTs) for the signal obtained from sensor L2 for one of the train passages. The point in time when a load transition occurs is represented by a dashed line in cyan. Each of the transformations were independently normalised from 0 to 1 (visualised with black for 0 and yellow for 1). (**a**) Acceleration signal of a single train passage. (**b**) Complex Gaussian CWT in frequency range of bridge. (**c**) Complex Gaussian CWT in frequency range of axles. (**d**) Gaussian CWT in frequency range of bridge. (**e**) Gaussian CWT in frequency range of axles. (**f**) Frequency B-Spline CWT in frequency range of bridge. (**g**) Frequency B-Spline CWT in frequency range of axles.

**Table 1.** Continuous wavelet transformation settings.

| **Wavelet** | **Figure** | **Lower Scale Limit** | **Upper Scale Limit** |
|---|---|---|---|
| First Order Complex | Figure 7b | 1 | 8 |
| Gaussian Derivative | Figure 7c | 8 | 50 |
| First Order Gaussian | Figure 7d | 0.6 | 6.5 |
| Derivative | Figure 7e | 6.5 | 35 |
| Default Frequency | Figure 7f | 1.5 | 10 |
| B-Spline [28] | Figure 7g | 10 | 40 |

We assume that it is nevertheless advantageous for the model to receive both pieces of information (influence of the bridge and of the axles) in order to be able to distinguish them better.

As a result, all 6 transformations were used in combination (Figure 7b–g). To create the final model inputs, each signal (per passage and per sensor, shown in Figure 3) was transformed according to our 6 settings. Afterwards, the transformations were independently normalised and stacked into a three-dimensional array $\mathbf{T}^{n_s \times n_f \times n_t}$ ($n_s$: number of samples, $n_f$: number of frequencies/scales, and $n_t$: number of transformations). Thus, each sensor functioned independently as a VAD, which means that for our ten sensors, our method can locate each axle ten times (ten time points for ten sensor coordinates).

### 2.3. Model Definition

Our approach for the Virtual Axle Detector (VAD) aimed to always evaluate entire passes in one step so that sufficient context is preserved before and after each axle. Therefore, in our case, the 60 s recordings were always combined into complete passages before evaluation. To ensure that passages of arbitrary length could be efficiently processed, a model with a flexible input length (in the time domain) was essential. Hence, we developed an FCN [29], which only used input size-independent layers such as convolution, pooling, or batch normalization. Our model was developed to output only a single value, between 0 and 1, for the same number of samples as those of the input. These output values represent the model's certainty for an axle at the $x$-ordinate of the respective sensor.

Our developed VAD model was based on the U-Net architecture, originally proposed by Ronneberger et al. [30], which was developed for semantic segmentation tasks. Here, the goal was to classify each pixel of the input image individually in order to preserve the resolution from the input. For the U-Net, the resolution of the input was halved 4 times (via max pooling) in the encoder path and then doubled 4 times (via transposed convolution) in the decoder path. In addition, the intermediate results before each pooling layer were appended to the intermediate results after the transposed convolution layer with the same resolution, after which they were processed together.

In our case, not each pixel but each sample had to be classified, thus reducing the resolution in the frequency domain to 1. We achieved this by increasing the resolution in the decoder path only in the time dimension (Figure 8), for which we used a transposed convolution layer with a kernel size of $3 \times 1$. Before the intermediate results from the encoder path could be appended to the intermediate results from the decoder path, its resolution and number of feature maps were adapted. Each purple arrow in Figure 8 consists of a reshape layer to reduce the frequency domain to the value of 1, and a convolution layer with a kernel size of $1 \times 1$ to adapt the number of feature maps.

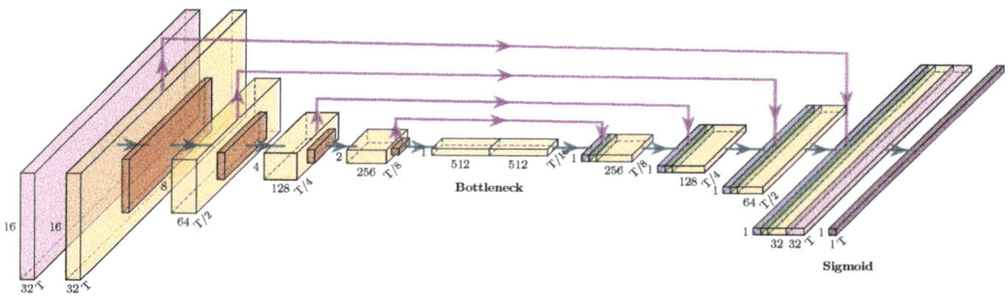


**Figure 8.** Definition of the Virtual Axle Detection model (VAD), with coloured boxes corresponding to the following layers: CB (light purple), RB (yellow), max pooling (red), concatenate (green), transposed convolution (blue), and reshaping skip connection (purple arrow). Dimensions of the output feature maps for the corresponding layer, with T samples at the bottom right, feature maps at the bottom, and frequencies at the left. The model dimensions for the input: 16 frequencies × 6 transforms × T samples; for the output: 1 × 1 pseudo-probabilities × T samples.

The convolution blocks (CBs) consist of a batch norm layer and a convolution layer with Rectified Linear Unit (ReLU) activation [31]. The CBs in Figure 8 have a 3 × 3 kernel size. The residual blocks (RB), originally proposed by He et al. [32], were implemented with 3 CBs in the filtering path and 1 CB in the skip connection. Here, the second CB in the filtering path had a 3 × 3 kernel size, while the other CBs had a 1 × 1 kernel size. The results of the filter path and the skip connection were added element-wise before further processing. Our model had 4 pooling steps as the U-Net [30]. We could therefore input transformed signals of any length (in the time domain) as long as they were divisible by 16; the resolution had to remain an integer after being halved 4 times. For lengths that were not multiples of 16, the signal was padded with zeros and thus extended by a maximum of 15 samples.

The last layer is a convolution layer consisting of a single kernel with the size of 3 × 3 and with sigmoid activation. Therefore, the resulting outputs could be interpreted as independent pseudo probabilities $p$, which indicate the predicted likeliness for a certain class per sample. The resulting model has an input size with an arbitrary number of samples (padded to a multiple of 16), an arbitrary number of signal transformations, and 16 frequencies, which were evenly spaced from the minimum to the maximum scale. The TensorFlow library [33] was used for the implementation of the model, and PlotNeuralNet [34] was used for its visualisation.

### 2.4. Loss Function

We defined the localisation task as a supervised classification problem instead of a regression problem in order to minimise complexity and maximise comprehensibility. We labelled each sample using one of the following classes: Axle at the same $x$-ordinate as the sensor (class 1) or not (class 0).

A common loss function for a binary classification task is Cross Entropy (CE), but for imbalanced data sets, Focal Loss (FL) has been shown to be more effective [35]. In our case, the total number of axles of a train is almost negligible compared to the total amount of samples of a passage. Hence, if the model predicts all values to be 0 (and cannot locate an axle), it will achieve an almost perfect loss for CE and will learn to ignore the axles. This brings us to the thesis that FL is necessary in order to achieve good results. The FL is defined as follows [35]:

$$\mathrm{FL}(p_t) = -1(1 - p_t)^{\gamma} \log(p_t), \tag{2}$$

where $p_t$ is defined as follows:

$$p_t = \begin{cases} p & \text{if } y = 1 \\ 1 - p & \text{otherwise.} \end{cases} \tag{3}$$

In the equation above, $p \in [0, 1]$ is the model's estimated probability for class 1, $y$ is the ground-truth class, and $\gamma$ is the focusing parameter. The equation of FL consists of $-\log(p_t)$, which is equal to the CE, and $(1 - p_t)$, which is a newly introduced modulating factor weighted by the focusing parameter $\gamma$. The larger the factor, the more significant the effect of the modulating factor is, and with a $\gamma$ of 0, the FL corresponds to the CE [35].

Due to the gamma value, the modulating factor was exponentially included in the equation. As a result, the loss became exponentially smaller, which gave a better prediction. For misclassified examples, the loss was unaffected compared to CE, which made misclassifications much more heavily weighted (a factor of 1000 and higher is possible [35]).

### 2.5. Evaluation Metrics

The loss function itself does not contain information about the number of correctly detected axles. Other metrics are needed to assess the overall performance of the VAD. Accuracy as a metric is also insufficient to draw a conclusion about the model's performance due to the imbalance in our data set. A prediction containing no axles at all would reach an accuracy of about 99% and would therefore not contain useful information. Precision and recall are suitable metrics for imbalanced data sets [31], but they only take into account binary results and not distance prediction and ground-truth. Due to the high sampling rate and the uncertainty of the labels described in Section 2, however, we wanted to recognise axle predictions within a few samples next to the ground-truth as correct and to measure the temporal error.

As already mentioned, the model output pseudo probabilities for each time step, whether an axle was on the same $x$ coordinate as the respective sensor of the input signal. However, these were continuous values that had to be converted into binary classes (0 for no axle and 1 for with axle). In addition, the model could not natively represent fuzziness and thus often output a large number of small spikes around the ground-truth axles. Thus, in order to obtain a definite point in time for the crossing of an axle over the sensor, despite the uncertainties of the model, the prediction had to be further processed (Figure 9). In order to keep only the meaningful peaks, they were classified using the find-peaks function from SciPy [36]. This function allowed for the use of additional logic for classifications that go beyond setting a threshold. For VAD, we fine-tuned the following parameters of the function: minimum height of the peak (0.25), minimum distance between two peaks (20 samples), and prominence of the peak compared to the surrounding points (0.15). We calculated the minimum distance $d$ between two peaks, with an assumed minimum wheel distance $\Delta w_{\text{min}} = 2\ \text{m}$ and the maximum velocity $v_{\text{max}} = 220\ \frac{\text{km}}{\text{h}}$, as follows:

$$d = \frac{\Delta w_{\text{min}} \cdot f_s}{v_{\text{max}}} \approx \frac{2\ \text{m} \cdot 600\ \frac{\text{samples}}{\text{s}}}{61.1\ \frac{\text{m}}{\text{s}}} \approx 20\ \text{samples} \tag{4}$$

A threshold was used to ensure that only predictions within a certain temporal error compared to the ground-truth would be considered correct. For example, the threshold could classify predicted axles as correct with a maximum temporal error of 30 milliseconds compared to the ground-truth. Depending on the application, its requirements may be decisive for the determination of the threshold. In general, it should be taken into account that good results cannot be expected with thresholds that are lower than the label and measurement accuracies. To avoid making assumptions that are too strict, we chose the largest reasonable threshold with 20 samples (Equation (4)) for the first evaluations. After having classified the peaks found as correct or incorrect, they were further evaluated using the following metrics: Precision, recall, and $F_1$ score. Precision describes the ratio of true positive predictions to positive predictions and thus allows for a statement regarding how many of the axles have been found. Recall describes the ratio of true positive predictions to false negative predictions and thus allows for a statement in relation to how many of the predicted axles are true. Since both of these metrics only describe a part of the problem, we used the $F_1$ score, which is the harmonic mean of precision and recall. Unlike the arithmetic mean, the harmonic mean strongly penalizes small values and thus ensures that axles are neither overlooked nor predicted arbitrarily often.

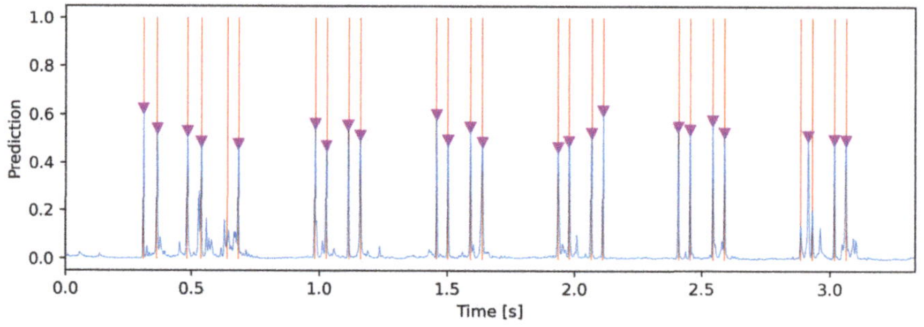


**Figure 9.** Exemplary output, with the model's output pseudo-probabilities represented by the blue lines, ground-truth by the red lines, and found peaks by the magenta triangles.

### 2.6. Optimization of $\gamma$

In order to find an optimal $\gamma$ value for the FL, we performed a parametric study with 150 epochs per run, 150 steps per epoch, and 16 samples per batch. We split the data set randomly, with 70% for training, 20% for validation, and 10% for testing. To ensure comparability, the same random state was used for every run. The selection criterion used for $\gamma$ was the $F_1$ score, because a high $F_1$ score indicates a high value for both recall and precision. The complete training logs and graphs for the determination of the $\gamma$ value are available online [37].

We thus confirmed our hypothesis that our data set is too unbalanced for standard loss functions such as Cross Entropy. The model training with small $\gamma$ values of 0 and 0.5 ended in dead ReLUs after 8 or 9 epochs and is therefore unusable. However, the modulation factor should also not be weighted too high to achieve the best performance. The relationship between $\gamma$, precision, and recall can be described as a trade-off between detecting too many axles and detecting too few axles (Figure 10).

The $\gamma$ values of 2, 2.5, and 3 achieved the highest $F_1$ score on the validation set. In order to decide which $\gamma$ value to use for the final evaluation, we trained the model with these $\gamma$ values in a second run for 300 epochs. In the second run, the $\gamma$ value of 2.5 achieved the highest $F_1$ score (Table 2) and was therefore kept for testing. Since the results of the $\gamma$ values were close to each other and the middle $\gamma$ value performed best, we assumed that the optimal value had been found. The complete training logs and graphs for the final models are available online [38] .

**Table 2.** The model's performance on the validation set, depending on the $\gamma$ value of FL with increased training length. Each of the precision and recall values was taken from the epoch with the highest $F_1$ value.

| $\gamma$ | $F_1$ | Precision | Recall |
|---|---|---|---|
| 3 | 0.9538 | 0.9477 | 0.9620 |
| 2.5 | 0.9544 | 0.9556 | 0.9542 |
| 2 | 0.9534 | 0.9559 | 0.9522 |

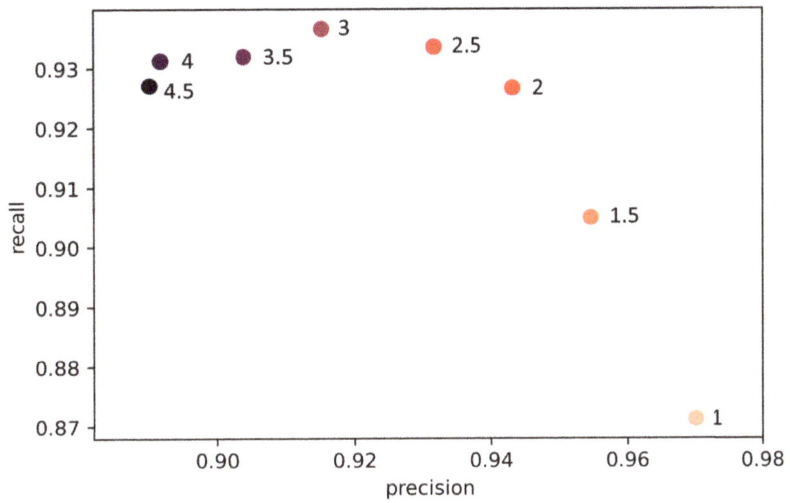


**Figure 10.** Relationship between $\gamma$, precision, and recall with median values of the training results on the validation set.

## 3. Results and Discussion

The test set consisted of 375 train passages with 13,480 axles in total. There were 10 acceleration sensors, for which the individual crossing times had to be determined, resulting in 134,800 times that had to be localised. On the test set, for a threshold of 20 samples, the VAD with a $\gamma$ value of 2.5 achieved an $F_1$ score of 0.938, a recall of 0.946, and a precision of 0.941. Thus, 126,449 out of 134,800 crossing times were localised correctly, with a maximum error of 0.033 s. On average, the predicted axle times had a temporal error of 1.16 samples (0.002 s) compared to the ground-truth, with a standard deviation of 3.06 samples (0.005 s).

Based on the distances between the sensors, we were able to convert the error from samples (temporal) into metres (spatial). In order to examine the spatial error more closely, we chose three threshold values:

- 200 cm as the minimum wheel distance;
- 37 cm as the maximum labelling error (Figure 6);
- 20 cm as the length of the wheel load measuring point.

The spatial errors for a threshold of 2 m were mostly at 0 cm, with an almost symmetrical distribution (Figure 11), thus indicating that there is no bias in the VAD. Most values were within a spatial error of 20 cm, and only a few values had an error higher than 25 cm. The maximum labelling error in the velocity range of most passages (30–60 $\frac{\mathrm{m}}{\mathrm{s}}$) was partially even above 25 cm (Figure 6).

We calculated the precision and recall per passage and sensor for each threshold in order to examine the distribution of the metrics in more detail, which resulted in 3750 values per threshold and metric (Figure 12). The differences in the results with thresholds of 20 cm and 37 cm were small, as even the 25% quantile stayed above 85% for both metrics (Figure 12). Precision and recall for a threshold of 200 cm were much better, with the 25% quantile remaining above 96%, while the mean spatial error greatly worsened by more than double the value compared to the other thresholds (Table 3). Therefore, we conclude that 37 cm is the optimal threshold value required to correctly evaluate the model's performance. In addition, we consider predictions with a spatial error above 37 cm as outliers. It should be possible for such outliers to be sorted out in post-processing by their comparison with known train configurations.

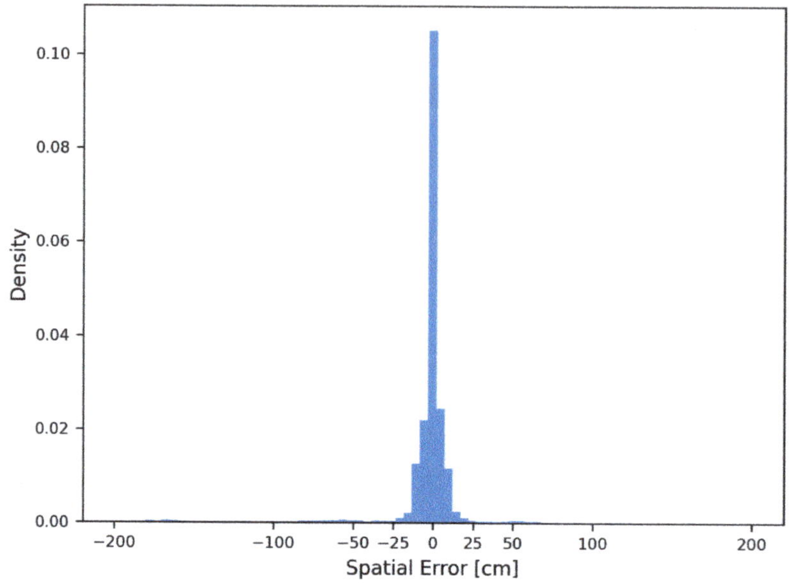


**Figure 11.** Differences between true and predicted axle positions for a threshold of 2 m.

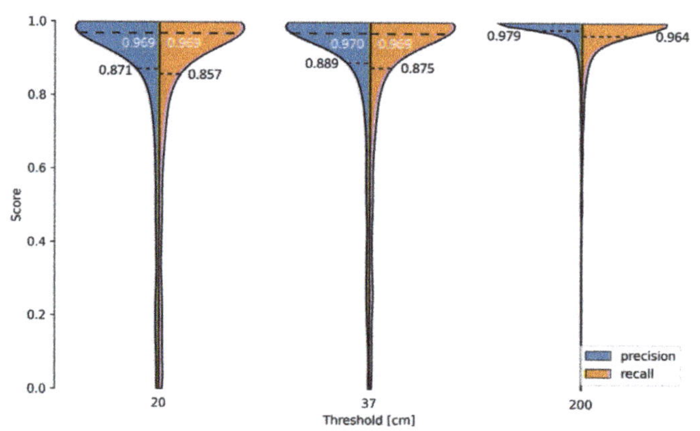


**Figure 12.** Precision and recall on test data set for different thresholds. Dotted lines with black text represent the 25% quantile, and dashed lines with white text represent the median, if not at 1.0.

**Table 3.** Influence of the threshold on mean spatial error, $F_1$, precision, and recall.

| Threshold (cm) | Mean (cm) | $F_1$ | Precision | Recall |
|---|---|---|---|---|
| 200 | 10.3 | 0.954 | 0.970 | 0.948 |
| 37 | 3.9 | 0.915 | 0.926 | 0.910 |
| 20 | 3.5 | 0.897 | 0.905 | 0.892 |

To investigate the influence of sensor placement, we determined the F1 score and the spatial accuracy per sensor. The spatial accuracy was calculated as follows:

$$spatial\ accuracy = 1 - \frac{spatial\ error}{200\ \text{cm}} \quad (5)$$

In this study, it was noticeable that sensor R3 performed significantly worse (Figure 13). In combination with a closer look at the measurement signal of R3, we came to the conclusion that the comparatively poor results were due to a degradation of the sensor. Since the remaining sensors performed comparably well, we concluded that the influence of the sensor placement was negligible.

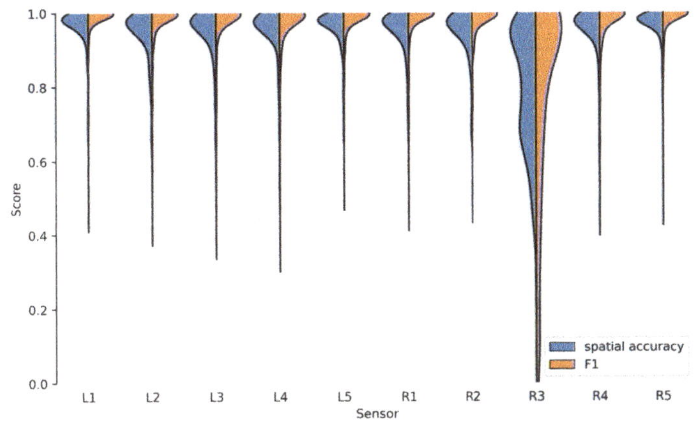


**Figure 13.** Spatial accuracy (Equation (5)) and $F_1$ score on test data set for each sensor (Figure 3).

The evaluation of the test data took 335 s for 375 passages and 10 sensors with the use of an NVIDIA RTX 3090. The model therefore needs 0.089 s per signal, and for our entire measurement setup, 0.89 s per passage. This would allow for the real-time application of the VAD and is a flexible trade-off between accuracy and computing speed due to the number of sensors used.

Compared to the work of Chatterjee et al. [24], who used FAD sensors and wavelets to detect more axles with the FAD, our model shows a comparable success rate in detecting axles. They were able to successfully evaluate 42/47 (about 89.4%) passages. The mean absolute spatial error was about 10.6 cm, which is about three times as much as that in our study. The achieved spatial accuracy in our study is still 1.4 times better compared to that obtained in a study using FAD sensors combined with an optimized mother wavelet and wavelet scale for the identification of axles [20]. Taking into account that we did not use FAD in our method, and that the velocities were about twice as high, this is a confirmation of our hypothesis that it is advantageous not to limit the analysis to certain mother wavelets and certain scales. In contrast to the method of Zhu et al. [22], due to our model architecture, the VAD can be applied at any point of the bridge. This allows for the use of common SHM measurement setups in axle detection without the need to attach additional sensors. The accuracies of the methods are similar. It should be noted that in all cases, the detection of car axles is compared with that of train axles.

## 4. Conclusions

We demonstrated that with our proposed method, no additional FADs or strain gauges on the main girders are required to realise a NOR-BWIM system. Instead, our method allows for accelerometers at any point of the structure to be used as VADs because our model can learn to account for the influence of the bridge structure and the sensor placement. As a result, the much more complex installation of strain gauges as well as track closures can be avoided.

We were also able to show that FCNs can detect axles using only acceleration measurements within a spatial accuracy of 37 cm, with a precision of 93% and a recall of 91%. The mean value of the absolute values of the spatial errors compared to the ground-truth here is about 3.9 cm. The results showed that the method can detect axles with spatial errors similar to the data used for labelling.

Even though our results show higher accuracy compared to other studies that used different methodologies, we assume that the accuracy of determining the vehicle configuration and velocity could be increased through the joint evaluation of several sensors, increased model complexity, improved signal transformation, or the use of different measured quantities such as strain and displacement. Enabling the method to be used with other measured quantities would also increase the amount of use cases.

Finally, the most important issue is the generalisability of the model. How efficiently the model can be used depends on whether it needs to be re-trained to be able to use the method, and if so, whether real or simulated data should be employed. Should retraining with real data be necessary, we propose to determine the axle position during the passages using vehicles with known axle configuration and DGPS.

**Author Contributions:** Conceptualization, S.R.L. and H.R.; data curation, S.R.L. and H.R.; formal analysis, M.M.R.; funding acquisition, S.R.L., H.B. and A.F.; investigation, S.R.L. and H.B.; methodology, H.R.; resources, J.S.; software, H.R., M.M.R. and L.S.; supervision, J.S.; validation, H.R.; visualization, S.R.L. and H.R.; writing—original draft, S.R.L., H.R. and M.M.R.; writing—review and editing, L.S., H.B., A.F. and J.S. All authors have read and agreed to the published version of the manuscript.

**Funding:** We acknowledge support from the Deutsche Forschungsgemeinschaft (DFG—German Research Foundation) and the Open Access Publishing Fund of the Technical University of Darmstadt. The research project ZEKISS (www.zekiss.de, accessed on 1 June 2022) was carried out in collaboration with the German railway company DB Netz AG, Wölfel Engineering GmbH, and GMG Ingenieurgesellschaft mbH. The project was funded by the mFund (mFund, 2020) and promoted by the Federal Ministry of Transport and Digital Infrastructure, grant number: 19F2123A. The research project DEEB-INFRA (www.deeb-infra.de, accessed on 1 June 2022) was carried out in collaboration with the sub-company DB Campus of Deutschen Bahn AG, AIT GmbH, Revotec zt GmbH, and iSEA Tec GmbH. It was funded by the mFund (mFund, 2020) and promoted by the Federal Ministry of Transport and Digital Infrastructure, grant number: 19F2139A.

**Data Availability Statement:** The data [25] as well as the source code [39] used in this paper are published and contain: (1) All measurement data; (2) Matlab code to label data and save as text files; (3) Python code for transformation, training, evaluation, and plotting. In addition, the training logs are also available for the determination of the gamma value [37] and for the final training [38].

**Conflicts of Interest:** The authors declare no conflict of interest. The funders had no role in the design of the study; in the collection, analyses, or interpretation of data; in the writing of the manuscript; or in the decision to publish the results.

## Abbreviations

The following abbreviations were used in this manuscript:

| | |
|---|---|
| BWIM | Bridge Weigh-In-Motion |
| CB | Convolution block |
| CE | Cross Entropy |
| CNN | Convolutional Neural Network |
| CWT | Continuous-Wavelet-Transformation |
| DGPS | Differential Global Positioning System |
| FAD | Free-of-axle-detector |
| FCN | Fully Convolutional Network |
| FL | Focal Loss |
| NOR | Nothing-on-road |

| | |
|---|---|
| RB | Residual block |
| ReLU | Rectified Linear Unit |
| SHM | Structural health monitoring |
| STFT | Short Time Fourier Transformation |
| VAD | Virtual Axle Detector |

## References


1. ASCE. Structurally Deficient Bridges | Bridge Infrastructure | ASCE's 2021 Infrastructure Report Card, 2021. Available online: https://infrastructurereportcard.org/cat-item/bridges-infrastructure/ (accessed on 28 June 2022).
2. Geißler, K. *Front Matter*; John Wiley & Sons, Ltd.: Hoboken, NJ, USA, 2014. [CrossRef]
3. Knapp, N. Brücken bei der Deutschen Bahn, 2019. Available online: https://www.deutschebahn.com/de/presse/suche_Medienpakete/medienpaket_bruecken-6854340 (accessed on 28 June 2022).
4. Chan, T.; Yu, L.; Law, S.; Yung, T. Moving Force Identification Studies, I: Theory. *J. Sound Vib.* **2001**, *247*, 59–76. [CrossRef]
5. Kouroussis, G.; Caucheteur, C.; Kinet, D.; Alexandrou, G.; Verlinden, O.; Moeyaert, V. Review of Trackside Monitoring Solutions: From Strain Gages to Optical Fibre Sensors. *Sensors* **2015**, *15*, 20115–20139. [CrossRef] [PubMed]
6. Firus, A.; Kemmler, R.; Berthold, H.; Lorenzen, S.; Schneider, J. A time domain method for reconstruction of pedestrian induced loads on vibrating structures. *Mech. Syst. Signal Process.* **2022**, *171*, 108887. [CrossRef]
7. Kazemi Amiri, A.; Bucher, C. A procedure for in situ wind load reconstruction from structural response only based on field testing data. *J. Wind. Eng. Ind. Aerodyn.* **2017**, *167*, 75–86. [CrossRef]
8. Hwang, J.; Kareem, A.; Kim, W. Estimation of modal loads using structural response. *J. Sound Vib.* **2009**, *326*, 522–539. [CrossRef]
9. Lourens, E.; Papadimitriou, C.; Gillijns, S.; Reynders, E.; De Roeck, G.; Lombaert, G. Joint input-response estimation for structural systems based on reduced-order models and vibration data from a limited number of sensors. *Mech. Syst. Signal Process.* **2012**, *29*, 310–327. [CrossRef]
10. Firus, A. A Contribution to Moving Force Identification in Bridge Dynamics. Ph.D. Thesis, Technische Universität, Darmstadt, Darmstadt, Germany, 2022. [CrossRef]
11. Lydon, M.; Robinson, D.; Taylor, S.E.; Amato, G.; Brien, E.J.O.; Uddin, N. Improved axle detection for bridge weigh-in-motion systems using fiber optic sensors. *J. Civ. Struct. Health Monit.* **2017**, *7*, 325–332. [CrossRef]
12. Wang, H.; Zhu, Q.; Li, J.; Mao, J.; Hu, S.; Zhao, X. Identification of moving train loads on railway bridge based on strain monitoring. *Smart Struct. Syst.* **2019**, *23*, 263–278. [CrossRef]
13. Yu, Y.; Cai, C.; Deng, L. State-of-the-art review on bridge weigh-in-motion technology. *Adv. Struct. Eng.* **2016**, *19*, 1514–1530. [CrossRef]
14. He, W.; Ling, T.; OBrien, E.J.; Deng, L. Virtual Axle Method for Bridge Weigh-in-Motion Systems Requiring No Axle Detector. *J. Bridge Eng.* **2019**, *24*, 04019086. [CrossRef]
15. Thater, G.; Chang, P.; Schelling, D.R.; Fu, C.C. Estimation of bridge static response and vehicle weights by frequency response analysis. *Can. J. Civ. Eng.* **1998**, *25*, 631–639. [CrossRef]
16. Zakharenko, M.; Frøseth, G.T.; Rönnquist, A. Train Classification Using a Weigh-in-Motion System and Associated Algorithms to Determine Fatigue Loads. *Sensors* **2022**, *22*, 1772. [CrossRef] [PubMed]
17. Bernas, M.; Płaczek, B.; Korski, W.; Loska, P.; Smyła, J.; Szymała, P. A Survey and Comparison of Low-Cost Sensing Technologies for Road Traffic Monitoring. *Sensors* **2018**, *18*, 3243. [CrossRef] [PubMed]
18. Yu, Y.; Cai, C.; Deng, L. Vehicle axle identification using wavelet analysis of bridge global responses. *J. Vib. Control.* **2017**, *23*, 2830–2840. [CrossRef]
19. O'Brien, E.J.; Hajializadeh, D.; Uddin, N.; Robinson, D.; Opitz, R. Strategies for Axle Detection in Bridge Weigh-in-Motion Systems. In Proceedings of the International Conference on Weigh-In-Motion, Dallas, TX, USA, 3–7 June 2012; pp. 79–88.
20. Zhao, H.; Tan, C.; OBrien, E.J.; Uddin, N.; Zhang, B. Wavelet-Based Optimum Identification of Vehicle Axles Using Bridge Measurements. *Appl. Sci.* **2020**, *10*, 7485. [CrossRef]
21. Kalhori, H.; Alamdari, M.M.; Zhu, X.; Samali, B.; Mustapha, S. Non-intrusive schemes for speed and axle identification in bridge-weigh-in-motion systems. *Meas. Sci. Technol.* **2017**, *28*, 025102. [CrossRef]
22. Zhu, Y.; Sekiya, H.; Okatani, T.; Yoshida, I.; Hirano, S. Acceleration-Based Deep Learning Method for Vehicle Monitoring. *IEEE Sensors J.* **2021**, *21*, 17154–17161. [CrossRef]
23. Daubechies, I. The wavelet transform, time-frequency localization and signal analysis. *IEEE Trans. Inf. Theory* **1990**, *36*, 961–1005. [CrossRef]
24. Chatterjee, P.; OBrien, E.; Li, Y.; González, A. Wavelet domain analysis for identification of vehicle axles from bridge measurements. *Comput. Struct.* **2006**, *84*, 1792–1801. [CrossRef]
25. Lorenzen, S.R.; Riedel, H.; Rupp, M.; Schmeiser, L.; Berthold, H.; Firus, A.; Schneider, J. Virtual Axle Detector based on Analysis of Bridge Acceleration Measurements by Fully Convolutional Network, *arXiv* **2022**, arXiv:2207.03758. https://doi.org/10.5281/zenodo.6782319.
26. Brunton, S.L.; Kutz, J.N. *Data-Driven Science and Engineering: Machine Learning, Dynamical Systems, and Control*; Cambridge University Press: Cambridge, UK, 2019. doi: 10.1017/9781108380690. [CrossRef]

27. Mallat, S. A theory for multiresolution signal decomposition: The wavelet representation. *IEEE Trans. Pattern Anal. Mach. Intell.* **1989**, *11*, 674–693. [CrossRef]
28. Lee, G.R.; Gommers, R.; Waselewski, F.; Wohlfahrt, K.; O'Leary, A. PyWavelets: A Python package for wavelet analysis. *J. Open Source Softw.* **2019**, *4*, 1237. [CrossRef]
29. Long, J.; Shelhamer, E.; Darrell, T. Fully convolutional networks for semantic segmentation. In Proceedings of the 2015 IEEE Conference on Computer Vision and Pattern Recognition (CVPR), Boston, MA, USA, 7–12 June 2015; pp. 3431–3440. [CrossRef]
30. Ronneberger, O.; Fischer, P.; Brox, T. U-Net: Convolutional Networks for Biomedical Image Segmentation. In Proceedings of the Medical Image Computing and Computer-Assisted Intervention—MICCAI 2015, Munich, Germany, 5–9 October 2015; Springer International Publishing: Cham, Switzerland, 2015; pp. 234–241. [CrossRef]
31. Géron, A. *Hands-On Machine Learning with Scikit-Learn, Keras, and TensorFlow: Concepts, Tools, and Techniques to Build Intelligent Systems*; O'Reilly UK Ltd.: Sebastopol, UK, 2019. ISBN: 9781492032649.
32. He, K.; Zhang, X.; Ren, S.; Sun, J. Deep Residual Learning for Image Recognition. In Proceedings of the 2016 IEEE Conference on Computer Vision and Pattern Recognition (CVPR), Las Vegas, NV, USA, 27–30 June 2016; pp. 770–778. [CrossRef]
33. Abadi, M.; Agarwal, A.; Barham, P.; Brevdo, E.; Chen, Z.; Citro, C.; Corrado, G.S.; Davis, A.; Dean, J.; Devin, M.; et al. TensorFlow: Large-Scale Machine Learning on Heterogeneous Systems, 2015. Available online: https://www.tensorflow.org/ (accessed on 11 August 2021).
34. Iqbal, H. *HarisIqbal88/PlotNeuralNet v1.0.0*; Zenodo: Geneve, Switzerland, 2018. [CrossRef]
35. Lin, T.Y.; Goyal, P.; Girshick, R.; He, K.; Dollár, P. Focal Loss for Dense Object Detection. In Proceedings of the 2017 IEEE International Conference on Computer Vision (ICCV), Venice, Italy, 22–29 October 2017; pp. 2999–3007. [CrossRef]
36. Virtanen, P.; Gommers, R.; Oliphant, T.E.; Haberland, M.; Reddy, T.; Cournapeau, D.; Burovski, E.; Peterson, P.; Weckesser, W.; Bright, J.; et al. SciPy 1.0: Fundamental Algorithms for Scientific Computing in Python. *Nat. Methods* **2020**, *17*, 261–272. [CrossRef] [PubMed]
37. Riedel, H. Training Logs for Determination of the Gamma Value. 2022. https://www.comet.com/imsdcomet/vader (accessed on 30 June 2022).
38. Riedel, H. Training Logs for the Final Models. 2022. https://www.comet.com/imsdcomet/vader2 (accessed on 30 June 2022).
39. Riedel, H.; Rupp, M. *VADer*; Zenodo: Geneva, Switzerland, 2022. [CrossRef]

*sensors*

MDPI

*Article*

# Characteristic Differences of Wind-Blown Sand Flow Field of Expressway Bridge and Subgrade and Their Implications on Expressway Design

**Shengbo Xie [1,*], Xian Zhang [1,2] and Yingjun Pang [3]**

[1] Key Laboratory of Desert and Desertification, Northwest Institute of Eco-Environment and Resources, Chinese Academy of Sciences, Lanzhou 730000, China; zhangxian@nieer.ac.cn
[2] University of Chinese Academy of Sciences, Beijing 100049, China
[3] Institute of Desertification Studies, Chinese Academy of Forestry, Beijing 100091, China; pangyingjun@caf.ac.cn
* Correspondence: xieshengbo@lzb.ac.cn

**Abstract:** Bridges and subgrades are the main route forms for expressways. The ideal form for passing through sandy areas remains unclear. This study aims to understand the differences in the influence of expressway bridges and subgrades on the near-surface blown sand environment and movement laws, such as the difference in wind speed and profile around the bridge and subgrade, the difference in wind flow-field characteristics, and the difference in sand transport rate, to provide a scientific basis for the selection of expressway route forms in sandy areas. Therefore, a wind tunnel test was carried out by making models of a highway bridge and subgrade and comparing the environmental effects of wind sand on them. The disturbance in the bridge to near-surface blown sand activities was less than that of the subgrade. The variation ranges of the wind speed of the bridge and its upwind and downwind directions were lower than those of the subgrade. However, the required distance to recover the wind speed downwind of the bridge was greater than that of the subgrade, resulting in the sand transport rate of the bridge being lower than that of the subgrade. The variation in the wind field of the subgrade was more drastic than that of the bridge, but the required distance to recover the wind field downwind of the bridge was greater than that of the subgrade. In the wind speed-weakening area upwind, the wind speed-weakening range and intensity of the bridge were smaller than those of the subgrade. In the wind speed-increasing area on the top of the model, the wind speed-increasing range and intensity of the bridge were smaller than those of the subgrade. In the wind-speed-weakening area downwind, the wind speed weakening range of the bridge was greater than that of the subgrade, and the wind speed-weakening intensity was smaller than that of the subgrade. This investigation has theoretical and practical significance for the selection of expressway route forms in sandy areas.

**Keywords:** expressway; bridge; subgrade; wind-blown sand flow field; sand transport

**Citation:** Xie, S.; Zhang, X.; Pang, Y. Characteristic Differences of Wind-Blown Sand Flow Field of Expressway Bridge and Subgrade and Their Implications on Expressway Design. *Sensors* **2022**, *22*, 3988. https://doi.org/10.3390/s22113988

Academic Editors: Araliya Mosleh, Diogo Ribeiro, Abdollah Malekjafarian, Maria D. Martínez-Rodrigo and Filippo Ubertini

Received: 22 March 2022
Accepted: 22 May 2022
Published: 24 May 2022



## 1. Introduction

Blown sand is the main natural disaster that threatens road driving safety in sandy areas [1,2]. In such areas, blown sand is also an important factor in highway engineering wiring, survey and design, route form selection, construction, operation, and maintenance [3]. Expressways have become a symbol of modern traffic because of their advantages, such as large transportation volume, low cost, fast speed, high traffic efficiency, flexible mobility, few traffic accidents, and a high degree of intensive use of land resources. In recent years, with the development of the social economy, the improvement of the urbanization level, and the progress of automobile industry technology, expressways have developed rapidly and have become the primary land transportation mode in sandy areas. Compared with ordinary highways, the hazards of wind-blown sand expressways have the following characteristics: First, the vehicle speed is very high, and a slight sand accumulation on the road surface causes traffic accidents. Second, the isolation belt in the middle of the expressway easily becomes an obstacle to wind-sand flow. To avoid sand accumulation there controlling the wind-sand flow on the road is necessary. Third, because the minimum width of expressway pavement is 26 m, its thickness is approximately three times that of ordinary highways, making the environmental impact of such expressways considerable. In sandy areas in particular, the disturbance of the blown sand activity is more intense after construction. Furthermore, the near-surface wind speed, wind field, and sand transport rate are significantly changed, as are the erosion, transport and accumulation conditions of wind-blown sand flow [4,5]. Therefore, preventing and controlling the harm caused by wind-blown sand by relying on the road surface to transport sand is difficult. Highway lines have three basic forms: subgrades, bridges, and tunnels. Among them, tunnels are less affected by blown sand [6], whereas the effects of the blown sand environment on subgrades and bridges are obvious [7,8]. These scientific problems need to be solved urgently for the construction of expressways in sandy areas. Determining the route form that is reasonable for passing through sandy areas to reduce, or even avoid, the harm of wind-blown sand is necessary. At present, the relevant studies mainly focus on the wind-blown sand hazards of traffic engineering and blown sand environmental monitoring. These focuses include wind-induced fatigue and asymmetric damage in bridges [9], buffeting response analysis of bridges [10], wind-blown sand along railway infrastructures and mitigation measures thereof [11], remote measurement of aeolian sand transport on sandy beaches and dunes [12], satellite monitoring of dust storms [13], the law of sand particle accumulation over railway subgrade [14], estimation methods and techniques of aeolian sand transport rate [15], sand dune ridge alignment effects on the surface [16], damage by wind-blown sand and its control measures along desert highways [17], wireless wind data acquisition systems at arid coastal foredunes [18], and wind speed forecasting in traffic control decision support systems [19]. However, these studies focus on the form of a single highway and railway line, and systematic studies are lacking on the environmental effects of blown sand in the form of subgrades and bridges. In particular, the optimization selection of the form of expressway lines in sandy areas has not yet been reported. This investigation selected the subgrade and bridge forms for a comparative study of the environmental effects of blown sand to understand the differences in the influence of expressway bridges and subgrades on the near-surface blown sand environment and movement laws, such as the difference in wind speed and profile around the bridge and subgrade, the difference in wind flow-field characteristics, and the difference in sand transport rate. The contribution of this investigation is to provide technical support for the survey and design of expressways in sandy areas and for the selection of route forms.

## 2. Research Methods

### 2.1. Models and Their Dimensions

The expressway pavement width is 26 m and the slope ratio is 1:1.75, following expressway technical standards. The subgrade and bridge models of the wind tunnel test were constructed at a 1:100 scale, based on previous studies [20,21]. The wind tunnel used for the test had a cross-section of 63 cm × 63 cm and a boundary layer thickness of 12–15 cm. Therefore, the bridge and subgrade models were both 8 cm in height and 62 cm in length, and their blockage ratios were 7.1% and 12.5%, respectively. The model sizes are shown in Figure 1.

**Figure 1.** Photos of model size and wind tunnel test of expressway bridge and subgrade.

### 2.2. Layout of Wind Tunnel Test

#### 2.2.1. Layout of Wind Speed Test in Wind Tunnel

A total of 10 observation points were used: −30H −25H, −20H, −15H, −10H, −5H, −3H, −1H, −0.5H, and −0H upwind of the bridge (H represents the model height, − represents upwind, + represents downwind, and −0H represents the slope foot of the bridge windward side). The observation point settings upwind of the subgrade were identical to those of the bridge. Five observation points were set under the bridge: the bottom of the windward side slope middle, bottom of the windward side slope shoulder, bridge bottom center, bottom of the leeward side slope shoulder, and bottom of the leeward side slope middle. Five observation points were set on the surface of the subgrade: the slope middle of the windward side, slope shoulder of the windward side, subgrade top center, slope shoulder of the leeward side, and slope middle of the leeward side. A total of 12 observation points were used: 0H (representing the slope foot of the bridge leeward side), 0.5H, 1H, 3H, 5H, 10H, 15H, 20H, 25H, 30H, 35H, and 40H downwind of the bridge. The observation point settings downwind of the subgrade were the same as those of the bridge (Figure 2). The wind profile of the wind tunnel test without a model is shown in Figure 3a. In total, 10 different heights of wind speed at each position were converted using pitot tubes that were placed at the bottom center of the wind tunnel. The measured heights were 0.6, 0.8, 1.3, 2.1, 8.3, 12.2, 16.4, 20.2, 24.2, and 28 cm. The wind speed was measured every 2 s 30 consecutive times, and the average value was obtained. The test wind speeds were set in five groups: 6, 9, 12, 15, and 18 m·s$^{-1}$.

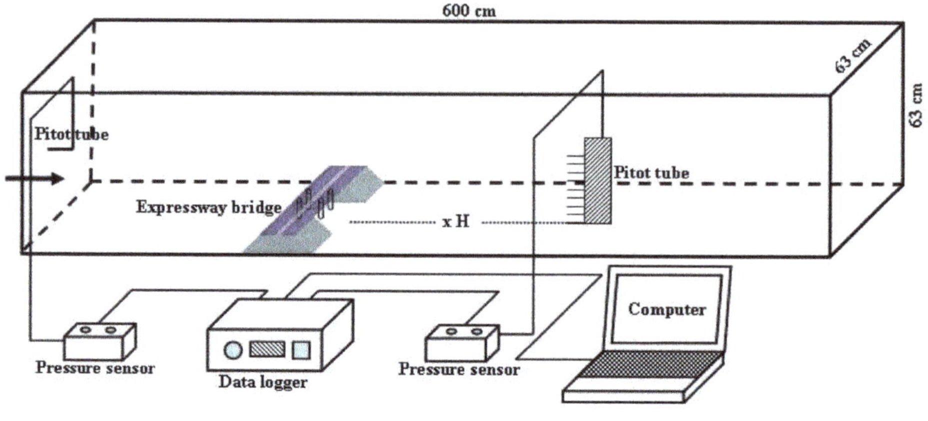


**Figure 2.** Wind speed experiment layout in wind tunnel.

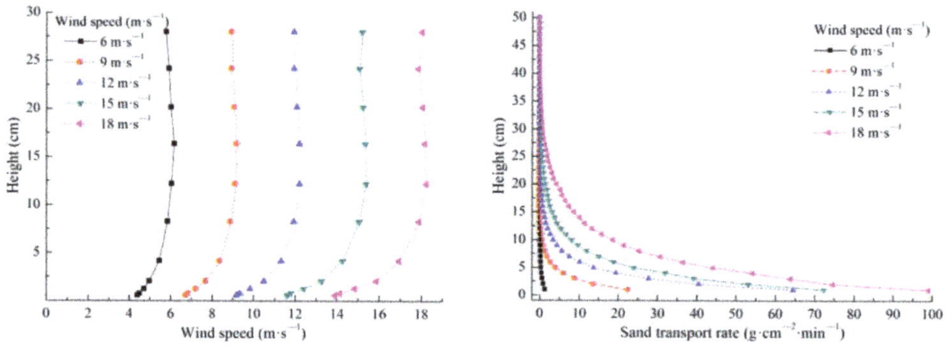


(**a**) Wind profile (**b**) Sand transport

**Figure 3.** Wind profile and sand transport of wind tunnel test without model.

2.2.2. Test Layout of Sand Transport in Wind Tunnel

The sandy bed surface was laid at the 10H distance upwind of the bridge. The length, width, and thickness of the sandy bed surface were 380, 63, and 5 cm, respectively. The sand used in the experiment was obtained from the original surface sand of the Kumtag Desert, and the grain size distribution curve of the sand is shown in Figure 4. The sand collector was set at the 10H distance downwind of the bridge. Every 1 cm height set up a set of sand collector mouths (the width and height of the sand collector mouths were 2 and 1 cm, respectively), for a total of 50 heights. The sand-moving wind speed was 5.0 $m \cdot s^{-1}$. Therefore, the test wind speed was set to 6, 9, 12, 15, and 18 $m \cdot s^{-1}$, for five groups in total. The sand transport in the wind tunnel test without a model is shown in Figure 3b. The test layout of the wind tunnel used to measure the sand transport of the subgrade was the same as that of the bridge (Figure 5).

Simulation experiments in wind tunnels should generally satisfy the principles of geometric similarity, kinematic similarity, and dynamic similarity [22,23]. In this wind tunnel experiment, the sand used was taken from the original surface sand of the desert, conforming to geometric similarity. The bridge and subgrade are models made according to equal-scale reduction, also conforming to geometric similarity. The bridge and subgrade models were placed within the boundary layer of the wind tunnel test section, the pitot tubes were arranged at the bottom center of the wind tunnel test section, and the measured initial wind speed profile was consistent with the distribution law in nature (Figure 3a). Thus, this experiment conforms to kinematic similarity. In the model, the sandy bed surface was laid upwind, and the sand collector was set downwind. The measured initial sand transport rate was consistent with the distribution law in nature (Figure 3b), indicating that this experiment conforms to dynamic similarity. Therefore, this investigation satisfies the principles of geometric, kinematic, and dynamic similarity of wind tunnel simulation experiments to the greatest extent.

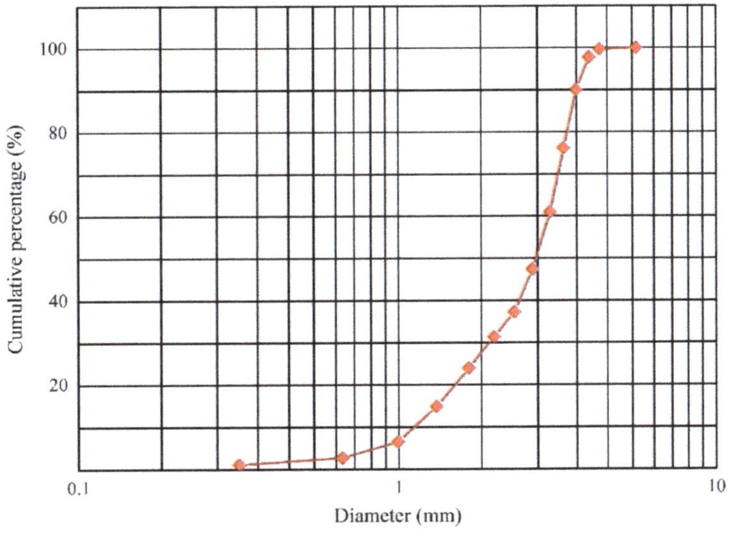


**Figure 4.** Grain size distribution curve of the test sand.

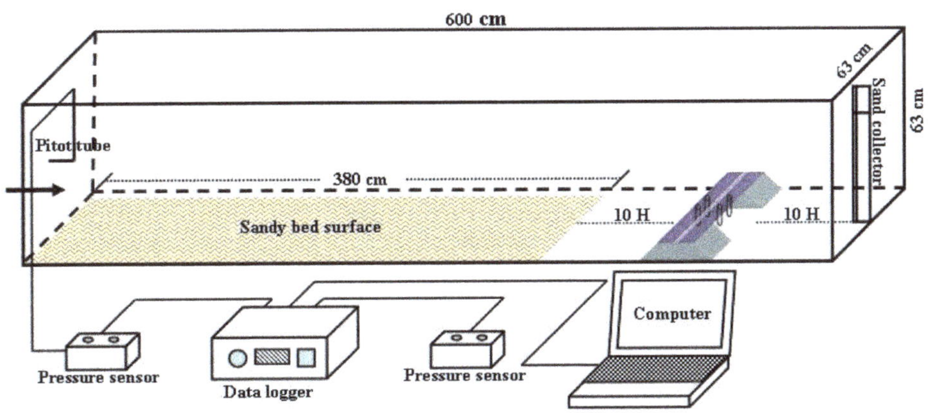


**Figure 5.** Test layout of sand transport in wind tunnel.

## 3. Test Results

### *3.1. Wind Speed at Each Observation Point*

According to the aforementioned layout of the wind speed experiment in a wind tunnel, the wind speed experiment results for the bridges and subgrades are shown in Figure 6.

(**a**)

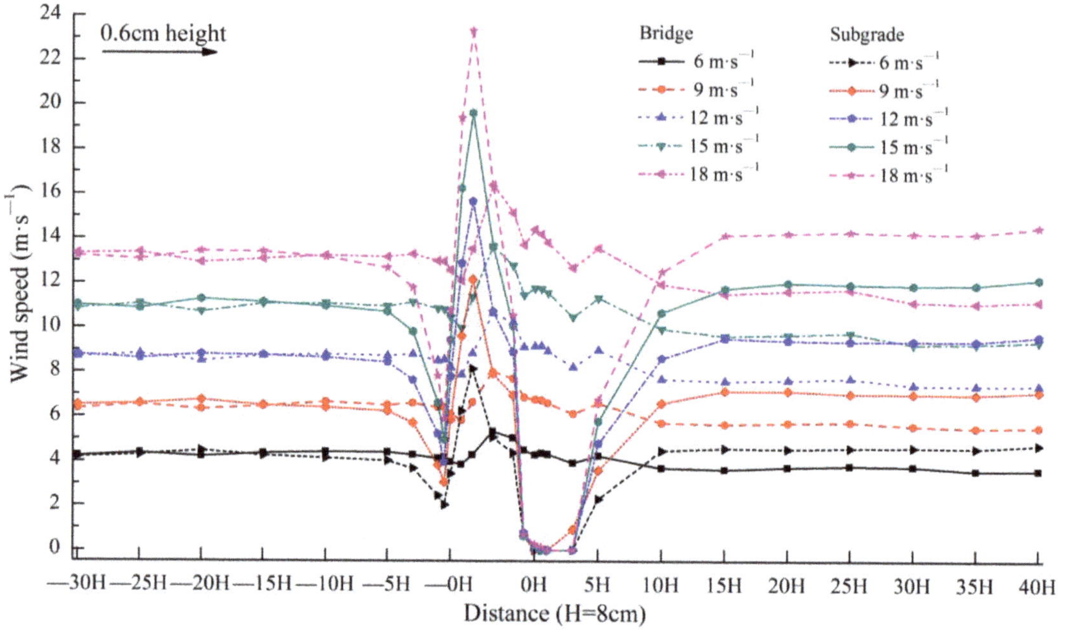


(**b**)

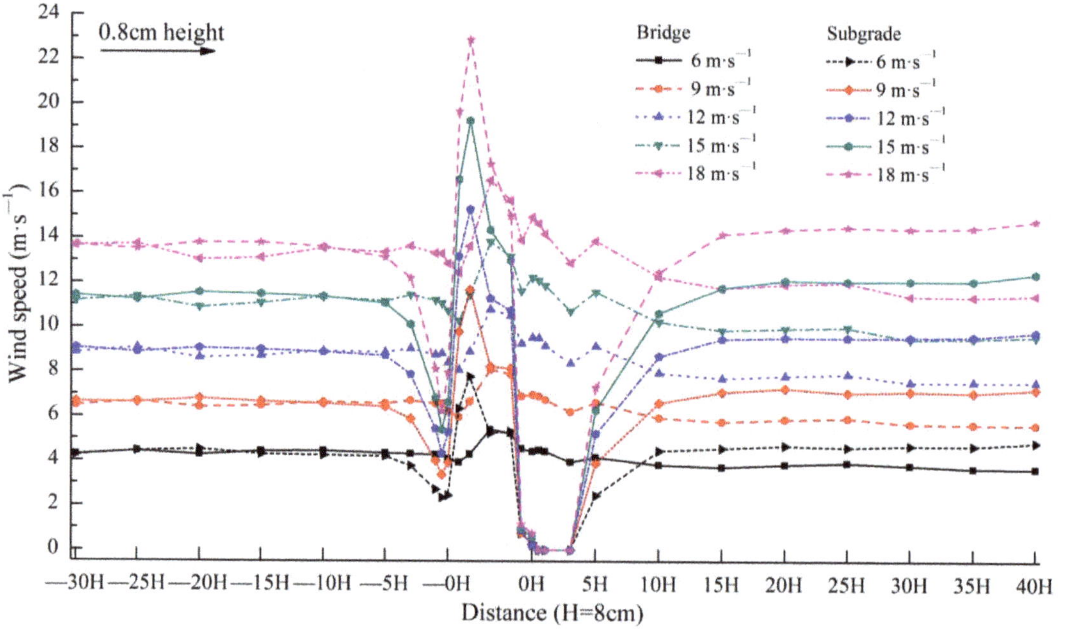


**Figure 6.** *Cont.*

(c)

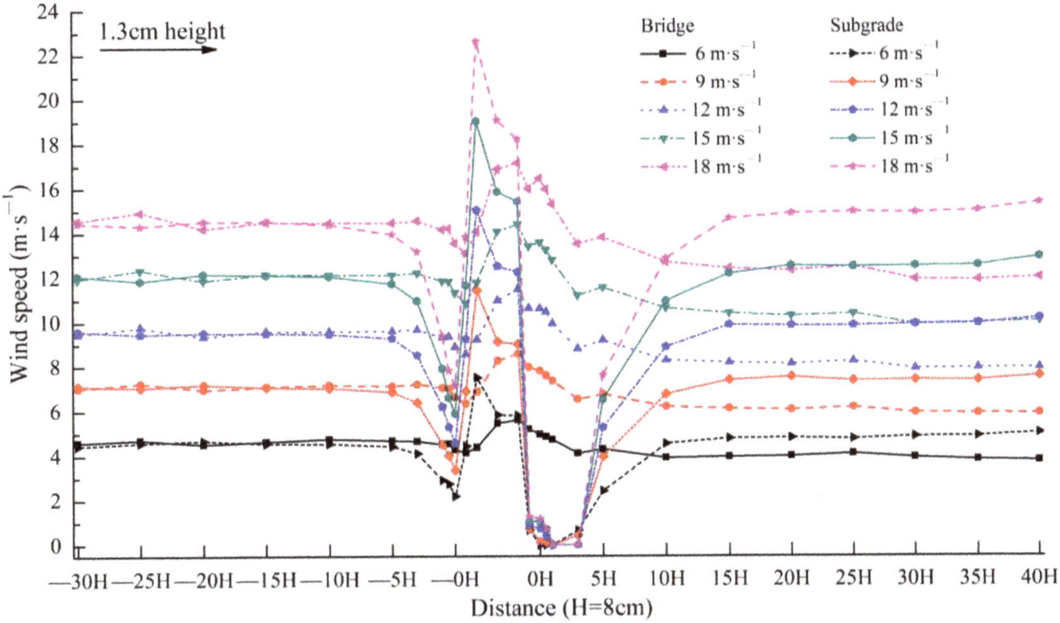


(d)

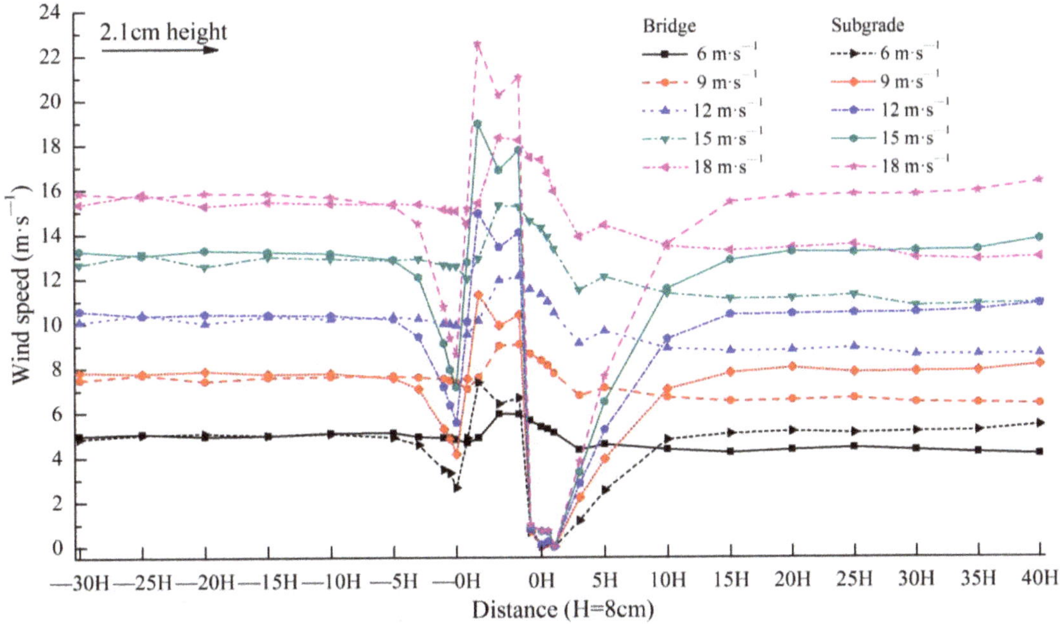


**Figure 6.** *Cont.*

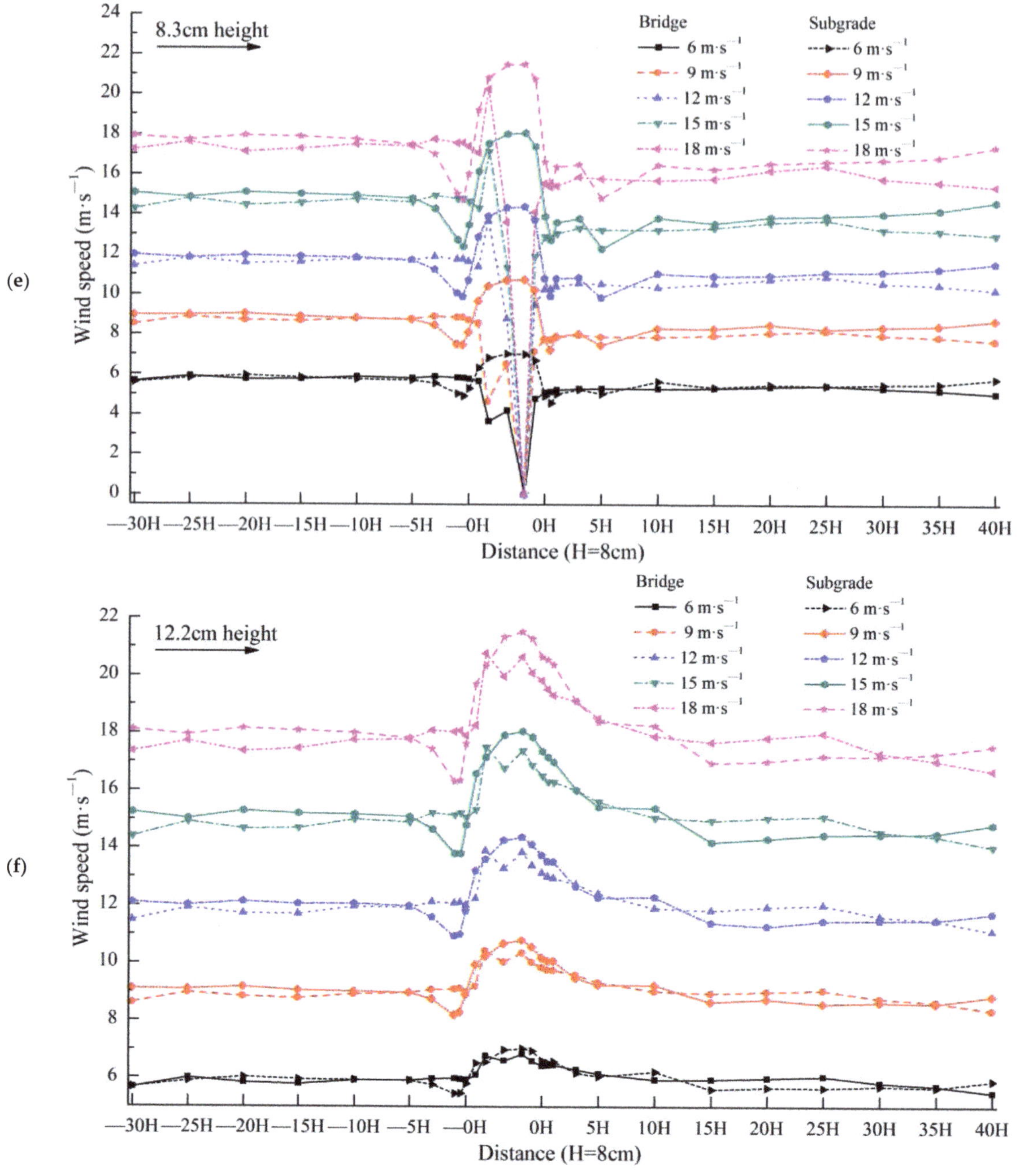


**Figure 6.** *Cont.*

(g)

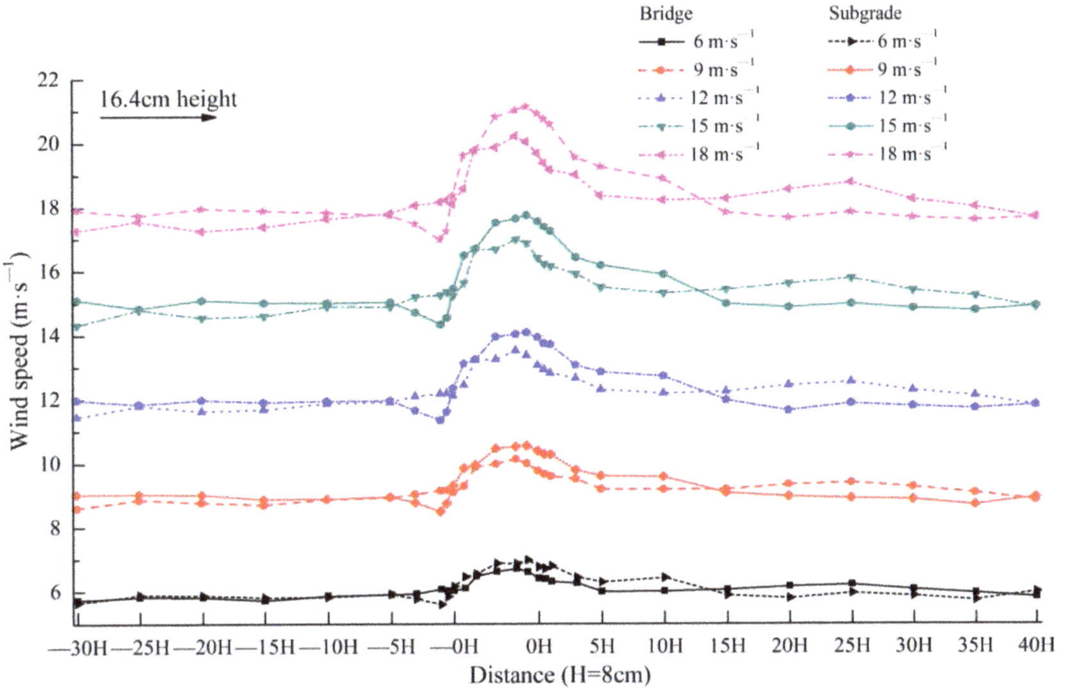


(h)

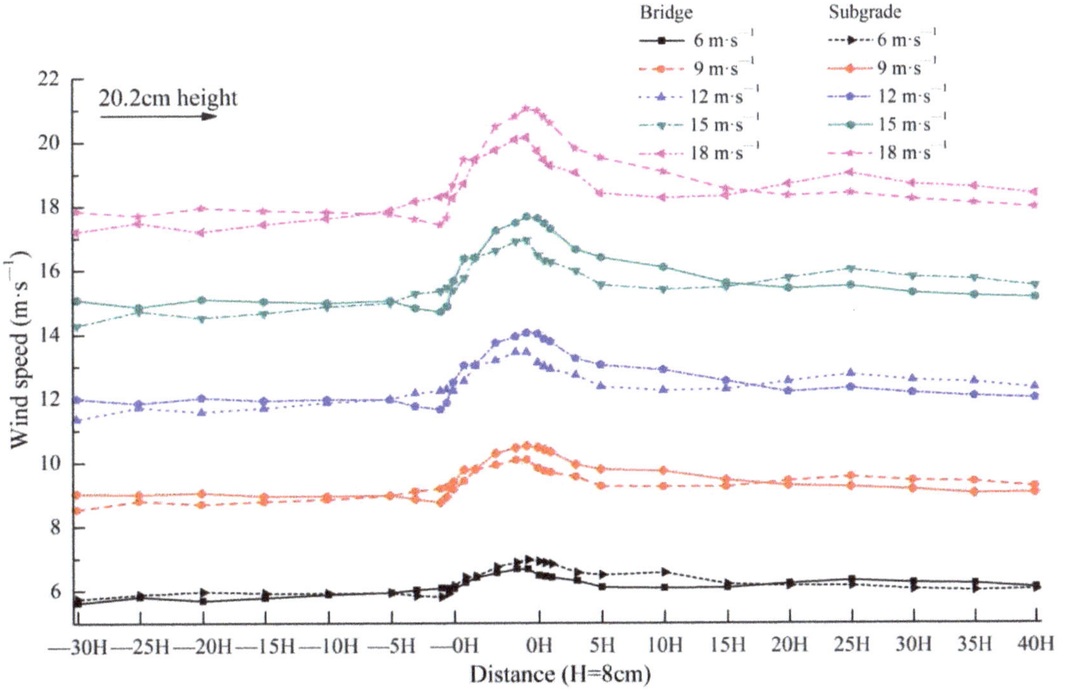


**Figure 6.** *Cont.*

(i)

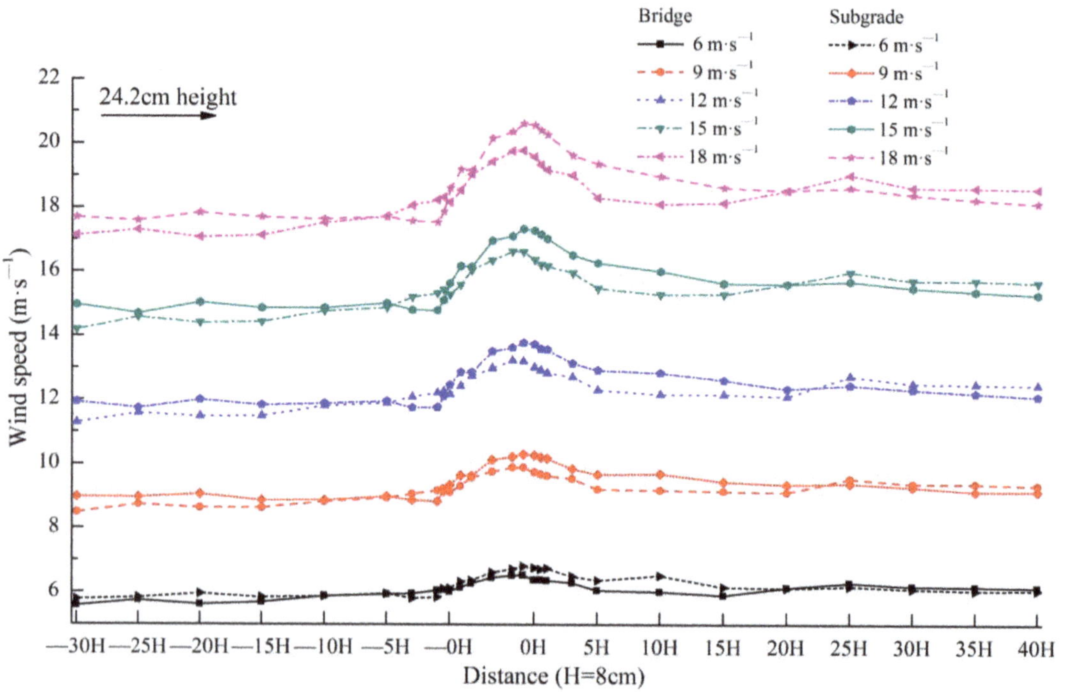


(j)

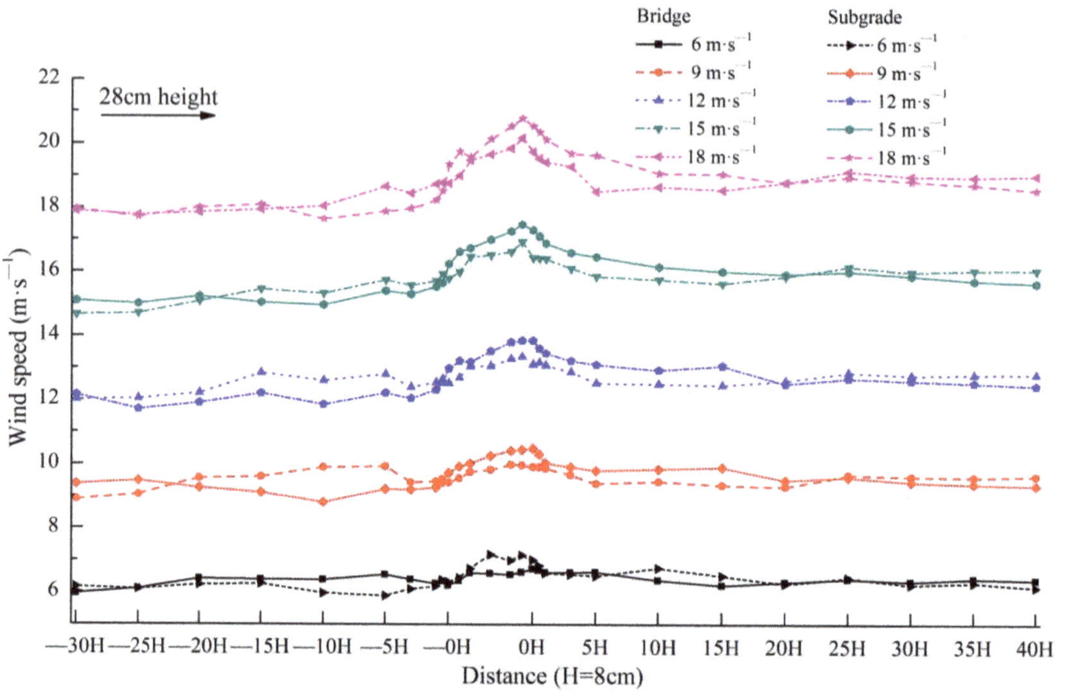


**Figure 6.** Wind speed differences between expressway bridge and subgrade.

When the height of the observation point was 0.6 cm (Figure 6a), the wind speeds of the bridge and the subgrade were nearly equal from −30H to −5H. From −5H to −0H, the wind speed of the bridge was significantly higher than that of the subgrade. From the slope middle of the windward side (the bridge is at the bottom of the windward side slope middle) to the center of the model (the bridge is the bottom center and the subgrade is the top center), the wind speed of the bridge was significantly lower than that of the subgrade. From the slope middle at the leeward side (the bridge is at the bottom of the leeward side slope middle) to 5H, the wind speed of the bridge was significantly higher than that of the subgrade. From 10H to 40H, the wind speed of the bridge was significantly lower than that of the subgrade. Notably, the wind speed of the subgrade recovered within this distance on the leeward side, but the wind speed of the bridge remained significantly lower than the corresponding wind speed on the windward side and did not recover. The variation range of bridge wind speed was significantly lower than that of the subgrade at the height of 0.6 cm.

When the height of the observation point was 0.8 cm (Figure 6b), the wind speed variations of the bridge and subgrade were very similar to those at 0.6 cm.

When the height of the observation point was 1.3 cm (Figure 6c), the wind speed variations of the bridge and subgrade were very similar to those at 0.6 cm.

When the height of the observation point was 2.1 cm (Figure 6d), the wind speed variations of the bridge and subgrade were very similar to those at 0.6 cm.

When the height of the observation point was 8.3 cm (Figure 6e), the wind speed of the bridge and that of the subgrade were nearly equal from −30H to −5H. From −3H to −0H, the wind speed of the bridge was significantly higher than that of the subgrade. From the slope middle of the windward side (the bridge is at the bottom of the windward side slope middle) to the slope middle of the leeward side (the bridge is at the bottom of the leeward side slope middle), the wind speed of the bridge is significantly lower than that of the subgrade. From 0H to 40H, the wind speed of the bridge and subgrade exhibited a minimal difference. The variation range of the bridge wind speed was greater than that of the subgrade at the height of 8.3 cm.

When the height of the observation point was 12.2 cm (Figure 6f), from −30H to −5H, the wind speed of the bridge and subgrade had a minimal overall difference. From −3H to −0H, the wind speed of the bridge was higher than that of the subgrade. From the center of the model (the bridge is the bottom center and the subgrade is the top center) to 1H, the wind speed of the bridge was lower than that of the subgrade. From 3H to 40H, the wind speed of the bridge and subgrade had minimal difference overall. The variation range of the bridge wind speed was lower than that of the subgrade at the height of 12.2 cm.

When the height of the observation point was 16.4 cm (Figure 6g), from −30H to −5H, the wind speed of the bridge and subgrade had a minimal overall difference. From −3H to −0.5H, the wind speed of the bridge is higher than that of the subgrade. From −0H to 10H, the wind speed of the bridge was lower than that of the subgrade. From 15H to 40H, the wind speed of the bridge and subgrade had minimal difference overall. The variation range of the bridge wind speed was lower than that of the subgrade at the height of 16.4 cm.

When the height of the observation point was 20.2 cm (Figure 6h), the wind speed variations in the bridge and subgrade were similar to those at 16.4 cm.

When the heights of the observation point were 24.2 and 28 cm (Figure 6i,j), although the wind speeds of the bridge and subgrade rose and fell with each other, they had a minimal difference overall. With the increase in height, the difference between the wind speeds of the bridge and subgrade decreased. The variations in their wind speeds became increasingly consistent, and the variation range of the wind speed became increasingly smaller. However, the variation range of the bridge wind speed was slightly lower than that of the subgrade wind speed.

### 3.2. Wind Flow Field

The Kriging interpolation method was used to draw the wind field map of the bridge and subgrade (wind speed contour map) based on the aforementioned wind speed test results. Figure 7 shows a wind-speed-weakening area between −3H upwind of the bridge to the slope shoulder of the windward side of the bridge. The figure also shows an obvious wind speed-weakening area between −5H upwind of the subgrade to the slope shoulder of the windward side of the subgrade. In these areas, the wind-speed-weakening range and intensity of the bridge are smaller than those of the subgrade. The wind-speed-increasing area was observed at the top of the bridge, as was an obvious increase in wind speed was at the top of the subgrade, particularly on the slope shoulder of the windward side. In those areas, the wind speed increase range and intensity of the bridge were smaller than those of the subgrade. A wind-speed-weakening area was found from the slope middle of the leeward side of the bridge to the 40H downwind, and an obvious wind-speed-weakening area was found from the slope shoulder of the leeward side of the subgrade to 15H downwind. In these areas, the wind-speed-weakening range of the bridge was greater than that of the subgrade, but the wind-speed-weakening intensity of the bridge was lower than that of the subgrade. These findings indicate that the overall range of wind speed variation of the bridge was lower than that of the subgrade, the wind field variation of the subgrade was more intense than that of the bridge, and the disturbance of the subgrade to the wind-blown sand environment was greater than that of the bridge.

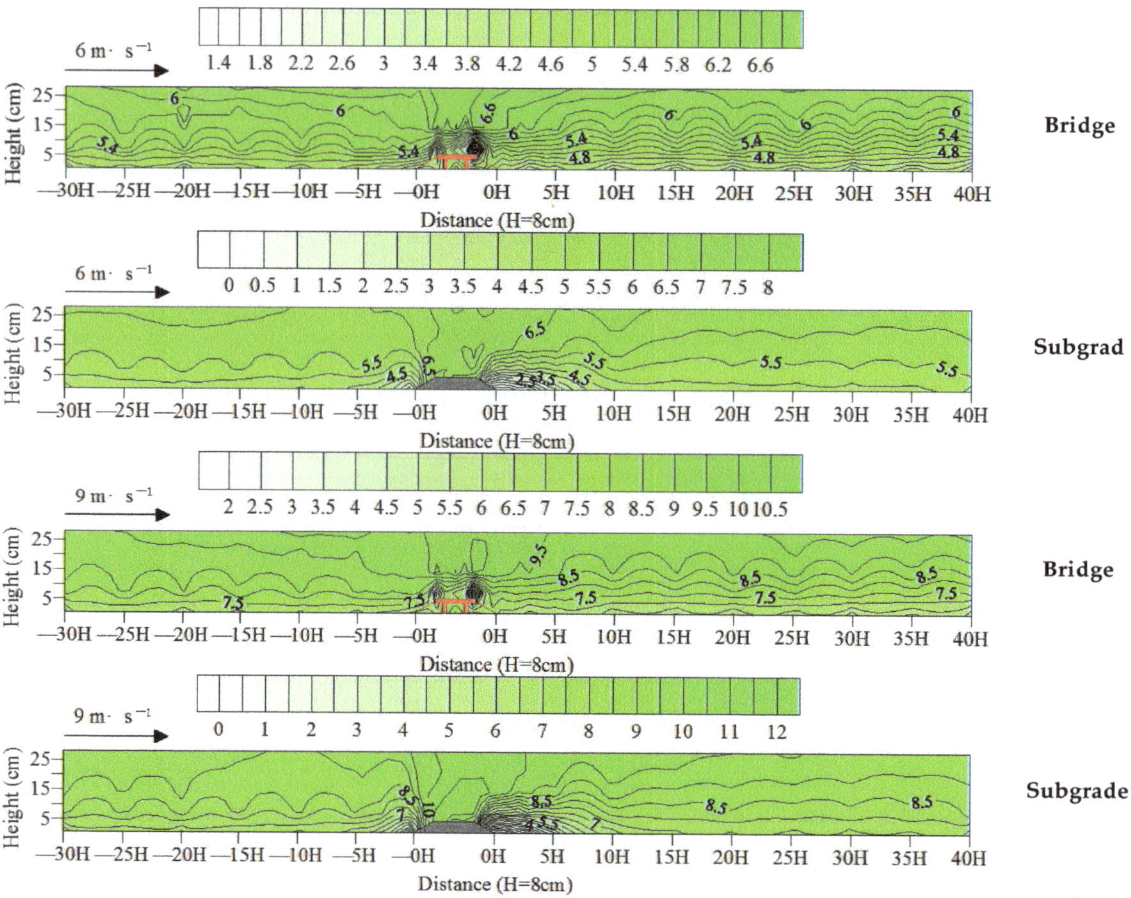


**Figure 7.** *Cont.*

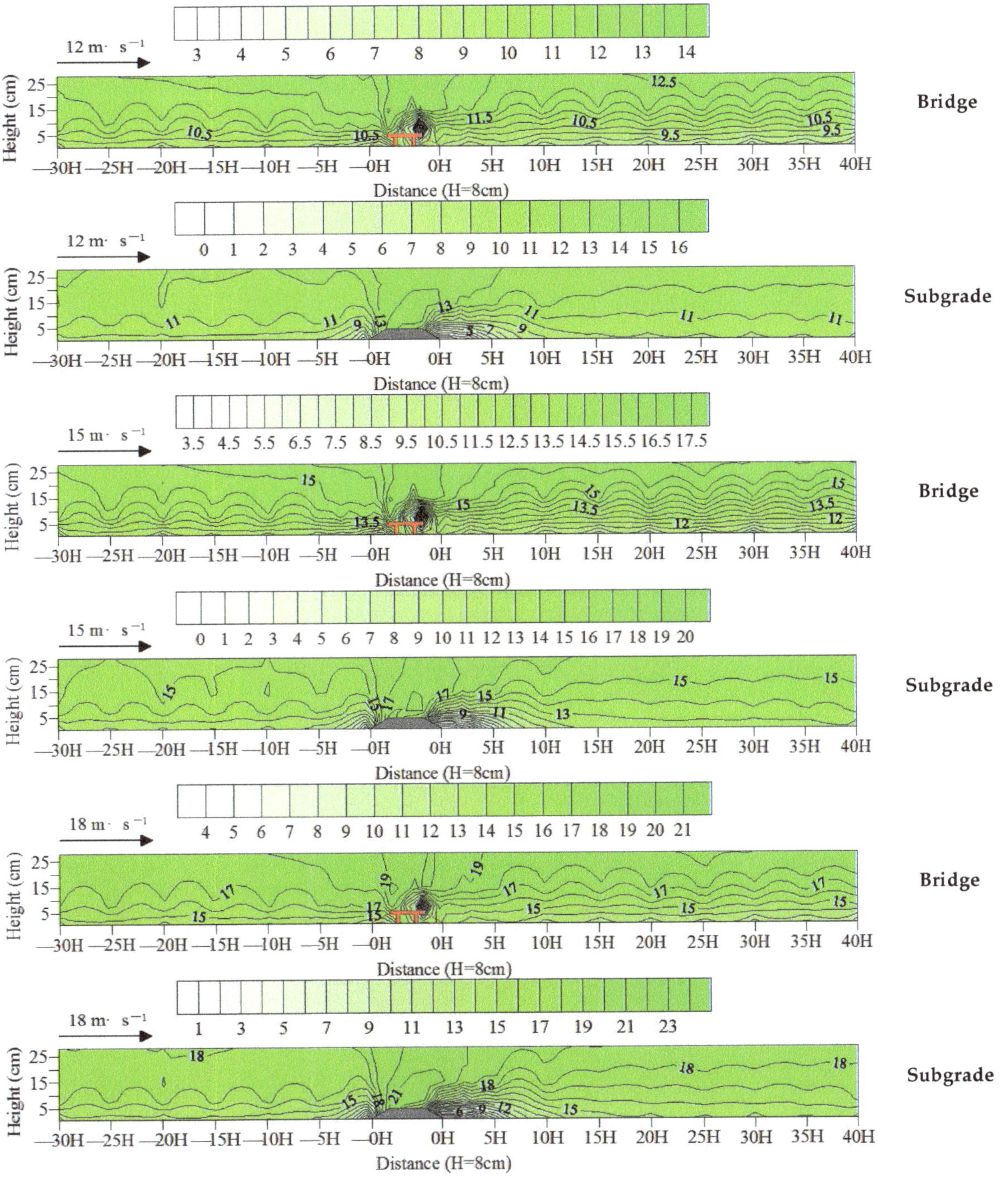


**Figure 7.** Wind field differences between expressway bridge and subgrade.

### 3.3. Sand Transport

The test results of the sand transport of bridges and subgrades are shown in Figure 8 and Table 1, based on the test layout of sand transport in the wind tunnel. The sand transport rates of the bridge for each height followed the distribution law of exponential functions, whereas the sand transport rates of the subgrade varying with height followed the distribution law of Gaussian functions. Under the experimental wind speeds of the five groups, the sand transport rates of the bridge were 7.56, 39.24, 141.05, 273.03, and 374.30 g·cm$^{-1}$·min$^{-1}$, and those of the subgrade were 9.56, 66.92, 164.16, 296.08, and 431.43 g·cm$^{-1}$·min$^{-1}$, respectively. The average sand transport rate of the bridge was 86.27% of that of the subgrade. The sand fluxes of the bridge were obviously lower than those of the subgrade, indicating that the passing rate of the wind-blown sand flow of the bridge was lower than that of the subgrade. Compared with the subgrade, the bridge blocks more wind-blown sand transport and causes more sand material to accumulate near the bridge.

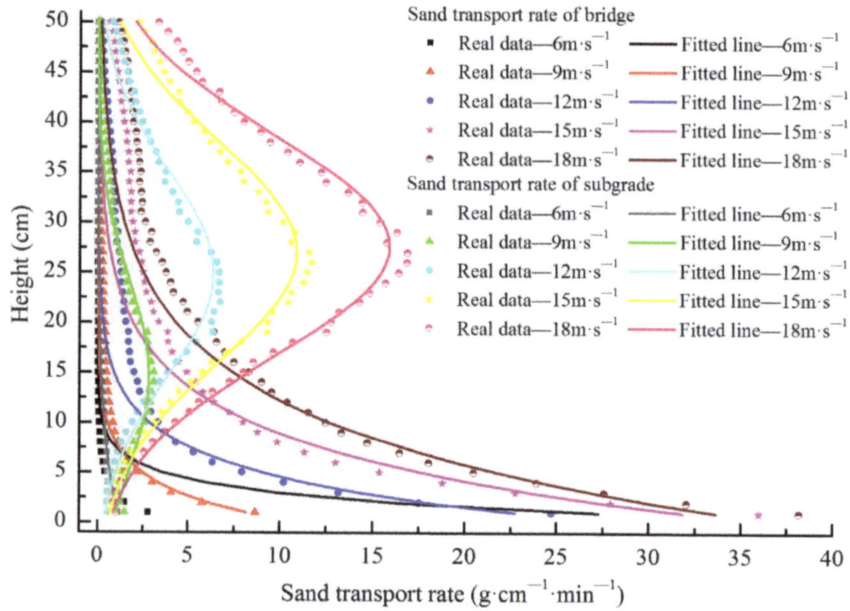


**Figure 8.** Sand transport rate of expressway bridge and subgrade.

**Table 1.** Fitting results of sand transport rate of expressway bridge and subgrade.

| Route Forms | Wind Speed (m·s$^{-1}$) | Fitting Function Type | Fitting Function Formula | a | b | c | $R^2$ |
|---|---|---|---|---|---|---|---|
| Bridge | 6 | Exponential | $y = a \times \exp(b \times x)$ | 45.30 | −0.51 | | 0.99 |
| Bridge | 9 | Exponential | $y = a \times \exp(b \times x)$ | 10.98 | −0.30 | | 0.98 |
| Bridge | 12 | Exponential | $y = a \times \exp(b \times x)$ | 28.60 | −0.23 | | 0.96 |
| Bridge | 15 | Exponential | $y = a \times \exp(b \times x)$ | 36.87 | −0.15 | | 0.96 |
| Bridge | 18 | Exponential | $y = a \times \exp(b \times x)$ | 37.48 | −0.11 | | 0.98 |
| Subgrade | 6 | Gaussian | $y = a \times \exp(-((x - b)/c)^2)$ | 999.70 | −118.30 | 45.69 | 0.94 |
| Subgrade | 9 | Gaussian | $y = a \times \exp(-((x - b)/c)^2)$ | 2.84 | 14.67 | 13.68 | 0.95 |
| Subgrade | 12 | Gaussian | $y = a \times \exp(-((x - b)/c)^2)$ | 6.34 | 24.88 | 14.65 | 0.95 |
| Subgrade | 15 | Gaussian | $y = a \times \exp(-((x - b)/c)^2)$ | 10.88 | 26.99 | 15.69 | 0.97 |
| Subgrade | 18 | Gaussian | $y = a \times \exp(-((x - b)/c)^2)$ | 15.89 | 27.64 | 15.68 | 0.98 |

## 4. Cause Analysis

In this test, the height of the model was set to 8 cm. Owing to the gap in the bridge, a minimal blocking effect was observed on the near-ground airflow below this height. When the airflow ran near the bridge, it passed through the gap at the bottom of the bridge. The disturbance received by the airflow was small. Therefore, the variation range of the wind speed was small, similar to the wind speed and flow field of a railway bridge [21].

Under the same experimental conditions, because no gap existed in the subgrade, it had a considerable blocking effect on the airflow near the ground. When the airflow ran in the −5H to −0H range upwind of the subgrade, it can only pass through the top of the subgrade with obstructions, resulting in considerable disturbance. Therefore, the wind speed decreased significantly. When the airflow ran to the windward side of the subgrade, the wind speed increased fleetingly with a climbing windward slope and reached the maximum value when it ran the shoulder of the windward slope. Thus, the wind speed increased significantly. When the airflow passed over the top of the subgrade, due to the terrain decrease of the leeward slope, the airflow dispersed rapidly and the wind speed dropped sharply. The wind speed dropped to the lowest value ($0\ m{\cdot}s^{-1}$ or close to $0\ m{\cdot}s^{-1}$) within 0H–3H downwind of the subgrade, and the wind speed dropped significantly. Then, as the distance increased, the influence of the subgrade on the airflow weakened and the wind speed recovered quickly. Consequently, the wind speed of the subgrade varied sharply, and the variation range was significantly higher than that of the bridge. Above 8.3 cm, with the increase in height, the influence of the bridge and subgrade became smaller and the airflow was less disturbed. Therefore, the wind speed difference between the bridge and subgrade became smaller, the variation trend of the wind speed was the same, and the variation range decreased.

Relevant studies show that sand transported by the wind accumulates around any type of obstacle [24,25], and the decrease in near-surface wind speed easily causes sand material accumulation, while the increase in wind speed easily causes blown sand flow erosion [26–29]. In the wind-speed-weakening area upwind, because the wind-speed-weakening range and intensity of the bridge were smaller than those of the subgrade, the range and intensity of sand material accumulation upwind of the bridge were smaller than those of the subgrade. In the wind-speed-increasing area at the top of the model, because the wind-speed-increasing range and intensity of the bridge were smaller than those of the subgrade, the erosion range and intensity of the wind-blown sand flow on the top of the bridge were smaller than those of the subgrade. In the wind-speed-weakening area downwind, because the wind-speed-weakening range of the bridge was greater than that of the subgrade, and the wind-speed-weakening intensity was smaller than that of the subgrade, the sand material accumulation range downwind of the bridge was larger than that of the subgrade. However, the accumulation intensity was smaller than that of the subgrade.

Relevant studies have shown that the influence of bridges on wind sand movement decreases with increasing height [21] and has no influence on the wind sand movement of the near-surface after reaching the threshold height. However, the opposite is true for the subgrade: the higher the subgrade, the stronger the disturbance to the wind-blown sand activity of the near-surface [30,31]. In this experiment, although the wind speed variation range of the bridge was less than that of the subgrade, the wind speed near the surface (below the model height of 8 cm) still did not recover within 40H downwind of the bridge. This lack of recovery resulted in the weakening of the driving force of wind-blown sand flow transport, a decrease in the passing rate, and more sand materials being intercepted near the bridge. Therefore, the sand transport rate of the bridge was lower than that of the subgrade. At the same time, these findings indicate that the required distance to recover the near-surface wind speed and its flow field downwind of the bridge is greater than that of the subgrade, causing the sand material accumulation range to also be greater than that of the subgrade.

## 5. Results Discussion

A comparison of the characteristics of the wind-blown sand environment of the expressway bridge and subgrade is shown in Table 2 through the test results and analysis. The disturbance of the bridge to the wind-blown sand environment was less than that of the subgrade in seven indices: the variation ranges of the wind speed, variation ranges of the flow field, wind-speed-weakening range and intensity in the wind-speed-weakening area upwind, wind-speed-increasing range and intensity in the wind-speed-increasing area on the top, and wind-speed-weakening intensity in the wind-speed-weakening area downwind. However, the disturbance of the bridge to the wind-blown sand environment was greater than that of the subgrade in several indices, such as the required distances to recover the wind speed and its flow field downwind and the wind-speed-weakening range in the wind-speed-weakening area downwind, thereby decreasing the passing rate of the wind-blown sand flow of the bridge and increasing the sand material accumulation. Therefore, from the perspective of prevention and control of wind-blown sand hazards, the wind-blown sand environment of the bridge was generally better than that of the subgrade.

**Table 2.** Comparison of characteristics of wind-blown sand environment of expressway bridge and subgrade.

| Environmental Indexes of Blown Sand | | Contrast | Advantage Item | Disadvantage Item |
|---|---|---|---|---|
| Wind speed | Variation range | bridge < subgrade | bridge | subgrade |
| | Required distance to recover the wind speed | bridge > subgrade | subgrade | bridge |
| Wind flow field | Variation range | bridge < subgrade | bridge | subgrade |
| | Required distance to recover the wind field | bridge > subgrade | subgrade | bridge |
| Wind-speed-weakening area upwind | Range | bridge < subgrade | bridge | subgrade |
| | Intensity | bridge < subgrade | bridge | subgrade |
| Wind-speed-increasing area on the top | Range | bridge < subgrade | bridge | subgrade |
| | Intensity | bridge < subgrade | bridge | subgrade |
| Wind-speed-weakening area downwind | Range | bridge > subgrade | subgrade | bridge |
| | Intensity | bridge < subgrade | bridge | subgrade |
| Passing rate of wind-blown sand flow (Average under the experimental wind speed of five groups) | Ratio (bridge/subgrade) | 0.8627 | subgrade | bridge |

According to the experimental results and their analysis, the following implications for practical engineering applications can be obtained. When surveying and designing expressways in sandy areas, if the construction cost is not considered, expressways through seriously blown sand areas should generally use the bridge form. However, when the downwind direction is limited by terrain such as river valleys or other special factors [32], the space is narrow, and the distance is limited, expressways should use the subgrade form. The subgrade height should also be lowered, and the bridge height should be raised as much as possible to reduce or even avoid sand disasters. Future work can focus on content such as the threshold distance of the downwind direction where the subgrade form should be adopted, environmental effects of wind-blown sand of bridges with different sizes (different heights and widths), and the threshold height of the bridge, which can avoid sand disasters.

## 6. Conclusions and Implications

At a height below 8.3 cm near the surface, the variation ranges of the wind speed of the bridge and its upwind and downwind directions were lower than those of the subgrade. However, the required distance to recover the wind speed downwind of the bridge was greater than that of the subgrade, resulting in the sand transport rate of the bridge being lower than that of the subgrade. Under the experimental wind speeds of the five groups, the average sand fluxes of the bridge was 86.27% of that of the subgrade. Above 8.3 cm, the wind speed difference between the bridge and subgrade became smaller, the variation trend of the wind speed was the same, and the variation range decreased.

The variation in the wind field of the subgrade was more drastic than that of the bridge, but the required distance to recover the wind field downwind of the subgrade was smaller than that of the bridge. In the wind speed-weakening area upwind, the wind speed-weakening range and intensity of the bridge were smaller than those of the subgrade. In the wind-speed-increasing area on the top of the model, the wind-speed-increasing range and intensity of the bridge were smaller than those of the subgrade. In the wind-speed-weakening area downwind, the wind-speed-weakening range of the bridge was greater than that of the subgrade, and the wind-speed-weakening intensity was smaller than that of the subgrade. From the perspective of prevention and control of wind-blown sand hazards, the wind-blown sand environment of the bridge was generally better than that of the subgrade. Therefore, expressways through seriously blown sand areas should prioritize the use of the bridge form.

**Author Contributions:** Investigation, S.X. and Y.P.; Writing—original draft, S.X.; Writing—review & editing, S.X. and X.Z. All authors have read and agreed to the published version of the manuscript.

**Funding:** This research project was funded by the National Natural Science Foundation of China (grant nos. 42077448 and 41877530) and the Youth Innovation Promotion Association CAS (member certification no. 2018459). This project was supported by the Ordos Science & Technology Plan (grant no. 2021EEDSCXQDFZ013).

**Institutional Review Board Statement:** Not applicable.

**Informed Consent Statement:** Not applicable.

**Data Availability Statement:** Not applicable.

**Conflicts of Interest:** The authors declare no conflict of interest.

## References

1. Bhattachan, A.; Okin, G.S.; Zhang, J.; Vimal, S.; Lettenmaier, D.P. Characterizing the role of wind and dust in traffic accidents in California. *GeoHealth* **2019**, *3*, 328–336. [CrossRef] [PubMed]
2. Li, J.R.; Kandakji, T.; Lee, J.A.; Tatarko, J.; Blackwell, J.; Gill, T.E.; Collins, J.D. Blowing dust and highway safety in the southwestern United States: Characteristics of dust emission "hotspots" and management implications. *Sci. Total Environ.* **2018**, *621*, 1023–1032. [CrossRef] [PubMed]
3. Horvat, M.; Bruno, L.; Khris, S.; Raffaele, L. Aerodynamic shape optimization of barriers for windblown sand mitigation using CFD analysis. *J. Wind Eng. Ind. Aerodyn.* **2020**, *197*, 104058. [CrossRef]
4. Hu, L.; Shan, Y.T.; Chen, R.H.; Guo, W.; Wang, Q.; Li, Z.B. A study of erosion control on expressway embankment sideslopes with three-dimensional net seeding on the Qinghai-Tibet Plateau. *Catena* **2016**, *147*, 463–468. [CrossRef]
5. Wang, C.; Li, S.Y.; Lei, J.Q.; Li, Z.N.; Chen, J. Effect of the W–beam central guardrails on wind–blown sand deposition on desert expressways in sandy regions. *J. Arid Land* **2020**, *12*, 154–165. [CrossRef]
6. Yan, M.; Wang, H.B.; Zuo, H.J.; Li, G.T. Wind tunnel simulation of an opencut tunnel airflow field along the Linhe–Ceke Railway, China. *Aeolian Res.* **2019**, *39*, 66–76. [CrossRef]
7. Li, S.H.; Li, C.; Yao, D.; Ge, X.D.; Zhang, G.P. Wind tunnel experiments for dynamic modeling and analysis of motion trajectories of wind–blown sands. *Eur. Phys. J. E* **2020**, *43*, 22. [CrossRef]
8. He, W.; Huang, N.; Xu, B.; Wang, W.B. Numerical simulation of wind–sand movement in the reversed flow region of a sand dune with a bridge built downstream. *Eur. Phys. J. E* **2018**, *41*, 53. [CrossRef]
9. Thalla, O.; Stiros, S.C. Wind-induced fatigue and asymmetric damage in a timber bridge. *Sensors* **2018**, *18*, 3867. [CrossRef]
10. Kim, S.; Jung, H.; Kong, M.J.; Lee, D.K.; An, Y.K. In-situ data-driven buffeting response analysis of a cable-stayed bridge. *Sensors* **2019**, *19*, 3048. [CrossRef]
11. Bruno, L.; Horvat, M.; Raffaele, L. Windblown sand along railway infrastructures: A review of challenges and mitigation measures. *J. Wind Eng. Ind. Aerodyn.* **2018**, *177*, 340–365. [CrossRef]
12. Pozzebon, A.; Cappelli, I.; Mecocci, A.; Bertoni, D.; Sarti, G.; Alquini, F. A wireless sensor network for the real-time remote measurement of aeolian sand transport on sandy beaches and dunes. *Sensors* **2018**, *18*, 820. [CrossRef] [PubMed]
13. Albarakat, R.; Lakshmi, V. Monitoring dust storms in Iraq using satellite data. *Sensors* **2019**, *19*, 3687. [CrossRef] [PubMed]
14. Huang, N.; Gong, K.; Xu, B.; Zhao, J.; Dun, H.C.; He, W.; Xin, G.W. Investigations into the law of sand particle accumulation over railway subgrade with wind-break wall. *Eur. Phys. J. E* **2019**, *42*, 145. [CrossRef] [PubMed]
15. Udo, K. New method for estimation of aeolian sand transport rate using ceramic sand flux sensor (UD-101). *Sensors* **2009**, *9*, 9058–9072. [CrossRef]
16. Govaerts, Y.M. Sand dune ridge alignment effects on surface BRF over the Libya-4 CEOS calibration site. *Sensors* **2015**, *15*, 3453–3470. [CrossRef]

17. Li, C.J.; Wang, Y.D.; Lei, J.Q.; Xu, X.W.; Wang, S.J.; Fan, J.L.; Li, S.Y. Damage by wind–blown sand and its control measures along the Taklimakan Desert Highway in China. *J. Arid Land* **2021**, *13*, 98–106. [CrossRef]
18. Antonio, C.D.; Jorge, C.; Manuel, V.; Eduardo, R.; Luis, H. A DIY low-cost wireless wind data acquisition system used to study an arid coastal foredune. *Sensors* **2020**, *20*, 1064.
19. Zdravko, K.; Bernard, Ž.; Biljana, M.B. FOCUSED–short-term wind speed forecast correction algorithm based on successive NWP forecasts for use in traffic control decision support systems. *Sensors* **2021**, *21*, 3405.
20. Xie, S.B.; Qu, J.J.; Zhang, K.C.; Han, Q.J.; Pang, Y.J. The mechanism of sand damage at the Fushaliang section of the Liuyuan–Golmud Expressway. *Aeolian Res.* **2021**, *48*, 100648. [CrossRef]
21. Xie, S.B.; Qu, J.J.; Han, Q.J.; Pang, Y.J. Wind dynamic environment and wind tunnel simulation experiment of bridge sand damage in Xierong section of Lhasa–Linzhi Railway. *Sustainability* **2020**, *12*, 5689. [CrossRef]
22. Tetsuya, K.; Yamagishi, Y.; Kimura, S.; Sato, K. Aerodynamic behavior of snowflakes on an uneven road surface during a snowstorm. *Open J. Fluid Dyn.* **2017**, *7*, 696–708. [CrossRef]
23. Dun, H.C.; Xin, G.W.; Huang, N.; Shi, G.T.; Zhang, J. Wind-tunnel studies on sand sedimentation around wind-break walls of Lanxin High-Speed Railway II and its prevention. *Appl. Sci.* **2021**, *11*, 5989. [CrossRef]
24. Raffaele, L.; Bruno, L. Windblown sand mitigation along railway megaprojects: A comparative study. *Struct. Eng. Int.* **2020**, *30*, 355–364. [CrossRef]
25. Lima, I.A.; Parteli, E.J.R.; Shao, Y.P.; Andrade, J.S.; Herrmann, H.J.; Araújo, A.D. CFD simulation of the wind field over a terrain with sand fences: Critical spacing for the wind shear velocity. *Aeolian Res.* **2020**, *43*, 100574. [CrossRef]
26. Liu, J.Q.; Kimura, R.; Miyawaki, M.; Kinugasab, T. Effects of plants with different shapes and coverage on the blown-sand flux and roughness length examined by wind tunnel experiments. *Catena* **2021**, *197*, 104976. [CrossRef]
27. Shen, Y.P.; Zhang, C.L.; Huang, X.Q.; Wang, X.S.; Cen, S.B. The effect of wind speed averaging time on sand transport estimates. *Catena* **2019**, *175*, 286–293. [CrossRef]
28. Jiang, Y.S.; Gao, Y.H.; Dong, Z.B.; Liu, B.L.; Zhao, L. Simulations of wind erosion along the Qinghai–Tibet Railway in north–central Tibet. *Aeolian Res.* **2018**, *32*, 192–201. [CrossRef]
29. Cheng, H.; He, J.J.; Xu, X.R.; Zou, X.Y.; Wu, Y.Q.; Liu, C.C.; Dong, Y.F.; Pan, M.H.; Wang, Y.Z.; Zhang, H.Y. Blown sand motion within the sand–control system in the southern section of the Taklimakan Desert Highway. *J. Arid Land* **2015**, *7*, 599–611. [CrossRef]
30. Xiao, J.H.; Yao, Z.Y.; Qu, J.J. Influence of Golmud–Lhasa section of Qinghai–Tibet Railway on blown sand transport. *Chin. Geogr. Sci.* **2015**, *25*, 39–50. [CrossRef]
31. Zhang, K.C.; Qu, J.J.; Han, Q.J.; Xie, S.B.; Kai, K.; Niu, Q.H.; An, Z.S. Wind tunnel simulation of windblown sand along China's Qinghai–Tibet Railway. *Land Degrad. Dev.* **2014**, *25*, 244–250. [CrossRef]
32. Draut, A. Effects of river regulation on aeolian landscapes, Colorado River, southwestern USA. *J. Geophys. Res. Earth Surf.* **2012**, *117*, 1–22. [CrossRef]

MDPI
St. Alban-Anlage 66
4052 Basel
Switzerland
www.mdpi.com

*Sensors* Editorial Office
E-mail: sensors@mdpi.com
www.mdpi.com/journal/sensors

www.ingramcontent.com/pod-product-compliance
Lightning Source LLC
Chambersburg PA
CBHW061516210926
51850CB00008B/1942
* 9 7 8 3 7 2 5 8 1 2 6 9 1 *